THE WADSWORTH INTERNATIONAL MATHEMATICS SERIES

Behzad, M., Gary Chartrand, and Linda Lesniak-Foster. ***Graphs and Digraphs.***

Cochran, James A. ***Applied Mathematics: Principles, Techniques, and Applications.***

Stromberg, Karl R. ***An Introduction to Classical Real Analysis.***

APPLIED MATHEMATICS

PRINCIPLES, TECHNIQUES, AND APPLICATIONS

JAMES A. COCHRAN

WASHINGTON STATE UNIVERSITY

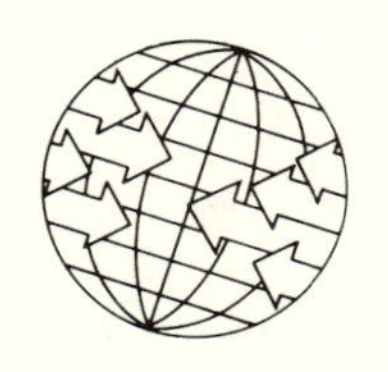

WADSWORTH INTERNATIONAL GROUP
a division of WADSWORTH, INC.
BELMONT, CALIFORNIA

"Curiosity and the urge to solve problems are the emotional hallmarks of our species; and the most characteristically human activities are mathematics, science, technology. . ."

Carl Sagan, *The Dragons of Eden*

This book is a copublishing project of Wadsworth International Group and Prindle, Weber & Schmidt.

Library of Congress Cataloging in Publication Data
Cochran, James Alan.
Applied mathematics.

Bibliography: p.
Includes index.
1. Mathematics—1961- I. Title.
QA37.2.C6 1982 510 82-7055
ISBN 0-534-98026-0 AACR2

Printed in the United States of America

Jacket image by NASA/Ames Research Center is a simulated visualization of compressible flow surrounding an air foil (a wing cross-section of an aircraft model).

Jacket and cover design by David Foss. Text design by Helen Walden. Text composition in Times Roman by Weimer Typesetting. Jacket printed by New England Book Components. Text printed and bound by Maple/Vail.

Preface

Much of both classical and modern mathematics is intimately intertwined with applications. Indeed, the need for timely answers to various practical problems over the years not only has engendered new uses for old mathematics but even, on occasion, has stimulated new mathematics itself. This, of course, does not mean that the mathematical "solution" of physical problems is typically a straightforward undertaking. As most readers realize, the opposite is often the case. Difficult "real-world" problems must be idealized and simplified, the resulting mathematical abstractions analyzed, answers determined, and results checked for relevance to the original physical situations. Considerable skill of various kinds and in varying degrees is needed at each stage of the process. But more than anything else, a certain level of mathematical knowledge and sophistication is absolutely essential.

This book represents an effort to impart some of that knowledge and sophistication. Since for many of us, mathematics really comes alive only as it is used in the analysis of physically meaningful problems, discussion of each major mathematical topic in this book is preceded by consideration of a relevant practical application of some substance. Indeed, the modelling of these diverse physical problems serves to motivate the mathematics that follows. Throughout the book there is also a healthy respect for "answers," and occasionally rather detailed analyses of various computational procedures and techniques for obtaining these "results" are presented. Thus, although mathematics is central to the text, the approach taken to the material is unique for a book at this level.

The idea for this text had its genesis at Bell Laboratories in New Jersey. The first draft of portions of the manuscript came into being, however, at Virginia Tech, where an extremely successful advanced course in applied mathematics was developed for engineering students during the mid-1970s. Since then, the presentation has been considerably honed by cogent review and classroom experience at Washington State University as part of a program designed for students who will ultimately pursue industrial and/or business careers.

Here then is a modestly sized advanced level text in applied mathematics that intermingles theory and practice. The book is not intended to be comprehensive in the style of the popular texts of Kreyszig or Wylie, nor is it directed to the same audience. Rather, the book is designed to be used in a 'second' course in applied mathematics where the emphasis may well be on depth instead of breadth. The topics presented have been deliberately selected for their relevance to *nonroutine* applications encountered in today's (and hopefully tomorrow's) world. For this reason, advanced topics such as stability theory, conformal mapping, generalized functions, integral equations, and asymptotics are included as are seemingly more elementary topics such as linear algebra, differential equations, and special functions.

The separate chapters stand alone for the most part, although there are some threads woven through the text, and later chapters tend to deal with more advanced material than earlier ones. This allows particularly eclectic instructors who are using this book as a text to adopt a modular approach, here and there replacing chapters with supplementary classroom material of their own devising. Many of the proofs of theorems that are included can also be skipped, without undue impact on continuity, if the course being taught is deemphasizing theoretical aspects of the material. Assignment of an appropriate number of the problems given at the end of each chapter, however, is a must. These have been carefully chosen to illustrate or amplify various portions of the text and constitute an extremely important component of the learning process. More difficult problems have been starred for ease in recognition.

Naturally, certain prerequisite knowledge is assumed on the part of the student. In general terms this includes a thorough grounding in the calculus through linear algebra and ordinary differential equations. A year-long post-calculus course in applied mathematics (perhaps using one of the substantial texts mentioned earlier) which incorporates exposure to partial differential equations, Fourier series, Laplace/Fourier transforms, vector analysis, and elementary complex variables is also taken for granted.

Acknowledgments

In an undertaking of this scope, there are numerous individuals who deserve recognition. Beresford Parlett, Richard Askey, Duane DeTemple, and Kenneth Shaw read and commented on selected chapters. Peter Duren, Raymond Roan, Charles Chui, Robert Muncaster, Gregory Kriegsmann, and Charles Oehring critiqued in depth substantial portions of the final manuscript. My wife Katherine helped with the indexing. Each has my grateful appreciation. Thanks also go to my several typists: Sylvia Davis, Vera Morgan, Gloria Hoover, Janet Chapman, and Pamela Wample. Without their consummate skill and assistance, especially that of Janet and Pam, the project would never have reached fruition.

James A. Cochran
Pullman, June 1982

Contents

Introduction

The first step toward understanding a given physical phenomenon through the use of mathematics involves formulation of the actual problem in mathematical terms, in other words, development of a "practical" mathematical model. This initial stage, which may involve the creative talents of more than one applied scientist, is a give-and-take affair between the complexities of nature, on the one hand, and the need for simplicity in the resulting mathematical formulation, on the other. It is also an evolutionary effort that may be redone a number of times. As physical knowledge increases, refined judgements on the importance of various underlying processes may suggest the inclusion or exclusion of certain components of the model. As mathematical knowledge expands, a more sophisticated model may be allowed or, contrarily, a simpler one dictated.

Following the formulation of the mathematical model comes the analysis of the associated mathematical problem(s). In this purely mathematical phase, questions of existence and uniqueness are posed and decided. Moreover, practical answers or data in one form or another are obtained. Sometimes this involves extensive numerical computation; on other occasions, delicate approximation techniques are needed. Only rarely, however, are the attendant mathematical problems solved completely and rigorously. More commonly, guided by physical intuition and mathematical know-how, the "applied problem solver" produces a rather varied collection of results, some qualitative and others quantitative. It is from these results that the utility of the mathematical model must be assessed.

Verifying the relevance of the mathematical model is an essential last step. In the ideal situation the output of the model is compared with known, or experimentally obtainable, physical data. In this way the loop is closed—the physical problem is both the beginning and the end. If the comparison is unfavorable, adjustments will have to be made. Indeed, in many instances a substantially revised formulation of the model may be necessary before specific scientific conclusions can be inferred with confidence. On the other hand, if the mathematical model is sound, then new physical insights and increased understanding of the phenomenon under investigation may come from a careful analysis of the model's properties.

It is appropriate to call this multistage process *applied mathematics* and those who labor in this fashion *applied mathematicians.** Viewed in this context, applied mathematics is the bridge that connects "pure" mathematics and science. The applied mathematician begins on the scientific side of the bridge with a physical problem of interest. Careful abstraction, idealization, and formulation of an associated mathematical model then permit him to cross to the other side. Buttressed by the extensive lore of the mathematical profession, he directs his effects toward the "solution" of the strictly mathematical problems to which his approach has led. Finally, equipped with new mathematical knowledge, the applied mathematician returns across the bridge to the scientific side, ready to compare results, draw conclusions, and (especially) communicate his findings.

To a great extent the chapters in this book have been fashioned in a manner reflecting this typical program of applied mathematical effort. Near the beginning of almost every chapter there is a problem (or problems) not unlike those currently being encountered in one or another branch of science or engineering. For the most part these are nontrivial (and sometimes rather sophisticated) physical applications. They are designed to present the flavor of what is common in today's real-life industrial, business, and research settings. During the consideration of each one of these problems, a mathematical model is developed and those specific mathematical principles and techniques which will be needed to "solve" the attendant mathematical problems are indicated. Within natural limits, enough detail regarding the interplay between physical laws and mathematical formulas is given so as to make the modelling process intelligible. We then move on to the principal portion of each chapter, discussing in some depth the pertinent body of applicable mathematics. It is here that due attention is given to the conditions under which various theoretical results or computational methods are applicable. Since formula proofs of the validity of most of these results and methods can be found in any number of other sources, they are generally omitted unless they serve a particular didactic purpose. Alternative approaches and procedures, however, are frequently compared and contrasted, and seemingly disparate scientific areas are mentioned in order to display their mathematical unity with the phenomenon under consideration. Each chapter closes with an analysis of results germane to the

---

*See Lin and Segel (1974), and other books and articles listed at the end of this section, for more detailed analyses of the scientific roles of applied mathematicians.

original scientific problem and a consideration of relevant conclusions. These summarizing sections generally include discussions of numerical data from computations and calculations typical of those which accompany most applied mathematical endeavors.

And now, on to Chapter 1. Indeed, *allez en avant, et la foi vous viendra* (d'Alembert).

References

Dettman, J.W. (1974): *Introduction to Linear Algebra and Differential Equations,* McGraw-Hill, New York; Section 5.2.

Hall, C.A. (1975): "Industrial Mathematics: A Course in Realism," *Amer. Math. Monthly, 82:* 651–659.

Klamkin, M.S. (1971): "On the Ideal Role of an Industrial Mathematician and Its Educational Implications," *Amer. Math. Monthly, 78:* 53–76.

Lin, C.C. and L.A. Segel (1974): *Mathematics Applied to Deterministic Problems in the Natural Sciences,* Macmillian, New York; especially Section 1.1 and Appendix 1.1.

1

Linear algebra and computation

1.1 An initial example from electrical circuits

Probably no other area of mathematics has been applied in such numerous and diverse contexts as the theory of matrices. In mechanics, electromagnetics, statistics, economics, operations research, the social sciences, and so on, the list of applications seems endless. By and large this is due to the utility of matrix structure and methodology in conceptualizing sometimes complicated relationships and in the orderly processing of otherwise tedious algebraic calculations and numerical manipulations. Let us illustrate this by considering a basic two-port electrical network as shown in Figure 1.1(a).

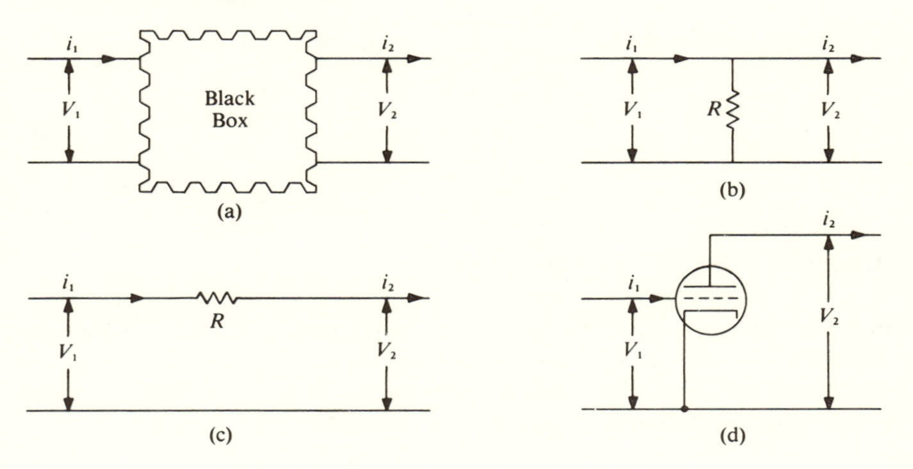

Figure 1.1 Two-port networks

If we assume that the "Black Box" contains only simple circuit elements such as resistors, capacitors, and vacuum tubes, the input and output (direct) currents and voltages can be related through the successive application of well-known fundamental physical principles. These key relationships are associated with the names of the German physicist Gustav Kirchhoff (1824–1887), who we will meet again in a later chapter, and his countryman Georg Ohm (1787–1854). Kirchhoff's laws state that the algebraic sums of both the currents flowing "into" any node as well as the voltage "drops" around any closed loop in an electrical circuit are zero. Ohm's law states that the voltage drop induced across a resistive element in the circuit is given by the product of the resistance of the element and the current flowing through that element. In the elementary cases of a shunt resistance, a series resistance, or a simple grounded-cathode vacuum tube with amplification μ and plate resistance r (see Figures 1.1b, c, d), these voltage and current laws of Ohm and Kirchhoff give rise to the following systems of equations written in matrix form:

$$\begin{bmatrix} v_1 \\ i_1 \end{bmatrix} = \begin{bmatrix} 1 & 0 \\ 1/R & 1 \end{bmatrix} \begin{bmatrix} v_2 \\ i_2 \end{bmatrix} \qquad \text{Shunt Resistance}$$

$$\begin{bmatrix} v_1 \\ i_1 \end{bmatrix} = \begin{bmatrix} 1 & R \\ 0 & 1 \end{bmatrix} \begin{bmatrix} v_2 \\ i_2 \end{bmatrix} \qquad \text{Series Resistance}$$

$$\begin{bmatrix} v_1 \\ i_1 \end{bmatrix} = \begin{bmatrix} -1/\mu & -r/\mu \\ 0 & 0 \end{bmatrix} \begin{bmatrix} v_2 \\ i_2 \end{bmatrix} \qquad \text{Vacuum Tube}$$

Employing these same empirical relations of Ohm and Kirchhoff, the more complicated two-port ladder network combination of shunt and series resistances exhibited in Figure 1.2 can be directly analyzed by considering the circuit as a

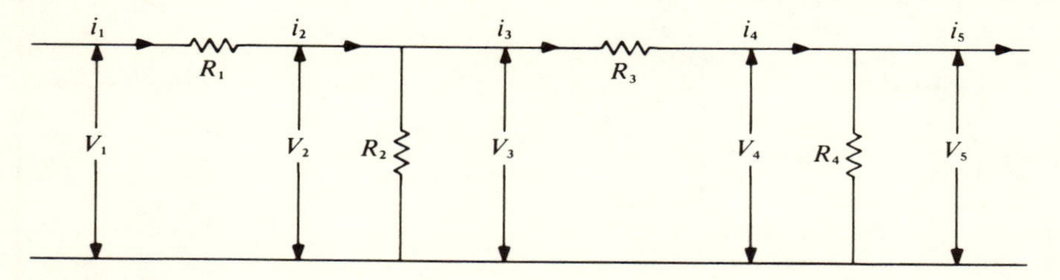

Figure 1.2 **Ladder network**

whole. Alternatively, since the output from each component stage of the network becomes the input for the next succeeding stage, the overall analysis may be streamlined considerably by using a matrix approach, which takes advantage of the *sequential* nature of the configuration. Indeed, if we let

$$\mathbf{x}_n \equiv \begin{bmatrix} v_n \\ i_n \end{bmatrix}, \quad A_n \equiv \begin{bmatrix} 1 & 0 \\ 1/R_n & 1 \end{bmatrix}, \quad B_n \equiv \begin{bmatrix} 1 & R_n \\ 0 & 1 \end{bmatrix}, \tag{1.1–1}$$

then in a systematic and natural manner we obtain

$$\mathbf{x}_1 = B_1 A_2 B_3 A_4 \mathbf{x}_5. \tag{1.1–2}$$

Specific numerical calculations with this compact expression do necessitate an understanding of matrix multiplication. On the other hand, the solution of a cumbersome set of equations in a possibly unorganized fashion has been avoided (see Problem 1 at the end of this chapter).

In the particular formulation just developed, the unique input associated with a given output may be readily determined. The inverse problem of finding $\mathbf{x}_5$ when $\mathbf{x}_1$ is given is only slightly more difficult, involving as it does merely the inversion of simple 2×2 matrices. Readers may be aware, though, that this is not the usual situation. The "inverse" (or reciprocal) problem is generally much more difficult than the "direct" problem, and sometimes in its analysis we find ourselves very quickly at the frontiers of our knowledge.*

1.2 Systems of linear algebraic equations

Our interest in this chapter is matrix theory, in general, and techniques for the numerical estimation of matrix eigenvalues, in particular. Before proceeding to a reasonably sophisticated practical example that will point us in the latter direction, however, it will be helpful to discuss in some detail the matrix inverse problem mentioned above. We do so here and in the next section since it is essential for those of us who are not trained specifically as numerical analysts to understand, at least in some broad sense, the implications that problem size and machine computation are going to have on our problem-solving efforts. As the size of the system increases, the task of finding the "solution" of a system of linear algebraic equations quickly becomes a far-from-elementary undertaking.

In order to fix our notation and provide a convenient survey of matrix fundamentals, we have gathered in Appendix 1 (at the end of the book) many of the basic definitions and properties that will be used in what follows. Some readers may find it helpful to review this material before continuing.

Although it is usually not considered in this fashion, the solution of the single equation $ax = b$, in which a and b are scalars and x is to be determined, actually devolves into three distinct cases:

(i) $a \neq 0$	Only one solution
(ii) $a = 0, b = 0$	Infinitely many solutions
(iii) $a = 0, b \neq 0$	No solution.

Using the notion of the *rank* $r(A)$ of a given matrix A (see Appendix 1), precisely analogous statements can be made concerning the solvability of the matrix equation $A\mathbf{x} = \mathbf{b}$. The following definition is helpful in formulating this result:

Definition 1 Given the arbitrary $m \times n$ coefficient matrix A, the **augmented matrix** $[A{:}b]$ is the $m \times (n+1)$ matrix formed from A by appending the column vector $\mathbf{b}$ as a new last column.

Then

Theorem:

Let A be the $m \times n$ coefficient matrix of the system of linear algebraic equations $A\mathbf{x} = \mathbf{b}$. The solvability of the matrix equation $A\mathbf{x} = \mathbf{b}$ is determined by the comparative values of $r(A)$ and of $r([A{:}b])$, namely

*For an interesting discussion of one class of inverse problems see Kac (1966); for an overview of a number of specific inverse problems arising in physics see Keller (1976).

(i) $r(A) = r([A{:}b]) = n$ Only one solution
(ii) $r(A) = r([A{:}b]) < n$ Infinitely many solutions
(iii) $r(A) < r([A{:}b])$ No solution. □

This classification scheme can be verified constructively by applying the so-called *elementary row operations* to $[A{:}b]$, namely

(I) interchange two rows,
(II) multiply every element in a row by a given nonzero scalar,
(III) replace the ith row by the sum of the ith row and j^{th} row (see, for example, Dettman (1974), pp. 50ff.).

From a theoretical point of view, it is worth noting that each of these operations can be accomplished on a given matrix by left-multiplication with an appropriate *elementary matrix* obtained from the identity I by the corresponding operation (see Problem 7). As examples we have

$$\begin{bmatrix} 3 & 4 \\ 1 & 2 \end{bmatrix} = \begin{bmatrix} 0 & 1 \\ 1 & 0 \end{bmatrix} \begin{bmatrix} 1 & 2 \\ 3 & 4 \end{bmatrix},$$

$$\begin{bmatrix} 2 & 4 \\ 3 & 4 \end{bmatrix} = \begin{bmatrix} 2 & 0 \\ 0 & 1 \end{bmatrix} \begin{bmatrix} 1 & 2 \\ 3 & 4 \end{bmatrix}, \quad \begin{bmatrix} 4 & 6 \\ 3 & 4 \end{bmatrix} = \begin{bmatrix} 1 & 1 \\ 0 & 1 \end{bmatrix} \begin{bmatrix} 1 & 2 \\ 3 & 4 \end{bmatrix}.$$

More importantly, however, these elementary row operations do not change the rank of the matrix (Dettman (1974), pp. 79f.). Moreover, a finite sequence of the operations can be put together systematically so as to reduce a given set of equations to a simplified form from which the solution(s) can be easily determined.

The most popular and efficient such method of reducing given systems of equations to simple equivalent systems is commonly termed *Gaussian elimination*. Readers should be familiar with the process, as applied below.

Example Consider the set of equations

$$\begin{aligned} x_1 + 2x_3 - x_4 &= 3 \\ x_2 + x_3 &= 5 \\ 3x_1 + 2x_2 - 2x_4 &= -1 \\ -x_3 + 4x_4 &= 13 \\ 2x_1 - x_3 + 3x_4 &= 11, \end{aligned}$$

which has the easily verified solution $x_1 = 1$, $x_2 = 2$, $x_3 = 3$, and $x_4 = 4$. Using the symbol $\sim$ to designate matrix equivalence and carrying out the obvious elementary row operations, we find

$$[A{:}b] \equiv \begin{bmatrix} 1 & 0 & 2 & -1 & 3 \\ 0 & 1 & 1 & 0 & 5 \\ 3 & 2 & 0 & -2 & -1 \\ 0 & 0 & -1 & 4 & 13 \\ 2 & 0 & -1 & 3 & 11 \end{bmatrix} \sim \begin{bmatrix} 1 & 0 & 2 & -1 & 3 \\ 0 & 1 & 1 & 0 & 5 \\ 0 & 2 & -6 & 1 & -10 \\ 0 & 0 & -1 & 4 & 13 \\ 0 & 0 & -5 & 5 & 5 \end{bmatrix}$$

$$\sim \begin{bmatrix} 1 & 0 & 2 & -1 & 3 \\ 0 & 1 & 1 & 0 & 5 \\ 0 & 0 & -8 & 1 & -20 \\ 0 & 0 & -1 & 4 & 13 \\ 0 & 0 & -5 & 5 & 5 \end{bmatrix} \sim \begin{bmatrix} 1 & 0 & 2 & -1 & 3 \\ 0 & 1 & 1 & 0 & 5 \\ 0 & 0 & -8 & 1 & -20 \\ 0 & 0 & 0 & 31/8 & 31/2 \\ 0 & 0 & 0 & 35/8 & 35/2 \end{bmatrix}$$

$$\sim \begin{bmatrix} 1 & 0 & 2 & -1 & 3 \\ 0 & 1 & 1 & 0 & 5 \\ 0 & 0 & -8 & 1 & -20 \\ 0 & 0 & 0 & 31/8 & 31/2 \\ 0 & 0 & 0 & 0 & 0 \end{bmatrix}$$

Gaussian elimination has reduced the original augmented coefficient matrix $[A:b]$ to the equivalent augmented matrix $[\tilde{A}:\tilde{b}]$. From the latter the successive unknowns are found by back-substitution, that is, determination of x_4 from the 4th equation, then x_3 from the 3rd equation, and so on. □

The results encountered in this specialized example suggest what happens in the more general setting. If $\tilde{b}_i \neq 0$ for some $i > r(\tilde{A})$, the original system of equations is inconsistent and there cannot be a solution. Note especially that in this case $r(A) = r(\tilde{A}) < r([\tilde{A}:\tilde{b}]) = r([A:b])$. On the other hand, if $r(A) = r([A:b])$ (denoted by r), then the nature of the solution set depends upon whether or not $r = n$. If $r = n$, the solution is unique; if $r < n$, then $x_1, x_2, \ldots, x_r$ can be written in terms of $x_{r+1}, x_{r+2}, \ldots, x_n$, so that there is obviously an $(n - r)$-fold infinity of solutions. Taken in its totality, this is essentially the state of affairs guaranteed by the "classification" theorem above.

1.3 Error management

Pivot selection and scaling

The leading nonzero row elements 1, 1, −8 and 31/8 encountered in the application of Gaussian elimination in the example of Section 1.2 are called *pivots*. These are the matrix elements (coefficients of unknowns) used to reduce elements in other rows (equations) to zero. Clearly a different set of pivots would have been obtained if the elimination process had been carried out in different order. Different pivots would have also resulted if the equations had been *prescaled* prior to beginning Gaussian elimination. In practice, the questions of whether or not to scale and which pivot selection procedure should be used to solve $A\mathbf{x} = \mathbf{b}$ are of considerable importance, especially if the system of equations is particularly large and the accuracy of the desired solution is of some concern. A good (although not foolproof) "rule of thumb" is the following:

(i) Scale the set of equations so that the largest elements in each row and column of A are of the same order (say unity for convenience);

(ii) Use *partial pivoting* (sometimes called *partial positioning for size*), eliminating the unknowns in their natural order $x_1, x_2, \ldots, x_n$, but at the kth stage ($k = 1, 2, \ldots, n-1$) interchanging rows first so that the coefficient of x_k of largest modulus in the remaining $(n - k + 1)$ equations becomes the pivot.

Applying Gaussian elimination with partial pivoting to the example of Section 1.2 would have led to the following sequence of equivalent matrices:

$$[A:b] \sim \begin{bmatrix} 3 & 2 & 0 & -2 & -1 \\ 0 & 1 & 1 & 0 & 5 \\ 1 & 0 & 2 & -1 & 3 \\ 0 & 0 & -1 & 4 & 13 \\ 2 & 0 & -1 & 3 & 11 \end{bmatrix} \sim \begin{bmatrix} 3 & 2 & 0 & -2 & -1 \\ 0 & 1 & 1 & 0 & 5 \\ 0 & -2/3 & 2 & -1/3 & 10/3 \\ 0 & 0 & -1 & 4 & 13 \\ 0 & -4/3 & -1 & 13/3 & 35/3 \end{bmatrix}$$

$$\sim \begin{bmatrix} 3 & 2 & 0 & -2 & -1 \\ 0 & -4/3 & -1 & 13/3 & 35/3 \\ 0 & -2/3 & 2 & -1/3 & 10/3 \\ 0 & 0 & -1 & 4 & 13 \\ 0 & 1 & 1 & 0 & 5 \end{bmatrix} \sim \begin{bmatrix} 3 & 2 & 0 & -2 & -1 \\ 0 & -4/3 & -1 & 13/3 & 35/3 \\ 0 & 0 & 5/2 & -5/2 & -5/2 \\ 0 & 0 & -1 & 4 & 13 \\ 0 & 0 & 1/4 & 13/4 & 55/4 \end{bmatrix}$$

$$\sim \begin{bmatrix} 3 & 2 & 0 & -2 & -1 \\ 0 & -4/3 & -1 & 13/3 & 35/3 \\ 0 & 0 & 5/2 & -5/2 & -5/2 \\ 0 & 0 & 0 & 3 & 12 \\ 0 & 0 & 0 & 7/2 & 14 \end{bmatrix} \sim \begin{bmatrix} 3 & 2 & 0 & -2 & -1 \\ 0 & -4/3 & -1 & 13/3 & 35/3 \\ 0 & 0 & 5/2 & -5/2 & -5/2 \\ 0 & 0 & 0 & 7/2 & 14 \\ 0 & 0 & 0 & 0 & 0 \end{bmatrix}.$$

Having shown the efficacy of pivoting strategies, it would be nice now to do the same for scaling procedures. As readers may be aware, however, for some matrices scaling has absolutely no effect on the round-off errors that accompany the process of Gaussian elimination performed in fixed-digit arithmetic. In other cases scaling may be necessary to avoid under/overflow in the calculation, but when introduced it may mask the actual sensitivity of the model to inaccuracies in the given input data. In short, scaling questions can be difficult and vexing. (See Noble and Daniel (1977), Steinberg (1974), or Forsythe and Moler (1967) for more complete discussions of this important topic.)

Iterative improvement

Numerical analysts often talk about *iteration on the residual* as a means of improving the accuracy of computed solutions to systems of linear algebraic equations. The technique is designed for those cases in which the inaccuracy of a proposed solution $\mathbf{x}_c$ of $A\mathbf{x} = \mathbf{b}$, as measured by the magnitude $|\mathbf{r}|$* of the residual vector

$$\mathbf{r} \equiv \mathbf{b} - A\mathbf{x}_c, \tag{1.3–1}$$

---

* $|\mathbf{r}|^2 \equiv \mathbf{r}^* \mathbf{r} = \sum_{i=1}^{n} |r_i|^2$

is unacceptably large. Sometimes the following simple iterative algorithm can be used to obtain progressively better approximations to the "true" solution $\mathbf{x}_t$ of the system.

Let

$$\mathbf{e} \equiv \mathbf{x}_t - \mathbf{x}_c. \tag{1.3–2}$$

Then

$$\begin{aligned} A\mathbf{e} &= A\mathbf{x}_t - A\mathbf{x}_c \\ &= \mathbf{b} - A\mathbf{x}_c \\ &= \mathbf{r}. \end{aligned} \tag{1.3–3}$$

The solution of the new system (1.3–3) will yield an estimate $\mathbf{e}_c$ of the true error $\mathbf{e}$ from which we can obtain an improved estimate of $\mathbf{x}_t$, namely

$$\tilde{\mathbf{x}}_c = \mathbf{x}_c + \mathbf{e}_c. \tag{1.3–4}$$

If $|\tilde{\mathbf{r}}| \equiv |\mathbf{b} - A\tilde{\mathbf{x}}_c|$ is still too large, the steps (1.3–2)–(1.3–4) can be repeated using $\tilde{\mathbf{x}}_c$ in place of $\mathbf{x}_c$ throughout. In fact, under favorable circumstances the procedure can be iterated until the values of $\mathbf{e}_c$ become comparable in size to the "noise level" of the calculation.

A few comments are in order about this method of iterative improvement:

(i) Each of the systems of equations we are interested in solving has the same coefficient matrix A. Therefore, if the multipliers* used at each step of the Gaussian elimination process are preserved, the complete procedure will not need to be repeated for each new right-hand side. For a square matrix A of order n the resultant saving is of the order of $(\frac{1}{3}n^3 - n^2)$ multiplication operations (Johnson and Riess (1977), pp. 26ff.).

(ii) On the other hand, without some elementary precautions, no improvement whatsoever is likely to result from applying the technique. This is due to the fact that although the residual $\mathbf{r}$ may not be so small that $\mathbf{x}_c$ constitutes an acceptable solution, it is nevertheless typically small enough so that computing it from (1.3–1) using *single-precision arithmetic* leads to a very high relative error. This high relative error in $\mathbf{r}$ then manifests itself in a meaningless $\mathbf{e}_c$. The way around this difficulty, of course, is to use higher-precision arithmetic, at least in the calculation of $\mathbf{r}$. Users must then evaluate whether this increased computing time (cost) is justified.

(iii) As we shall see below, the size of $|\mathbf{r}|$ is not always an accurate a priori indicator of the error $\mathbf{e}$. What is important is the size of $\mathbf{e}_c$ relative to $\mathbf{x}_c$.

---

*Remember that elementary row operations can be accomplished using matrix multiplication (see Problem 7).

Ill-conditioning

In the case of a nonsingular square matrix A, equations (1.3–2) and (1.3–3) above can be rewritten as

$$\mathbf{x}_t = \mathbf{x}_c + A^{-1}\mathbf{r}.$$

If the residual $\mathbf{r}$ is small and the largest element in A^{-1} is of modest size, then we would expect the error between the true and computed solutions of $A\mathbf{x} = \mathbf{b}$ to be small also. On the other hand, if A has been scaled (normalized) to order unity and yet A^{-1} contains some very large elements, $\mathbf{x}_c$ can be far from $\mathbf{x}_t$ even though $\mathbf{r}$ is small.

This lack of correlation between small residuals and small errors in the computed solution is a sign of *ill-conditioning*. From a practical point of view, it means that the effect of round-off errors in the computation process as well as any original (empirical) errors in the coefficients of A and $\mathbf{b}$ may be multiplied many-fold by the time the 'solution' is reached. For example, compare the following three problems and their exact solutions:

$$\text{(a)} \quad \begin{bmatrix} 2 & 4 \\ 2 & 4.0001 \end{bmatrix} \begin{bmatrix} x_1 \\ x_2 \end{bmatrix} = \begin{bmatrix} 6 \\ 6.0001 \end{bmatrix}, \quad \begin{bmatrix} x_1 \\ x_2 \end{bmatrix} = \begin{bmatrix} 1 \\ 1 \end{bmatrix};$$

$$\text{(b)} \quad \begin{bmatrix} 2 & 4 \\ 2 & 4.0001 \end{bmatrix} \begin{bmatrix} x_1 \\ x_2 \end{bmatrix} = \begin{bmatrix} 6 \\ 6 \end{bmatrix}, \quad \begin{bmatrix} x_1 \\ x_2 \end{bmatrix} = \begin{bmatrix} 3 \\ 0 \end{bmatrix};$$

$$\text{(c)} \quad \begin{bmatrix} 2 & 4 \\ 2 & 3.9999 \end{bmatrix} \begin{bmatrix} x_1 \\ x_2 \end{bmatrix} = \begin{bmatrix} 6 \\ 6.0001 \end{bmatrix}, \quad \begin{bmatrix} x_1 \\ x_2 \end{bmatrix} = \begin{bmatrix} 5 \\ -1 \end{bmatrix}.$$

The data varies from one case to the next by at most 10^{-4}; the solutions, however, are completely different.

The reason for this disconcerting state of affairs is that the coefficient matrices in each of these examples is *nearly* singular. In other words, in each case $\det A$ is much smaller than the individual elements of A. If we recall that for a square coefficient matrix A of order n,

$$\det A = \prod_{i=1}^{n} p_i$$

where p_i is the ith pivot, the appearance of small pivots in the Gaussian elimination process (especially when coupled with pivots of more reasonable size) is at least a warning sign that the set of equations we are solving may be ill-conditioned.* The test is not foolproof, however, as Problem 10 at the end of this chapter demonstrates.

Final comments

The earlier remarks of this section should have suggested to the reader that the solution of linear systems of algebraic equations, although perhaps discussed in

---

*If A is actually singular, some of the small pivots that occur in the Gaussian elimination process may have the values they have (rather than zero) due to numerical round-off errors.

high school algebra or collegiate calculus courses as fairly straightforward undertakings, can be quite the opposite if computational techniques are needed, as they are when the number of unknowns and/or equations is large. Possible ill-conditioning in the set of equations needs to be recognized and, if feasible, circumvented by reformulating the original problem (see the pertinent and practical discussion of Noble and Daniel (1977) in this regard). In view of the ever-present round-off error, however, even when the equations are well-conditioned, the calculation of solutions as accurately as the input data warrants must be done carefully and efficiently.* As a practical matter then, many an applied scientist prefers to make use of the linear algebraic equation solver package available in his local computing center (see Dongarra et al (1979) for a discussion of the new and very efficient LINPACK subroutines). In this way the applied scientist is free from the worry about pivots, row interchanges, and the like, while bringing the efforts of knowledgeable numerical analysts and computer scientists to bear on the problem.

1.4 Mechanical vibrations and resonance

A much more complicated problem in matrix theory and one of substantial practical importance is the well-known matrix (or *algebraic) eigenvalue problem*. Here the task is to find scalars λ and nonzero (column) vectors $\mathbf{x}$ (termed *eigenvalues* and *eigenvectors*, respectively) such that

$$A\mathbf{x} = \lambda\mathbf{x} \tag{1.4–1}$$

for a given square matrix A. Interestingly, nature is replete with examples wherein the modelling process leads to an equation of the form (1.4–1). To illustrate this fact, we consider one particularly instructive situation involving the torsional vibrations of a system of elastically coupled disks.

Admittedly, as is true with most of the other practical problems discussed in the book, this is a distinctly nontrivial example. We will try to give enough background, however, so that those readers initially unfamiliar with this particular area of physical applications will eventually become at least moderately comfortable with the mathematical modelling that has been carried out. (For further information on the physics involved see any undergraduate text in mechanics; additional discussion of this application is also to be found in Wylie (1975), pp. 629 ff. and in Rasof (1973), pp. 151 ff.)

Let a number of individually uniform (circular) disks be mounted on a shaft, which is rigidly attached to a wall as depicted in Figure 1.3. We assume that each disk is firmly affixed to the shaft, that the shaft (although perhaps not of identical construction throughout) is uniform between any two disks, and that the only allowable motion is in an angular direction measured by angular displacement

*For matrices that are especially large ($n > 100$ say) or sparse (most elements zero), iterative procedures such as Gauss-Seidel may offer some computational advantages over direct methods. Interested readers may wish to consult references such as Ralston and Rabinowitz (1978) or Forsythe and Moler (1967) for discussions of these techniques.

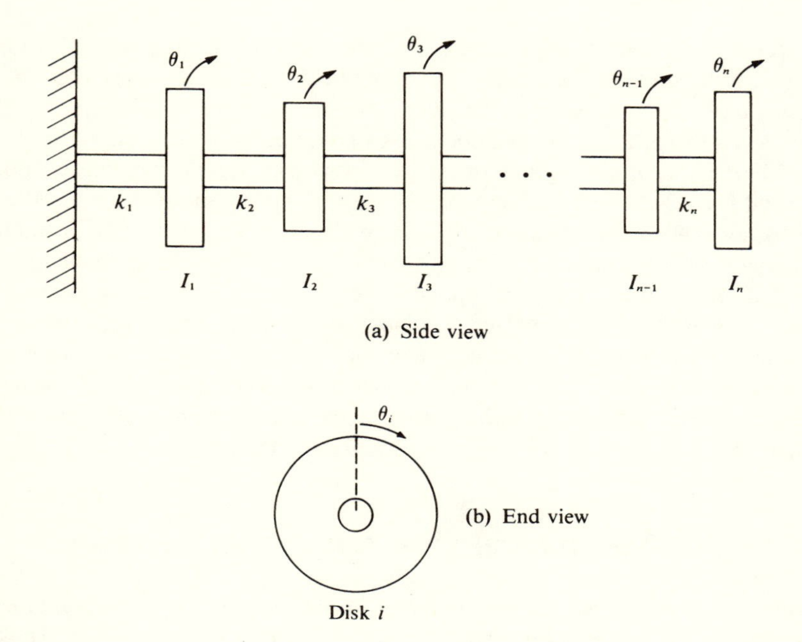

Figure 1.3 **A system of elastically coupled disks**

θ from some zero reference angle. We are interested in analyzing the rotational motions or vibrations that the various disks can undergo, as functions of time, under the influence of the inertias of the individual disks and the torsion (or twisting) of the separate portions of the shaft.

As in most problems of classical mechanics, expressions for the kinetic energy T and potential energy V of the system under investigation are central to our understanding of the physical situation. Akin to the familiar formula $T = \frac{1}{2} m\dot{x}^2$ for the kinetic energy associated with a single (point) mass m moving rectilinearly with velocity $\dot{x} \equiv dx/dt$, we have the relationship $T = \frac{1}{2} I\dot{\theta}^2$ for a single disk rotating with angular velocity $\dot{\theta} \equiv d\theta/dt$. In this expression I is the so-called "polar" moment of inertia: the moment of inertia of the disk about an infinitesimal perpendicular shaft through its geometrical center. In comparable fashion, an appropriate relationship for the potential energy of a single disk situated on a twisted shaft would be $V = \frac{1}{2} k\theta^2$; where k is the rotational flexibility or modulus (ratio of applied moment to resulting angular rotation) of the shaft. This formula is the precise analogue of the expression $V = \frac{1}{2} kx^2$; known as Hooke's law, for the potential energy associated with a spring of stiffness (modulus) k compressed a distance x within its elastic regime, and is valid in the case of small rotations.

If we assume that the shaft contributes to the system dynamics only through its torsion, the kinetic energy of the entire disk/shaft configuration can be obtained by adding up the individual disk energies. Thus we find

$$T = \frac{1}{2} \sum_{i=1}^{n} I_i \, \dot{\theta}_i^2, \qquad \textbf{(1.4–2)}$$

where I_i and $\theta_i(t)$ are the polar moment of inertia and the angular displacement of the ith disk, respectively. Similarly, for the potential energy we have

$$V = \frac{1}{2} k_1 \theta_1{}^2 + \frac{1}{2} \sum_{i=2}^{n} k_i (\theta_i - \theta_{i-1})^2. \qquad \textbf{(1.4–3)}$$

Here k_i is the rotational modulus of that portion of the shaft between the $(i - 1)$st and ith disks, and we have used the fact that it is only the *net* angle of rotation between the disks that contributes to their vibration.

With expressions for the kinetic and potential energies now in hand, it is a simple matter to derive the equations of motion for the disk/shaft system we are analyzing. We do this using an approach first suggested by the French applied mathematician Joseph Lagrange (1736–1813). As we shall see in Chapter 4, as far as the mathematics is concerned, the key equations of Lagrange are nothing more than a direct result of the calculus of variations applied to a fundamental principle of mechanics.* For our present purposes, it suffices to note that, in the absence of either external forces or (internal) damping, our dynamical system is what is called conservative, and Lagrange's equations of motion for the resulting vibrations have the form

$$\frac{d}{dt}\left(\frac{\partial L}{\partial \dot{\theta}_i}\right) - \frac{\partial L}{\partial \theta_i} = 0 \qquad i = 1, 2, \cdots, n.$$

In these equations, $L \equiv T - V$ is the so-called *Lagrangian function* and partial derivatives have been written with ∂ instead of d.

Substituting (1.4–2), (1.4–3) then gives

$$I_i \ddot{\theta}_i + k_i(\theta_i - \theta_{i-1}) - k_{i+1}(\theta_{i+1} - \theta_i) = 0 \qquad i = 1, 2, \cdots, n,$$

---

*From the point of view of classical dynamics, Lagrange's equations are based upon energy conservation and are an alternate (but often more convenient) method of describing the motion of a given mechanical system. For rectilinear motion of a single (point) mass m constrained by a spring with stiffness (modulus) k following Hooke's law we would have

$$T = \tfrac{1}{2} m\dot{x}^2, \qquad V = \tfrac{1}{2} kx^2, \qquad L = \tfrac{1}{2} m\dot{x}^2 - \tfrac{1}{2} kx^2$$

and

$$\begin{aligned} 0 &= \frac{d}{dt}\left(\frac{\partial L}{\partial \dot{x}}\right) - \frac{\partial L}{\partial x} \\ &= m\ddot{x} + kx. \end{aligned}$$

We recognize this last expression as the familiar form of Newton's Second Law of Motion for this problem.

where, since they have no meaning, we take $\theta_0 = 0$ and $k_{n+1} = 0$. In matrix form these coupled equations look like

$$\begin{bmatrix} I_1\ddot{\theta}_1 \\ I_2\ddot{\theta}_2 \\ \cdot \\ \cdot \\ \cdot \\ I_n\ddot{\theta}_n \end{bmatrix} + \begin{bmatrix} k_1+k_2 & -k_2 & 0 & \cdots & 0 \\ -k_2 & k_2+k_3 & -k_3 & \cdots & 0 \\ \cdot & \cdot & \cdot & & \cdot \\ \cdot & \cdot & \cdot & & \cdot \\ \cdot & \cdot & \cdot & & \cdot \\ 0 & 0 & 0 & \cdots & k_n \end{bmatrix} \begin{bmatrix} \theta_1 \\ \theta_2 \\ \cdot \\ \cdot \\ \cdot \\ \theta_n \end{bmatrix} = \mathbf{0},$$

and constitute a linear system of differential equations for the unknown angular displacements $\theta_i(t)$ with a real symmetric coefficient matrix.

The special case when the disks are identical and equally spaced, and the shaft is uniform throughout, is of particular interest. For this configuration, $I_1 = I_2 = \ldots = I_n \equiv I$, $k_1 = k_2 = \ldots = k_n \equiv k$, and the above set of linear equations simplifies substantially. If we then ask whether the uniform system supports any pure resonant vibrations, that is, whether there are any *real* (natural) frequencies ω for which solutions of the form

$$\theta_i(t) = x_i \sin \omega t \qquad i = 1, 2, \cdots, n$$

exist, then we are led, after rearranging terms, to the set of equations

$$\begin{bmatrix} 2 & -1 & 0 & \cdot & \cdot & \cdot & 0 \\ -1 & 2 & -1 & \cdot & \cdot & \cdot & 0 \\ 0 & -1 & 2 & \cdot & \cdot & \cdot & 0 \\ \cdot & \cdot & \cdot & & & \cdot & \cdot \\ \cdot & \cdot & \cdot & & & 2 & -1 \\ 0 & 0 & 0 & \cdot & \cdot & -1 & 1 \end{bmatrix} \begin{bmatrix} x_1 \\ x_2 \\ x_3 \\ \cdot \\ \cdot \\ x_n \end{bmatrix} = \frac{I\omega^2}{k} \begin{bmatrix} x_1 \\ x_2 \\ x_3 \\ \cdot \\ \cdot \\ x_n \end{bmatrix}. \qquad \textbf{(1.4–4)}$$

In this expression $\mathbf{x}$ and $\lambda \equiv I\omega^2/k$ are unknowns, to be determined. A search for resonant solutions (normal modes) of this particular mechanical system wherein all the disks oscillate with the same frequency, therefore, brings us to a matrix problem of the promised form (1.4–1). To complete our analysis we need to solve the algebraic eigenvalue problem for this special matrix A.

For later reference we note that when (1.4–4) is written in the form $A\mathbf{x} = \lambda\mathbf{x}$, the matrix A turns out to symmetric. Indeed, since

$$\mathbf{x}^*A\mathbf{x} = |x_1|^2 + \sum_{i=2}^{n} |x_{i-1} - x_i|^2,$$

which is strictly positive unless $\mathbf{x} \equiv \mathbf{0}$, A belongs to that special subclass of symmetric (hermitian) matrices, the *positive definite* matrices. This ensures that the eigenvalues $\lambda = I\omega^2/k$ of A are positive, and hence the natural frequencies ω of this uniform configuration are indeed real. □

1.5 Matrix eigenvalues/eigenvectors and their calculation

Our concern now is with determining the eigenvalues and eigenvectors of the matrix A in equation (1.4–4). However, we want to place our discussion in a general context that will transcend the rather special character of our particular matrix. Thus, we will sample both theoretical results and computational techniques, intermingling practical remarks here and there with the numerical output obtained from specific calculations. Readers who desire greater depth on any given topic should consult some of the texts listed at the end of the chapter, particularly the monumental work by Wilkinson (1965) or the recent effort of Parlett (1980).

Matrix theoretic results

We recall that, given a square matrix A of order n, its eigenvalues $\lambda_1, \lambda_2, \ldots, \lambda_n$ are the complex numbers that are the solutions of the *characteristic equation*

$$\det(A - \lambda I) = 0. \qquad \textbf{(1.5–1)}$$

Some of these complex numbers may be equal to one another. Associated with each distinct eigenvalue λ_i, however, there is at least one eigenvector satisfying $A\mathbf{x} = \lambda_i\mathbf{x}$. In the case of either hermitian or real symmetric matrices, moreover, there are a number of other theoretical results concerning eigenvalues and eigenvectors that can help to guide our investigation. For example:

Property 1
If A is hermitian, then its eigenvalues are real.

Proof: Let $A\mathbf{x} = \lambda\mathbf{x}$ with $|\mathbf{x}| \neq 0$. Then

$$\begin{aligned} \lambda\mathbf{x}^*\mathbf{x} &= \mathbf{x}^*(\lambda\mathbf{x}) = \mathbf{x}^*A\mathbf{x} \\ &= \mathbf{x}^*A^*\mathbf{x} = (A\mathbf{x})^*\mathbf{x} = \bar{\lambda}\mathbf{x}^*\mathbf{x}. \end{aligned}$$ □

Property 2
If A is positive definite, then its eigenvalues λ are all positive.

Proof: $\mathbf{x}^*A\mathbf{x}$ positive in the proof of Property 1 implies that $\lambda > 0$. □

Property 3
If A is real symmetric, then its eigenvectors may be taken to be real.

Definition 1 A column vector $\mathbf{x}$ is said to be **normalized** (or **of unit length**) if $|\mathbf{x}|^2 \equiv \mathbf{x}^*\mathbf{x} = 1$. Two column vectors $\mathbf{x}$ and $\mathbf{y}$ with the same number of elements are termed **orthogonal** if $\mathbf{x}^*\mathbf{y} = 0$. More generally, the set of vectors $\mathbf{x}_1, \mathbf{x}_2, \ldots, \mathbf{x}_n$ is designated an **orthogonal** set if $\mathbf{x}_i$ and $\mathbf{x}_j$ are orthogonal for $i \neq j$, and an **orthonormal** set if also each $\mathbf{x}_i$ is normalized.

Property 4

If A is hermitian, then its eigenvectors may be taken to be orthonormal.

Proof: Assume that $A\mathbf{x} = \lambda\mathbf{x}$, $A\mathbf{y} = \mu\mathbf{y}$ with both $|\mathbf{x}| \neq 0$, $|\mathbf{y}| \neq 0$ and λ, μ real. Then

$$\begin{aligned}\mu\mathbf{x}^*\mathbf{y} &= \mathbf{x}^*A\mathbf{y} = \mathbf{x}^*A^*\mathbf{y} \\ &= (\mathbf{y}^*A\mathbf{x})^* = \lambda(\mathbf{y}^*\mathbf{x})^* \\ &= \lambda\mathbf{x}^*\mathbf{y}.\end{aligned}$$

If $\lambda \neq \mu$, the orthogonality follows immediately. The normalization is then attained, as needed, by scaling. If $\lambda = \mu$, then we may assume that $\mathbf{x}$ and $\mathbf{y}$ are distinct (linearly independent) and thus they can be rendered orthonormal using the well-known Gram-Schmidt procedure. □

Property 4 ensures that the distinct eigenvectors of hermitian matrices can be transformed into an orthonormal set. The corollaries of the following result confirm that a hermitian matrix of order n has a full set of (that is, precisely n) distinct eigenvectors. The result itself is included since it shows how the more general case must be approached. This theorem will also be useful in its own right in our subsequent analysis.

Schur's Theorem:

Given an arbitrary square matrix A, there exists a unitary matrix (transformation) Q such that

$$Q^*AQ = T,$$

where T is upper triangular.

Proof: We proceed by induction on the order n of the matrix A. If $n = 1$, the result is obvious since a matrix of order 1 is automatically triangular. Assuming then that A has order n and the desired result is valid for all matrices of order $n - 1$, we let $\mathbf{x}$ be a normalized eigenvector of A corresponding to some eigenvalue λ. Form the orthogonal set $\{\mathbf{x}, \mathbf{y}_1, \mathbf{y}_2, \ldots, \mathbf{y}_{n-1}\}$, that is, complete $\mathbf{x}$ to an orthogonal basis for n-space, and let $\tilde{Q}$ be the (unitary) matrix whose columns are these vectors. (Recall that a unitary matrix Q is such that $Q^*Q = I = QQ^*$.) It follows that

$$\tilde{A} \equiv \tilde{Q}^* A\tilde{Q}$$

$$= \begin{bmatrix} \mathbf{x}^* \\ \mathbf{y}_1^* \\ \cdot \\ \cdot \\ \cdot \\ \mathbf{y}_{n-1}^* \end{bmatrix} A\,[\mathbf{x}, \mathbf{y}_1, \ldots, \mathbf{y}_{n-1}]$$

$$= \begin{bmatrix} \lambda & \mathbf{u}^* \\ \mathbf{0} & B \end{bmatrix},$$

where $\mathbf{u}$ is a column vector with $(n-1)$ elements and B is a matrix of order $n-1$.

By virtue of the induction hypothesis, there exists a unitary matrix P of order $n-1$ such that $P^*BP = S$ where S is upper triangular. As a consequence, if we define Q by the matrix product

$$Q \equiv \tilde{Q} \begin{bmatrix} 1 & \mathbf{0}^* \\ \mathbf{0} & P \end{bmatrix},$$

we find

$$Q^*AQ = \begin{bmatrix} 1 & \mathbf{0}^* \\ \mathbf{0} & P^* \end{bmatrix} \tilde{A} \begin{bmatrix} 1 & \mathbf{0}^* \\ \mathbf{0} & P \end{bmatrix}$$

$$= \begin{bmatrix} \lambda & \mathbf{u}^*P \\ \mathbf{0} & S \end{bmatrix}.$$

Since Q is unitary and this last matrix is upper triangular, the proof is complete. □

Corollary 1:
If A is hermitian, then there exists a unitary matrix Q such that $Q^*AQ = D$ where D is a diagonal matrix with the eigenvalues of A as its diagonal elements.

Corollary 2:
If A is hermitian, then $AQ = QD$ and the columns of Q constitute a full orthonormal set of eigenvectors of A.

Power iteration

As stated earlier, the eigenvalues of a given square matrix are the roots of the characteristic equation (1.5–1). Since direct calculation of the coefficients in this equation could involve on the order of $n!$ multiplications, where n is the order of the matrix, for large matrices it is highly impractical to try to determine these polynomials directly, to say nothing of finding their roots (essentially an ill-posed problem). An indirect method such as Krylov's (see Ralston and Rabinowitz (1978) or Johnson and Riess (1977), for example) can be used to reduce the number of operations to the order of n^3, but even this approach is generally inferior to an iterative procedure.

Over the years one of the more widely used iterative techniques for eigenvalue estimation has been the *Power Method*. Although the method is applicable whenever the matrix of interest has a full set of linearly independent eigenvectors (and sometimes even more generally), we describe the technique for the case of hermitian matrices with a *dominant* maximum eigenvalue.

Let λ_i, $\mathbf{x}_i$ $(i = 1, 2, \ldots, n)$ be the eigenvalues and eigenvectors, respectively, of a given hermitian matrix A of order n. Assume the eigenvalues of A are such that one of them (call it λ_1) is strictly greater in magnitude than all the others. Choose an initial column vector $\mathbf{v}_0$ and assume it has a component in the direction of the (unknown) eigenvector $\mathbf{x}_1$, that is,

$$\mathbf{v}_0 = \sum_{i=1}^{n} \alpha_i \mathbf{x}_i \qquad \text{with } \alpha_1 \neq 0.$$

(We will see in a moment what happens, in practice, if $\mathbf{v}_0$ is orthogonal to $\mathbf{x}_1$.) Finally define

$$\mathbf{v}_k \equiv A^k \mathbf{v}_0 \qquad k = 1, 2, \cdots,$$

and note that, in view of the nature of the $\mathbf{x}_i$, $\mathbf{v}_k$ may be alternatively expressed as

$$\begin{aligned} \mathbf{v}_k &= \sum_{i=1}^{n} \alpha_i \lambda_i^k \mathbf{x}_i \\ &= \lambda_1^k \left[\alpha_1 \mathbf{x}_1 + \sum_{i=2}^{n} \alpha_i (\lambda_i/\lambda_1)^k \mathbf{x}_i\right]. \end{aligned} \tag{1.5–2}$$

In the limit of large k (numerous iterations) we have

$$\lim_{k\to\infty} \left(\frac{\mathbf{v}_k}{\lambda_1{}^k}\right) = \alpha_1 \mathbf{x}_1,$$

since $|\lambda_1| > |\lambda_i|$, $i = 2, \ldots, n$. Moreover, if $\mathbf{y}$ is any other vector that also has a nonvanishing component in the direction of $\mathbf{x}_1$, then

$$\lim_{k\to\infty} \left(\frac{\mathbf{y}^* \mathbf{v}_{k+1}}{\mathbf{y}^* \mathbf{v}_k}\right) = \lambda_1. \tag{1.5–3}$$

These last two relations constitute the essence of this iterative procedure.

Example For illustrative purposes, the power method is applied to the matrix A of Section 1.4 for the case $n = 3$, namely when

$$A \equiv \begin{bmatrix} 2 & -1 & 0 \\ -1 & 2 & -1 \\ 0 & -1 & 1 \end{bmatrix}. \qquad \textbf{1.5–4)}$$

Calculations have been performed on a hand calculator with ten-digit readout, and the four most significant digits have been tabulated below. Two sets of results are presented based upon different choices of the initial vector $\mathbf{v}_o$. In each case $\mathbf{v}_k$ has been scaled so that its third component equals unity, and the estimates for λ_1 were determined from the ratio in (1.5–3) using the convenient selection of $\mathbf{y}$ as the column vector with the element 1 in the position corresponding to the *maximum* component of $\mathbf{v}_k$ and zeros elsewhere.*

Table 1: Power iteration with $\mathbf{v}_0^* = (1, -1, 1)$

k	$\mathbf{v}_k^*$ (normalized)	λ_1
0	(1.000, −1.000, 1.000)	—
1	(1.500, −2.000, 1.000)	4.000
2	(1.667, −2.167, 1.000)	3.250
3	(1.737, −2.211, 1.000)	3.231
4	(1.770, −2.230, 1.000)	3.238
5	(1.787, −2.239, 1.000)	3.243
6	(1.795, −2.243, 1.000)	3.245
7	(1.798, −2.245, 1.000)	3.246
8	(1.800, −2.246, 1.000)	3.247
∞	(1.802, −2.247, 1.000)	3.247

Table 2: Power iteration with $\mathbf{v}_0^* = (-0.55493896, 0, 1.00000000)$

k	$\mathbf{v}_k^*$ (normalized)	λ_1
0	(−0.555, 0 , 1.000)	—
1	(−1.110, −0.445, 1.000)	−1.110
2	(−1.228, −0.540, 1.000)	1.599
3	(−1.244, −0.553, 1.000)	1.560
4	(−1.246, −0.555, 1.000)	1.556
5	(−1.246, −0.555, 1.000)	1.555
6	(−1.245, −0.556, 1.000)	1.554
10	(−1.217, −0.572, 1.000)	1.543
15	(−0.386, −1.033, 1.000)	2.287
20	(1.619, −2.145, 1.000)	3.193
25	(1.797, −2.244, 1.000)	3.246
∞	(1.802, −2.247, 1.000)	3.247

□

From (1.5–2) it is clear that the *rate* of convergence of the power method is dependent upon the speed with which the ratios $(\lambda_i/\lambda_1)^k$ tend to zero. However, as the tabular data of the above example shows, the number of iterations necessary to achieve a given level of accuracy is also strongly influenced by the choice of the initial vector $\mathbf{v}_o$. Indeed, in the second case, starting the iterative procedure with a column vector that is nearly orthogonal to the dominant eigenvector $\mathbf{x}_1$ of A induces

*Such a choice simplifies enormously the calculation of the estimates for λ. In the general case $\mathbf{y}$ will typically be unchanging with k after at most a few iterations.

a rather quick initial convergence to the 'wrong' values (which turn out to be the eigenpair λ_2, $\mathbf{x}_2$). But fortuitously here, the ever-present round-off error in the calculation finally comes along and *assists* the interations slowly toward the 'proper' results.

Therefore, in applications of the power method, physical considerations should be used to suggest a good first guess for $\mathbf{v}_0$. Combining this with judicious shift-of-origin techniques and other acceleration of convergence schemes can help make this iterative procedure a reasonable method for accurate determination of the dominant eigenvalue and eigenvector. Of course, matrix deflation techniques will have to be employed before the lower-order eigenpairs can be calculated in analogous fashion (see Wilkinson (1965), Parlett (1980), or Burden, Faires, and Reynolds (1981), for example).

LR and QR

The ease of application and the general success of the power method notwithstanding, there are numerous modern problems for which the procedure is inappropriate. Usually these are applications in which it is important to obtain rather accurate values for the lower-order as well as the higher-order eigenvalues and eigenvectors of a matrix of at least modestly (although not overly) large size. In such cases, the accumulation of round-off error attendant to a succession of power iterations and matrix deflations prohibits their use.*

Fortunately, there are a number of other procedures available that do not suffer from this defect. *Jacobi's Method* (see Isaacson and Keller (1966) or Ralston and Rabinowitz (1978)) is one such technique that has been employed on real symmetric matrices with good success for many years. In this method, iterative approximations are obtained for the orthogonal (real unitary) matrix which diagonalizes the given symmetric matrix (recall Schur's Theorem and its corollaries). The eigenvalues and eigenvectors are then virtually read off.

There are two other iterative procedures of more recent vintage, however, that deserve special consideration. They are the *LR Transformation Method* due to Rutishauser and the *QR Transformation Method* originated by Francis and Kublanovskaya. These eigenvalue determination techniques are known to converge for a wide class of matrices that includes, as a subset, symmetric (hermitian) matrices. In addition, the QR method is extremely stable numerically (as is the LR procedure when the given matrix is positive definite) and thus leads to excellent eigenvalue estimates.

The basis of these two methods is the successive decomposition, into a particular form, of a sequence of specially constructed similar matrices. If we start with the matrix A, the LR procedure involves the following stages. (Readers be-

---

*The standard power iteration process (without origin shifts) also breaks down when there are a number of unequal eigenvalues of the same, or nearly the same, modulus (Ralston and Rabinowitz (1978), pp. 494ff.).

ware: this does *not* describe how the procedure should be computationally implemented):

$$
\begin{aligned}
A \equiv A_1 &= L_1R_1 \\
R_1L_1 \equiv A_2 &= L_2R_2 \\
&\cdot \\
&\cdot \\
&\cdot \\
R_{k-1}L_{k-1} \equiv A_k &= L_kR_k,
\end{aligned}
$$

and so on, where L_k, R_k ($k = 1, 2, \ldots$) denote lower (left) and upper (right) triangular matrices, respectively (hence the mnemonic designation). For the QR method the comparable relations are

$$
\begin{aligned}
A \equiv A_1 &= Q_1R_1 \\
R_1Q_1 \equiv A_2 &= Q_2R_2 \\
&\cdot \\
&\cdot \\
&\cdot \\
R_{k-1}Q_{k-1} \equiv A_k &= Q_kR_k,
\end{aligned}
$$

where R_k is as above but the Q_k are orthogonal (unitary). In either case, if the given original matrix A is nonsingular, all of the A_k are similar ($A_{k+1} = R_kA_kR_k^{-1}$) and therefore they have the same eigenvalues. (Actually all the A_k have the same eigenvalues even when A is singular.) In fact, when these eigenvalues are real and distinct, A_k becomes triangular as $k \to \infty$, and the diagonal of A_∞ contains the desired eigenvalues in decreasing order of magnitude from left to right.*

In applying either the LR or QR process to a given practical problem, all that is needed at each stage are the two component matrices L_k, R_k or Q_k, R_k calculated in the previous stage. However, the efficient and accurate determination of these component matrices is a distinctly nontrivial undertaking, and the inexperienced applied mathematician is well-advised either to consult more knowledgeable colleagues or to use the sophisticated computer routines for these procedures that have been prepared by experts. In these modern eigenvalue packages, the original matrix is typically reduced to *tridiagonal* ($a_{ij} = 0$ for $|i - j| > 1$) or *Hessenberg* ($a_{ij} = 0$ for $i - j > 1$, or for $i - j < 1$) form before initiating the iterative process since this saves valuable computer time (order n^2 versus order n^3

---

*For the LR method in this case, A_∞ is lower triangular; in the QR it is upper triangular. In the more general situation it becomes "nearly" triangular. See Parlett (1964) for details as well as an excellent overview of these two procedures. The classic by Wilkinson (1965) and the handbook by Wilkinson and Reinsch (1971) are other fine references.

operations per iteration).* In this regard the EISPACK collection of matrix eigen-system computational subroutines should be mentioned (see Smith, et al.(1976) and Garbow, et al.(1977)). Utilizing this tremendously high quality and thoroughly reliable package of certified FORTRAN IV programs, even the novice can quickly find the eigenvalues (and eigenvectors, if so desired) of small matrices ($n < 100$).

Example To suggest the nature of LR/QR calculations, the LR transformation method has been applied to the matrix A of interest given in (1.5–4).[†] The results are tabulated to three decimal places with dash marks denoting repeated values.

Table 3: Elements of the triangular component matrices

k	q_{1k}	ϵ_{1k}	q_{2k}	ϵ_{2k}	q_{3k}
1	2.000	0.500	1.500	0.667	0.333
2	2.500	0.300	1.867	0.119	0.214
3	2.800	0.200	1.786	0.014	0.200
4	3.000	0.119	1.681	0.002	0.198
5	3.119	0.064	1.619	0.000	—
6	3.183	0.033	1.586	—	—
7	3.216	0.016	1.570	—	—
8	3.232	0.008	1.562	—	—
9	3.240	0.004	1.558	—	—
10	3.244	0.002	1.556	—	—
11	3.246	0.001	1.555	—	—
12	3.247	0.000	—	—	—

In this example, $A_k = L_k R_k$ where $L_k = R_k'$ and

$$R_k = \begin{bmatrix} \sqrt{q_{1k}} & -\sqrt{\epsilon_{1k}} & 0 \\ 0 & \sqrt{q_{2k}} & -\sqrt{\epsilon_{2k}} \\ 0 & 0 & \sqrt{q_{3k}} \end{bmatrix}$$

with

$$q_{j,k+1} = q_{jk} + \epsilon_{jk} - \epsilon_{j-1,k+1}$$

and

$$\epsilon_{j,k+1} = \epsilon_{jk}\, q_{j+1,k} \big/ q_{j,k+1}$$

*Householder's method is a very stable technique for reducing symmetric matrices to tridiagonal (nonsymmetric to Hessenberg) form. The process requires $(n-2)$ square roots and on the order of $\frac{2}{3} n^3$ multiplications for a symmetric (or skew-symmetric) matrix of order n. The number of multiplications is increased approximately by a factor 2 in the nonsymmetric case. On the other hand, there do exist alternative methods based upon Gaussian elimination that obviate the need for any square roots at all (Ralston and Rabinowitz (1978) Section 10.4.2, contains the details).

[†]Note that this matrix is already in tridiagonal form.

for $j = 1, 2, 3$; $k = 1, 2, \ldots$ $(\epsilon_{0,k+1} \equiv 0 \equiv \epsilon_{3k})$. As is clear from these tabular results, after some twelve iterations, R_k, L_k and hence A_k have each become diagonal (to three decimals), and all three eigenvalues have been determined. □

Inverse iteration

For a complete solution of the algebraic eigenvalue/eigenvector problem, iterative routines such as LR and QR, which effectively provide only the eigenvalues of a given matrix, must be combined with appropriate eigenvector estimation procedures. The most powerful and accurate of such methods for computing matrix eigenvectors is called *Inverse Iteration.*

Assume λ_i, $\mathbf{x}_i$ $(i = 1, 2, \ldots n)$ are the eigenvalues and eigenvectors of a given matrix A of order n. If $\tilde{\lambda}$ is an approximate value for one of the λ_i, then

$$(A - \tilde{\lambda} I)\mathbf{x}_i = (\lambda_i - \tilde{\lambda})\mathbf{x}_i.$$

This suggests the iteration defined by

$$(A - \tilde{\lambda} I)\mathbf{v}_{k+1} = \mathbf{u}_k, \qquad \mathbf{u}_{k+1} = \mathbf{v}_{k+1}\big/\|\mathbf{v}_{k+1}\|_\infty, \tag{1.5–5}$$

where the second step is taken merely to normalize $\mathbf{v}_k$, relative to its maximum element, after each iteration.

The inverse iteration method given by the relations (1.5–5) converges extremely rapidly for all but the very worst choices of initial vectors $\mathbf{u}_0$ (see the numerical examples below). This may be surprising in view of the fact that $A - \tilde{\lambda} I$ is nearly singular (recall our earlier discussion regarding ill-conditioned systems). In this case, however, the ill-conditioning works to our advantage; the errors are indeed enormous, but they are essentially proportional to the eigenvector we seek unless there is more than one eigenvalue of A near $\tilde{\lambda}$.* Wilkinson (1965) (especially pp. 321ff, 619ff) considers these matters in detail.

Example In order to provide some representative numerical evidence, we have applied the inverse iteration procedure to the matrix A of (1.5–4) using $\tilde{\lambda} = 3.247$ (other cases are left to the problems). For comparison purposes, the choices of initial vector are the same as were employed above in the subsection on the power method. Classical Gaussian elimination with partial pivoting was used, and all calculations were again made on a hand calculator with ten-digit readout. The results given in the tables have been rounded to three decimal digits. The rapidity of numerical convergence exhibited in these sets of tabular data, as compared to the earlier results, underscores the importance of so-called shift-of-origin techniques (where $A - \tilde{\lambda} I$ is used with appropriately chosen $\tilde{\lambda}$ rather than just A) in accelerating the general power iteration procedure.

---

*Apart from round-off errors, the procedure given by (1.5–5) is equivalent to the power method using $(A - \tilde{\lambda} I)^{-1}$. The rate of convergence of this latter process, of course, depends upon

$$|\lambda_i - \tilde{\lambda}|\Big/\min_{j \neq i}|\lambda_j - \tilde{\lambda}|.$$

Table 4: Inverse iteration with
$\mathbf{u}_o^* = (1, -1, 1)$

k	$\mathbf{u}_k^*$ (normalized)*
0	(1.000, −1.000, 1.000)
1	(1.802, −2.247, 1.000)

Table 5: Inverse iteration with
$\mathbf{u}_o^* = (-0.55493896, 0, 1.00000000)$

k	$\mathbf{u}_k^*$ (normalized)
0	(−0.555, 0 , 1.000)
1	(−0.071, −0.745, 1.000)
2	(1.802, −2.247, 1.000)

□

Before leaving this subsection it is worth noting that inverse iteration with tridiagonal matrices is extremely economical, with the number of operations needed in each iteration proportional to the *first* power of the order of the given matrix (Ralston and Rabinowitz (1978), p. 499). For matrices in Hessenberg form the number of operations is approximately $\frac{1}{2} n^2$, compared with n^3 for a full matrix. If the tridiagonal or Hessenberg matrix has been obtained by reduction of a full matrix (via Householder's method, say) subsidiary calculations will, of course, be needed to compute the eigenvectors of the original matrix from those of the reduced matrix. These are standard features of modern matrix eigensystem computational routines such as EISPACK (see Smith, et al (1976)).

Accuracy checks

A reasonable measure of the accuracy of a computed eigenvector is the magnitude of the residual

$$\mathbf{r}_i = A\mathbf{x}_i - \lambda_i \mathbf{x}_i. \tag{1.5–6}$$

If the given matrix is hermitian and several eigenvectors have been determined, then an orthogonality check will provide a further indication of the accuracy level.

There are a number of comparable checks that can be made on the accuracy of the eigenvalues. These follow from the well-known invariance, under similarity transformations, of both the determinant and the trace of a matrix. To be specific,

Definition The **trace** of a square matrix A is given by the sum of its diagonal elements, that is,

$$\operatorname{tr} A \equiv \sum_{i=1}^{n} a_{ii}.$$

*For comparison with the results of Tables 1, 2 the $\mathbf{u}_k$ have been renormalized so as to have unit third component.

Property
If $P^{-1}AP = B$ for some nonsingular matrix P, then $\det B = \det A$ and $\operatorname{tr} B = \operatorname{tr} A$.

Proof: See Problem 21. □

With this result in hand, we can easily establish the following.

Theorem:

Let $\lambda_1, \lambda_2, \ldots, \lambda_n$ be the eigenvalues of an arbitrary square matrix A of order n. Then

$$\text{(i)} \prod_{i=1}^{n} \lambda_i = \det A,$$

$$\text{(ii)} \sum_{i=1}^{n} \lambda_i = \operatorname{tr} A, \qquad \textbf{(1.5–7)}$$

$$\text{(iii)} \sum_{i=1}^{n} |\lambda_i|^2 \leq \operatorname{tr} AA^* \equiv \sum_{i,j=1}^{n} |a_{ij}|^2.$$

Equality holds in (iii) if A is hermitian.

Proof: By virtue of Schur's Theorem, A is unitarily equivalent to a triangular matrix T. Moreover, the diagonal elements of T must be the eigenvalues of A. The relations (i) and (ii) are a ready consequence, therefore, of the above property. To establish (iii) we merely note that

$$\begin{aligned}\sum_{i=1}^{n} |\lambda_i|^2 \leq \operatorname{tr} TT^* &= \operatorname{tr} (Q^*AQ\,Q^*A^*Q) \\ &= \operatorname{tr} (Q^*AA^*Q) \\ &= \operatorname{tr} AA^*.\end{aligned}$$

Equality holds when A is hermitian since T then is diagonal. □

Example These several checks may be carried out on the matrix A of (1.5–4), which has occupied our attention in this section. Using the values $\lambda_1 \doteq 3.247$, $\lambda_2 \doteq 1.555$, $\lambda_3 \doteq 0.198$ and $\mathbf{x}_1^* \doteq (1.802, -2.247, 1.000)$, we find from (1.5–6) and (1.5–7) to three decimals,

$$|\mathbf{r}_1| = 0.000$$

$$\prod_{i=1}^{3} \lambda_i = 1.000$$

$$\sum_{i=1}^{3} \lambda_i = 5.000$$

$$\sum_{i=1}^{3} \lambda_i^2 = 13.000.$$

The latter values are to be compared with det $A = 1$, tr $A = 5$, and tr $AA^* = 13$ as calculated directly from the matrix A itself. □

1.6 A summary for the three-disk problem

As far as the uniform 3-disk configuration of Section 1.4 is concerned, our analysis is complete. The original differential equations for the angular displacements $\theta_i(t)$ had the form

$$I\begin{bmatrix} \ddot{\theta}_1 \\ \ddot{\theta}_2 \\ \ddot{\theta}_3 \end{bmatrix} + k\begin{bmatrix} 2 & -1 & 0 \\ -1 & 2 & -1 \\ 0 & -1 & 1 \end{bmatrix}\begin{bmatrix} \theta_1 \\ \theta_2 \\ \theta_3 \end{bmatrix} = \mathbf{0}$$

where I and k are the polar moment of inertia of each disk and the rotational modulus of the shaft, respectively. The search for pure resonant vibrations wherein

$$\theta_i(t) = x_i \sin \omega t \qquad i = 1, 2, 3$$

led us to the matrix eigenvalue problem

$$A\mathbf{x} = \lambda\mathbf{x} \tag{1.6–1}$$

where A is the symmetric 3×3 matrix above and $\lambda \equiv I\omega^2/k$. The application of various eigenvalue/eigenvector approximation techniques suggested in Section 1.5 then provided rather accurate numerical estimates for the desired quantities (see also Problem 19). If k and I are specified, the natural frequencies and normal modes of the resulting mechanical vibrating system clearly can be determined easily and with confidence.

Other applications that lead to the algebraic eigenvalue problem with matrices A of modest size can be investigated in comparable fashion. Analogous calculations involving matrices that are so "large" that only part of them can be held at any one time in the high-speed storage of a given computing system, however, necessitate rather different considerations. We leave these to the experts (see Parlett (1980) for example).

References

Burden, R. L., J. D. Faires, and A. C. Reynolds (1981): *Numerical Analysis*, 2nd ed., Prindle, Weber, & Schmidt, Boston; Chapters 6, 8.

Dettman, J. W. (1974): *Introduction to Linear Algebra and Differential Equations*, McGraw-Hill, New York; Chapters 2, 4.

Dongarra, J. J., J. R. Bunch, C. B. Moler, and G. W. Stewart (1979): *LINPACK Users' Guide*, Soc. for Indust. and Appl. Math., Philadelphia.

Forsythe, G. E., and C. B. Moler (1967): *Computer Solution of Linear Algebraic Systems*, Prentice-Hall, Englewood Cliffs, N.J.

Garbow, B. S., J. M. Boyle, J. J. Dongarra, and C. B. Moler (1977): "Matrix Eigensystem Routines—EISPACK Guide Extension," *Lecture Notes in Computer Science*, Vol. 51, Springer-Verlag, New York.

Gourlay, A. R., and G. A. Watson (1973): *Computational Methods for Matrix Eigenproblems*, Wiley, New York.

Isaacson, E., and H. B. Keller (1966): *Analysis of Numerical Methods*, Wiley, New York.

Johnson, L. W. and R. D. Riess (1977): *Numerical Analysis*, Addison-Wesley, Reading, Mass.; Chapters 2, 3.

Kac, M. (1966): "Can One Hear the Shape of a Drum?" *Amer. Math. Monthly*, *73* (no. 4, part II): 1–24.

Keller, J. B. (1976): "Inverse Problems," *Amer. Math. Monthly*, *83*: 107–118.

Noble, B. and J. W. Daniel (1977): *Applied Linear Algebra*, 2nd ed., Prentice-Hall, Englewood Cliffs, N.J.

Parlett, B. (1964): "The Development and Use of Methods of LR Type," *SIAM Review, 6*: 275–295.

——— (1980): *The Symmetric Eigenvalue Problem*, Prentice-Hall, Englewood Cliffs, N.J.

Ralston, A. and P. Rabinowitz (1978): *A First Course in Numerical Analysis*, 2nd ed., McGraw-Hill, New York; Chapters 9, 10.

Rasof, B. (1973): "Applications of Linear Algebra—A Mechanical Vibrating System," *Proceedings of the Summer Conference for College Teachers on Applied Mathematics, Univ. of Missouri, 1971*, CUPM, Berkeley, Calif.

Smith, B. T., J. M. Boyle, B. S. Garbow, Y. Ikebe, V. C. Klema, and C. B. Moler (1976): "Matrix Eigensystem Routines—EISPACK Guide," *Lecture Notes in Computer Science,* Vol. 6, 2nd ed., Springer-Verlag, New York.

Steinberg, D. I. (1974): *Computational Matrix Algebra*, McGraw-Hill, New York.

Wilkinson, J. H. (1965): *The Algebraic Eigenvalue Problem*, Oxford Univ. Press, London.

———, and C. Reinsch (1971): *Handbook for Automatic Computation: Linear Algebra*, Vol. II, Springer-Verlag, New York.

Wylie, C. R. (1975): *Advanced Engineering Mathematics*, 4th ed., McGraw-Hill, New York; Chapters 3, 5, 11, 12.

Problems

Sections 1.1, 1.2

1. Assume Ohm's law ($v = iR$) for voltage drops across resistances and Kirchhoff's law ($\Sigma\, i_n = 0$) relating currents at a circuit node.
   (a) Directly analyze the electrical ladder network of Figure 1.2 on p. 2 by applying Ohm's law across each resistance and

Kirchhoff's law at every nodal junction, respectively. Express both the input currents and voltages in terms of the comparable output quantities and vice versa.

(b) Compare these results with those obtained using equations (1.1–1), (1.1–2).

2. Give an example showing that matrix multiplication is not, in general, commutative.

3. Verify that the transpose of the product of two matrices is given by $(AB)' = B'A'$.

★4. Let $\mathbf{x}$, $\mathbf{y}$ be arbitrary column vectors.
(a) Demonstrate that if $\mathbf{y}^*A\mathbf{x} = 0$ for all such $\mathbf{x}$, $\mathbf{y}$, then $A = \mathbf{0}$.
(b) Verify that A is hermitian if and only if $\mathbf{x}^*A\mathbf{x}$ is real for all such $\mathbf{x}$.

5. (a) Show that $\det A = \pm 1$ whenever A is orthogonal.
(b) State and prove an analogous result for unitary matrices.

6. (a) Determine the rank of the matrix A given by

$$A = \begin{bmatrix} 4 & -1 & 2 & 6 \\ -1 & 5 & -1 & -3 \\ 3 & 4 & 1 & 3 \end{bmatrix}.$$

(b) For which column vectors $\mathbf{b}$ can the matrix equation $A\mathbf{x} = \mathbf{b}$ be solved with this A?

7. Let A be an $m \times n$ matrix. Show that each of the three *elementary row operations* defined in Section 1.2 can be accomplished on A by left-multiplication with a matrix P, where P is formed by performing that operation on corresponding rows of the identity matrix I of order m. In each case, show also that P is nonsingular.

★8. (a) Verify that an arbitrary nonsingular matrix A can be reduced to (transformed into) the identity matrix I by a finite number of elementary row operations.
(b) Show that A^{-1} can be determined by performing the same operations in the same order on the identity matrix.
(c) Use this procedure to invert the matrix given by equation (1.5–4).

Section 1.3

9. Compare the solutions of the system

$$\begin{bmatrix} 10^{-4} & 1 & 1 \\ 10^{-1} & 2 & 1 \\ 1 & 10^{-4} & 0 \end{bmatrix} \begin{bmatrix} x_1 \\ x_2 \\ x_3 \end{bmatrix} = \begin{bmatrix} 1.0 \\ 1.1 \\ 1.0 \end{bmatrix}$$

obtained using 3-digit floating-decimal arithmetic (that is, 3 digits plus exponent) and Gaussian elimination with
(a) partial pivoting; (b) no pivoting strategy.

10. Discuss the conditioning (in the sense of the considerations of Section 1.3) of the system

$$\begin{bmatrix} 2 & 1 & 1 \\ 1 & \epsilon & \epsilon \\ 1 & \epsilon & -\epsilon \end{bmatrix} \begin{bmatrix} x_1 \\ x_2 \\ x_3 \end{bmatrix} = \begin{bmatrix} 1 \\ 2\epsilon \\ \epsilon \end{bmatrix}$$

as a function of the parameter ϵ.

Section 1.4

11. (a) Discuss completely the small *longitudinal* vibrations of the system of identical masses and springs depicted in Figure 1.4.

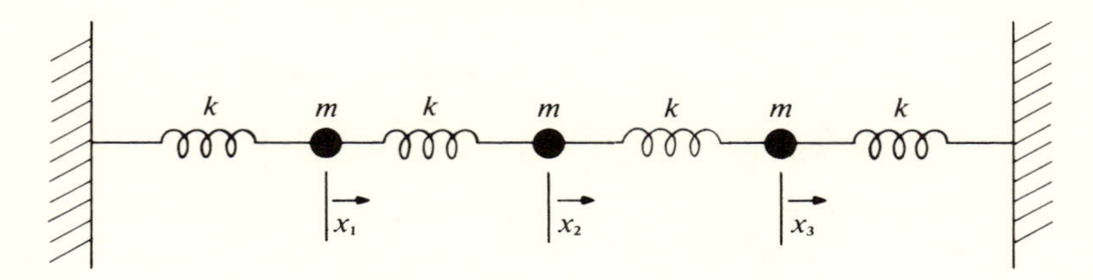

Figure 1.4

(b) What does the relevant eigenvalue problem become if the end spring and the wall are removed on one side? (Assume, of course, that gravity plays no role.)

12. Show that the problem

$$x_1 - 2x_2 = \lambda x_1$$
$$x_1 - x_2 = \lambda x_2$$

does not possess real nontrivial solutions for any values of λ.

Section 1.5

13. Determine the eigenvalues and normalized eigenvectors of the hermitian matrix

$$A = \begin{bmatrix} 9 & 2 + 2i \\ 2 - 2i & 2 \end{bmatrix}.$$

★14. Use Schur's Theorem on unitary equivalence (p. 14) to show that a normal matrix is

(a) hermitian if and only if its eigenvalues are real;

(b) unitary if and only if each of its eigenvalues has absolute value equal to one.

15. Find eigenvectors for the matrix

$$A = \begin{bmatrix} 1 & \alpha \\ 0 & 1 \end{bmatrix}$$

and discuss why there is not a full set if $\alpha \neq 0$.

16. Determine the orthogonal transformation that diagonalizes the matrix

$$A = \begin{bmatrix} 2 & -\tfrac{1}{2} & 0 \\ -\tfrac{1}{2} & 2 & -\tfrac{1}{2} \\ 0 & -\tfrac{1}{2} & 2 \end{bmatrix}.$$

17. Invert the matrix A given by equation (1.5–4) and then employ the power method to find the dominant eigenvalue of A^{-1}. (This will provide an estimate of the eigenvalue $\lambda_3 \doteq 0.198$ of A.)

18. When Householder's method is applied to the inverse of the matrix A given by equation (1.5–4), the matrix

$$B = \begin{bmatrix} 1 & -\sqrt{2} & 0 \\ -\sqrt{2} & 9/2 & -1/2 \\ 0 & -1/2 & 1/2 \end{bmatrix}$$

is obtained. Use the LR procedure on this new matrix in order to determine estimates of its characteristic values, accurate to three decimal places. (Since the matrix B is tridiagonal, the simplified approach employed in Section 1.5 on A itself is applicable.)

19. Use inverse iteration to determine eigenvectors associated with the lower-order eigenvalues of the matrix A of (1.5–4). Check to see whether the eigenvectors so obtained are indeed orthogonal to the dominant eigenvector given in the text.

20. Invert the matrix A given by equation (1.5–4) and then employ inverse iteration to determine eigenvectors associated with the various eigenvalues of A^{-1}. How do these eigenvectors compare with those calculated in Problem 19?

21. Demonstrate that if A and B are similar matrices (that is, $P^{-1}AP = B$ for some nonsingular matrix P), then

(a) $\det B = \det A$; (b) $\operatorname{tr} B = \operatorname{tr} A$.

22. Consider the real symmetric matrix

$$C = \begin{bmatrix} 2 & 1 & -1 & 0 \\ 1 & 3 & 4 & 2 \\ -1 & 4 & 1 & 2 \\ 0 & 2 & 2 & 1 \end{bmatrix}.$$

(a) Determine whether the matrix C is positive definite.
(b) Find the sum and the product of all of the eigenvalues of C.

2

Eigenvalue problems for differential equations

2.1 Shallow water waves

One of the most interesting fields for mathematical applications is the area of fluid mechanics. Describing the motion of fluids in various physical situations has attracted the best efforts of scores of mathematicians over the years, and has been central to important investigations in such modern fields as plasma physics and biomathematics (see Boyd and Sanderson (1969) and Lighthill (1975), for example).

Greenspan (1973) has proposed *discrete* modelling procedures for a variety of problems in applied mechanics, including ones dealing with fluids. These models involve large systems of algebraic equations to which some of the considerations of Chapter 1 apply. Although pursuing such an approach would be extremely interesting, we choose instead to develop a *continuous* model for the fluid mechanics application that will motivate our efforts in this chapter. This will lead us again to eigenvalue problems, but this time for differential equations rather than matrices. Since there are both similarities and differences between the two classes of eigenvalue problems, we will make comparisons where appropriate.

In what follows, we will be discussing the motion of liquids constrained to move at shallow depths. A typical classical application (see Lamb (1945) or Stoker (1957)), might concern the so-called "tidal" oscillations associated with the water waves present in small inland channels connected to larger bodies of water such as bays or oceans. In another context, our results have relevance to the sloshing of liquids in carrying containers and transport vessels, for example.

For our purposes, we will regard a liquid to be an *ideal* or *perfect* fluid, that is, a fluid in which there are no shearing forces. Within such a fluid, if we analyze the internal forces acting on an infinitesimal tetrahedron, the force acting on each face is perpendicular to that face. In the limit, the quotient of the force by the area of the face becomes a quantity, called the *pressure*, which is independent of the orientation of the original face. Thus, an ideal fluid can be described in terms of a pressure function p, which is a point function of position and time and is the same in all directions. As a further simplification, we assume that our liquids are *incompressible*; that is, they have constant density.

As a practical problem of interest, let us consider a liquid of density ρ constrained to move in a straight channel of gradually varying cross-section $S = S(x)$ as depicted in Figure 2.1. For convenience we choose the coordinate system so that the x-axis is parallel to the length of the channel with the z-axis upward and presume that the fluid motion takes place only in these two directions. We let z_o be the ordinate of the free liquid surface in the undisturbed state and $z_o + \eta(x, t)$ be the free surface ordinate in general. We also assume that gravity is the only external force acting on the liquid.

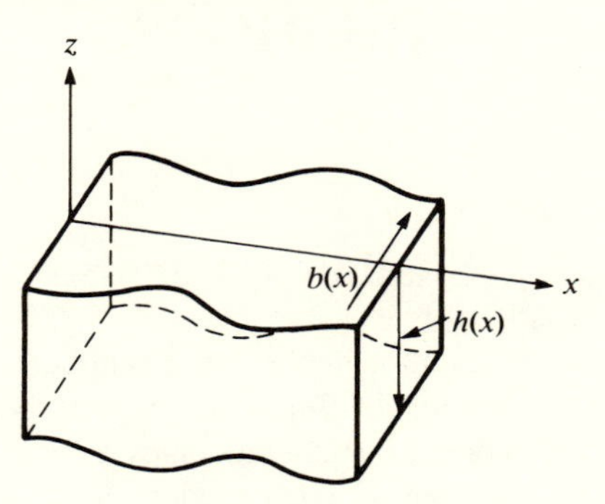

Figure 2.1 Shallow water channel of varying cross-section $S = bh$

The equations of motion governing the flow of our channeled liquid are particular cases of the equations of motion for general fluid flows. These more general equations of fluid dynamics can be derived in a number of ways and are a consequence solely of Newton's law of *conservation of momentum*. In vector form the general equations may be expressed as

$$\frac{d\mathbf{v}}{dt} = -\frac{1}{\rho}\operatorname{grad} p + \mathbf{f} \qquad \textbf{(2.1–1)}$$

where $\mathbf{v}$, p, and $\mathbf{f}$ are the vector velocity, the internal pressure, and the externally induced force per unit mass, respectively, acting at any point in the fluid. As expected, the relations equate, on a point-by-point basis, the local vector acceleration with the local total force per unit mass.

The equations (2.1–1) need to be specialized to our shallow-water configuration. We begin by making the reasonable assumption that the left-hand side of (2.1–1) is negligibly small in the vertical direction. This is equivalent to assuming that, as in the hydrostatic case, the pressure $p(x,z,t)$ at any point within the fluid is a function only of the distance of that point below the free liquid surface. If we shift the pressure term to the left-hand side and integrate the vertical component of (2.1–1) with respect to z, we are then led to

$$p(x,z,t) = p_o + [z_o + \eta(x,t) - z]\, g\rho.$$

(Remember, gravity is the only external force). In this expression, p_o is the uniform (assumed constant) external pressure acting on the fluid at the free surface and g is the gravitational acceleration constant. When we differentiate this relationship with respect to x, we obtain as our first really useful result

$$\frac{\partial p}{\partial x} = g\rho \frac{\partial \eta}{\partial x}. \tag{2.1–2}$$

As announced earlier, this last relation is representative of conservation of momentum in the vertical direction.

Before continuing, we note that the right-hand side of (2.1–2) is independent of z. This implies that the horizontal forces and hence the horizontal accelerations of the fluid particles situated in any plane perpendicular to the x-direction are identical. In other words, if we denote by u the x-component of the fluid velocity, then u is also independent of z and is therefore a function of x and t alone.

For our second governing equation, we will use the horizontal component of the vector equation (2.1-1). Since now $u = u(x,t) \equiv dx/dt$, when we employ the chain rule this component becomes

$$\frac{\partial u}{\partial t} + u\frac{\partial u}{\partial x} = -\frac{1}{\rho}\frac{\partial p}{\partial x}. \tag{2.1–3}^*$$

As in Section 1.4, we will be primarily interested here in "small oscillations." We can then make further simplifications. For example, u and $\partial u/\partial x$ both generally turn out to be small quantities, and we can neglect the mixed term in (2.1–3) in favor of the linear terms. When we combine this modified result with (2.1–2), we then have

$$\frac{\partial u}{\partial t} = -g\frac{\partial \eta}{\partial x}.$$

For convenience we can rewrite this in terms of $\xi \equiv \int^t u(x,\tau)d\tau$, where the time integral represents fluid displaced past the cross-sectional plane at x up to the time t:

$$\frac{\partial^2 \xi}{\partial t^2} = -g\frac{\partial \eta}{\partial x}. \tag{2.1–4}$$

To complete our mathematical description of shallow-water theory we require an expression for the kinematic relationship between the velocity of the fluid and its density, or, since the fluid is assumed incompressible, between the velocity and geometrical parameters such as the cross-sectional area $S(x)$. This "equation of continuity" can be derived in several ways and is a consequence of the basic principle of *conservation of mass* for the fluid. We will proceed from first principles. If we decompose S into the product of a *mean* channel depth function $h(x)$ and a breadth (at the free surface) function $b(x)$ (so that $S = bh$), then mass

---

*This equation of motion, which represents conservation of momentum in the horizontal direction, is in so-called "Eulerian" form. Alternatively, a "Lagrangian" approach could have been taken in which the motion of individual fluid particles is the central construct. The names attached to these representations are those of Leonhard Euler (1707–1783) and Joseph Louis Lagrange (1736–1813), who made fundamental contributions to the mathematical foundations of fluid mechanics. Interestingly, both forms of the basic fluid dynamical equations are in fact due to Euler.

conservation applied to the infinitesimal volume of height η, width b, and length δx requires

$$\eta b\ \delta x = S(x)\ \xi(x,t) - S(x + \delta x)\ \xi(x + \delta x,t)$$

or, in the limit,

$$\eta = -\frac{1}{b}\frac{\partial}{\partial x}(bh\xi). \tag{2.1–5}$$

Here, in keeping with our small oscillation assumptions, we have neglected η and $\partial\eta/\partial x$ relative to b and S.

Either of the two dependent variables can be eliminated between (2.1–4) and (2.1–5). Accordingly, there are alternative forms for the continuity equation:

$$\frac{\partial^2\xi}{\partial t^2} = \frac{g}{b}\frac{\partial^2}{\partial x^2}(bh\xi),$$

or **(2.1–6)**

$$\frac{\partial^2\eta}{\partial t^2} = \frac{g}{b}\frac{\partial}{\partial x}\left(bh\frac{\partial\eta}{\partial x}\right).$$

If the channel breadth b and depth h are both constant, it follows that η and ξ (and hence u) satisfy the one-dimensional *linear wave equation*

$$\frac{\partial^2 f}{\partial x^2} = \frac{1}{c^2}\frac{\partial^2 f}{\partial t^2}.$$

It may be noted that in this special case, the speed c at which the disturbance propagates is given by $\sqrt{gh}$.

As might be expected, the features of those solutions of the general differential equations (2.1–6) that exhibit periodic time-wise behavior, that is, those that represent harmonic motion of the form

$$\eta(x,t) = y(x)\sin(\omega t + \theta),$$

are of especial interest. (Compare with the normal-mode analysis of Section 1.4.) Under these conditions, the second of the equations (2.1–6) gives rise to

$$(bhy')' + (b\omega^2/g)y = 0, \tag{2.1–7}$$

where $'$ denotes differentiation with respect to x. Eventually we shall make various physically meaningful choices for b and h and impose other realistic constraints on the solutions of (2.1–7), and these auxiliary conditions naturally restrict the values of the "frequency" ω that may be associated with the solutions. In much the same way that the practical problem of Section 1.4 brought us to the eigenvalue problem for matrices, therefore, our consideration of water waves in shallow channels has led us to an analogous eigenvalue problem for ordinary differential equations.

2.2 Sturm-Liouville problems

The motivational example of the previous section has provided a rather different approach to eigenvalue problems. More familiar to most readers undoubtedly would have been considerations based upon the application of separation of variables to well-known partial differential equations. The end result is the same, however. We now have an extremely general class of ordinary differential equations whose solutions will help us better to understand various physical phenomena.

Before we investigate a number of interesting special situations associated with the relation (2.1–7), it is appropriate to provide some background on general eigenvalue problems embodying ordinary differential equations. To simplify our discussion we restrict attention to second-order homogeneous linear equations and assume that the differential operator in question has the *self-adjoint* form

$$Ly \equiv (py')' - qy \qquad a < x < b. \tag{2.2–1}$$*

In our considerations it will be essential that $p(x)$ be nonvanishing for $a \leq x \leq b$. Consequently, we suppose that $p(x)$ is positive, and we assume that $p'(x)$ and $q(x)$ are real-valued continuous functions on the bounded interval $[a,b]$.

In view of the above definition we find that for *any two functions* $u(x)$ and $v(x)$,

$$\begin{aligned} uLv - vLu &= [p(uv' - vu')]' \\ &= [p(x)\, W(u,v)]' \end{aligned}$$

where $W(u,v) \equiv uv' - vu'$ is the Wronskian of u and v. It readily follows then that

$$\int_a^b (uLv - vLu)dx = p(x)\, W(u,v) \Big|_a^b . \tag{2.2–2}$$

For the self-adjoint operator (2.2.1), the value of this fundamental integral depends therefore, solely upon the right-hand side of this expression, that is, *only* on the *boundary values* satisfied by u, v, and p at the endpoints of the interval $[a,b]$.

Since boundary conditions are thus going to be important in our subsequent analyses, we need to agree upon some additional terminology:

Definition 1 A pair of relationships among the values of y and y' at a and b of the form

$$\begin{aligned} Ay(a) + By'(a) + Cy(b) + Dy'(b) &= 0 \\ Ey(a) + Fy'(a) + Gy(b) + Hy'(b) &= 0, \end{aligned} \tag{2.2–3}$$

where $A, B, \ldots, H$ are specified real constants such that the rank of the matrix $\begin{bmatrix} A & B & C & D \\ E & F & G & H \end{bmatrix}$ is 2, is called a set of **general mixed homogeneous** boundary conditions. If $C = D = 0 = E = F$, the boundary conditions are said to be **unmixed** or **separated**.

[Note that the rank condition merely ensures that the two boundary conditions are distinct.]

---

*Every second-order linear differential equation can be formally rewritten so as to have the form $Ly = 0$; see Problem 2. Problem 1 considers the rationale for the terminology "self-adjoint."

Definition 2 Boundary conditions of general mixed homogeneous type are termed **regular** (or **self-adjoint**) if, for all u, v that satisfy them,

$$p(x)\, W(u,v) \Big|_a^b = 0.$$

(Problem 3 contains an alternative form for this definition).

Two often-encountered types of boundary conditions are easily seen to be regular:

Property 1
Unmixed (separated) boundary conditions are regular.

Property 2
Periodic boundary conditions of the form $y(b) = y(a)$ and $y'(b) = y'(a)$ also are regular provided $p(b) = p(a)$.

The study of *boundary value* (*eigenvalue*) *problems* associated with the self-adjoint operator (2.2–1) and regular boundary conditions has a rich heritage. The classical theory is due to Jacques Sturm (1803–1855) and Joseph Liouville (1809–1882). In recognition of this, the following definition is often used:

Definition 3 A boundary value problem consisting of a differential equation with the self-adjoint operator (2.2–1) and *regular* boundary conditions is called a standard **self-adjoint** or **Sturm-Liouville** problem.

Readers who have studied functional analysis will have noted that our self-adjoint differential operator, by itself, is at variance with the usual definition of an abstract self-adjoint operator. However, if we introduce the notion of scalar product $(f,g) \equiv \int_a^b f(x)\, g(x)\, dx$, then for any functions* u and v that satisfy regular boundary conditions at $x = a, b$ it follows that $(u,Lv) = (Lu,v)$, which agrees with the customary characterization of self-adjoint operators on inner product spaces.

In the remainder of this chapter we assume that all the boundary conditions encountered are regular and all the boundary-value problems under consideration are of the Sturm-Liouville type.

2.3 Basic comparison theory and qualitative results

Our interest in Sturm-Liouville problems was originally stimulated by equation (2.1–7). If we associate bh with p, let $q = 0$, and write λ for ω^2 and r for b/g, this earlier relation could now be reexpressed using (2.2–1) as

$$Ly + \lambda\, r\, y = 0.$$

---

*If a function is to be operated upon by the differential operator L, it is assumed that it is twice continuously differentiable. Indeed, we will consider as solutions of given boundary-value problems only those functions having this smoothness.

This is the standard form for the differential equation in a Sturm-Liouville eigenvalue problem. In order to complete the problem description, however, a set of regular boundary conditions is needed. Since the eigenvalue problems of practical interest commonly entail unmixed (separated) boundary conditions, we will concentrate on this important case. In other words, the Sturm-Liouville problem with which we will concern ourselves in what follows is assumed to have the special form

$$Ly + \lambda r y = (py')' + (\lambda r - q)y = 0 \qquad a < x < b \tag{2.3–1a}$$

$$\begin{aligned} Ay(a) - By'(a) &= 0 \\ Gy(b) + Hy'(b) &= 0, \end{aligned} \tag{2.3–1b}$$

where A and B are real constants, not both zero, and the same is true for G and H. (The sign in front of the constant B has been changed from (2.2–3) in order to facilitate the statement of some later results.)

The reader is reminded that $p'(x)$, $q(x)$ and $r(x)$ in (2.3–1a) are assumed to be real-valued continuous functions on the bounded interval $[a,b]$. For reasons that will soon be apparent, we will also suppose that $r(x) > 0$ on $[a,b]$; this condition is usually satisfied in applied problems.

Example The Sturm-Liouville problem (2.3–1) is a generalization of the familiar problem

$$\begin{aligned} y'' + \lambda y &= 0 \qquad 0 < x < \pi \\ y(0) &= 0 \\ y(\pi) &= 0. \end{aligned}$$

Solutions of the differential equation that satisfy the first boundary condition have the form $Y(x,\lambda) = C \sin(\sqrt{\lambda}\, x)$ where C is constant. Unless λ is one of the "special" values n^2 for some integer n, these solutions will satisfy the second boundary condition only if $C = 0$. □

The situation encountered in the above example also obtains in the more general case. As we see in the next section, for most values of λ in (2.3–1) only the trivial solution of the differential equation also satisfies both boundary conditions. An *eigenvalue* λ is then defined to be a complex number with the property that there exists a *non*trivial solution of (2.3–1a) satisfying (2.3–1b). The solution y is called an *eigenfunction* corresponding to the eigenvalue λ.

Our considerations will eventually lead us to analyze these eigensolutions of the boundary-value problem (2.3–1) for various choices of the functions p, q, r and constants a, b, A, B, G, H. However, as Sturm and others have shown, much can be deduced about the qualitative behavior of the solutions independent of specific properties of p, q, r, etc. For example, if u and v are eigenfunctions of (2.3-1) associated with the eigenvalues λ and μ respectively, then

$$(pu')' + (\lambda r - q)u = 0$$

and

$$(pv')' + (\mu r - q)v = 0.$$

Multiplying the first equation by v, the second by u, and subtracting gives rise to the expression

$$\begin{aligned}(\lambda - \mu)ruv &= uLv - vLu \\ &= [p(x)\, W(u,v)]'.\end{aligned}$$

Making use of the boundary conditions (2.3–1b), the reader may verify that the following ensue immediately:

Property 1
If $\lambda = \mu$, then $W(u,v) = 0$, so u and v are necessarily linearly dependent on $[a,b]$.*

This means that to each distinct eigenvalue of (2.3–1) there corresponds only a single eigenfunction, unique up to a multiplicative constant. The terminology "the eigenvalues of (2.3–1) are *simple*" is often used to describe this behavior. (This property of eigenvalue simplicity is occasionally not the case in other regular boundary-value problems, for example those with periodic boundary conditions.)

Property 2
If $\lambda \neq \mu$, then $\int_a^b r\, uv\, dx = 0$ and u and v are linearly independent.

The integral of the product of two functions that appears in this second property is a generalization of the inner (scalar) product usually encountered in Fourier series and mentioned earlier. The function $r(x)$ is called the *weight function* of the inner product; and if it is positive we are assured that for real $u(x)$

$$\int_a^b r\, u^2\, dx = 0 \quad \text{if and only if} \quad u \equiv 0.$$

As with the matrix-theoretic situation, we say that a real-valued function is *normalized* if $\int_a^b r\, u^2\, dx = 1$; and that two functions are *orthogonal* (with respect to the weight function r) if their (generalized) inner product vanishes. Property 2 can thus be reexpressed in the form

Property 2′
If $\lambda \neq \mu$, then $\int_a^b r\, u\, v\, dx = 0$, so u and v are not only linearly independent but orthogonal on $[a,b]$ with respect to the weight function $r(x)$.

---

*It is a fact presumably familiar to the reader that the vanishing of the Wronskian of two solutions of the same linear homogeneous equation implies their linear dependence.

As a last result we note it can be easily established (see Problem 5) that

Property 3
Every eigenvalue λ of (2.3–1) is real and the eigenfunction associated with it may be chosen to be real also. (We will make such a choice throughout the rest of this chapter. Compare with Property 3 near the beginning of Section 1.5.)

Sturm-type comparisons

The three properties above provide a foundation for our considerations of qualitative behavior. To proceed we observe that, in view of the positivity of r, the orthogonality of eigenfunctions expressed in Property 2′ implies that these solutions of (2.3–1) cannot be entirely of one sign in $[a, b]$, that is, they must have zeros there. Perhaps surprisingly, a careful analysis of the properties of these zeros of the eigenfunctions is going to lead us to a rather rich harvest of results.

We begin our investigation by considering the initial-value problem (IVP) comprised of the equation (2.3–1a) for fixed real λ and the *first* of the conditions (2.3–1b). In fact, for purposes of comparison, we consider two such IVPs in which all of the coefficients and constants satisfy our standard hypotheses:

$$\begin{aligned} (pu')' + (\lambda r - q)u &= 0 \\ \alpha u(a) - \beta u'(a) &= 0 \end{aligned} \qquad \textbf{(2.3–2)}$$

$$\begin{aligned} (Pv')' + (\lambda r - Q)v &= 0 \\ Av(a) - Bv'(a) &= 0. \end{aligned} \qquad \textbf{(2.3–3)}$$

Fundamental theory for ordinary differential equations (Birkhoff and Rota (1978) or Ince (1953) or Coddington and Levinson (1955), for example) ensures the existence of nontrivial twice-continuously differentiable real solutions of these problems. Most importantly, these solutions depend continuously upon the various coefficients and constants as well as the real parameter λ. As regards the zeros of these solutions, we have

Theorem 1:

Each nontrivial solution of (2.3–2) or (2.3–3) has a finite number of zeros in $[a,b]$ and these zeros are all simple, that is, if $u(c) = 0$ then $u'(c) \neq 0$.

Theorem 2:

The zeros of real linearly independent solutions of the same equation separate one another. In other words, between any two consecutive zeros in $[a,b]$ of one real solution there lies a zero of any other real linearly independent solution of the same equation.

Proof: See Problem 7. □

A slightly deeper result in the spirit of Sturm's original considerations is:

Theorem 3:

Let $u(x)$, $v(x)$ be real nontrivial solutions of the differential equations in (2.3–2), (2.3–3), respectively, where $P(x) = p(x)$ but $Q(x) \leq q(x)$ on $[a,b]$. Assume that u is such that $u(c) = 0$ and $u(d) = 0$, where $a \leq c < d \leq b$. Then $v(x)$ vanishes at least once in $(c,d]$. Furthermore, if $v(c) = 0$, $v(x)$ has at least one zero in the open interval (c,d) unless $Q = q$ throughout.

Proof: Without loss of generality we may suppose that c and d are consecutive zeros of u with u positive in between. Thus $u'(c) > 0$ and $u'(d) < 0$. Multiplying the differential equation of (2.3–2) by v, that of (2.3–3) by u, subtracting the resulting equations and integrating over $[c,d]$ then leads to

$$\int_c^d (q - Q)\, uv\, dx = -p(x)\, W(u,v)\Big|_c^d$$
$$= p(d)\, u'(d)\, v(d) - p(c)\, u'(c)\, v(c).$$

If v is always of one sign in $(c,d]$, the signs of the two sides of this equality are at variance due to the behavior of u and the positivity of p; thus we have a contradiction. □

A simple modification, made by Picone in 1909, allows the same conclusion to be drawn for the more general case wherein both $P(x) \leq p(x)$ and $Q(x) \leq q(x)$ on $[a,b]$.

Results such as Theorem 3 and the cited extension may be described in a casual way as inferring that "solutions of the Sturm-Liouville equation (2.3–1a) *oscillate more rapidly* as p and/or q decrease." This terminology has the desirable feature of illustrating how these results generalize the well-known oscillation properties of the solutions of a familiar constant-coefficient differential equation such as $my'' + ky = 0$. For this example, as $m > 0$ decreases and/or k increases (recall the minus sign in front of q), the solutions exhibit increasing numbers of zeros in any fixed interval. Not only that, but each individual zero appears to migrate to the left. With a little effort, we can establish the following rather precise result, which shows that this behavior also occurs in the more general setting:

Theorem 4:

Let $u(x)$, $v(x)$ be real nontrivial solutions of (2.3–2), (2.3–3), respectively, where $P(x) \leq p(x)$, $Q(x) \leq q(x)$ on $[a,b]$ and either $\beta = 0$ or $P(a)\, A/B \leq p(a)\, \alpha/\beta$ with $\beta B \neq 0$. If $u(x)$ has m zeros in $(a,b]$, then $v(x)$ has *at least* m zeros in $(a,b]$. Moreoever, if the zeros are enumerated according to increasing algebraic size, each zero of v in $(a,b]$ lies to the *left* of the zero of u with corresponding index.

Proof: We begin by enumerating the m zeros of u in $(a,b]$, namely $a < x_1 < x_2 < \ldots < x_m \leq b$, and let c and d designate any two consecutive zeros. If $v \neq 0$ in $[c,d]$ the Picone identity (see Problem 8)

$$\int_c^d \{q - Q)u^2 + (p - P)(u')^2 + P[W(u,v)/v]^2\}\, dx = u^2(pu'/u - Pv'/v)\Big|_c^d \qquad \textbf{(2.3–4)}$$

leads to a contradiction. Just as in Theorem 3, then, we find that between each pair of these zeros of u there is at least one zero of v. To complete the verification we only need demonstrate that $v(x)$ also vanishes in the interval $(a,x_1]$. This is easily accomplished, however, by setting $c = a$, $d = x_1$ in (2.3–4) and reasoning as before. □

Example As an example of the application of Theorem 4, consider the two functions $u = \sin \gamma\pi x$ and $v = P_n(x)$ where P_n is the familiar Legendre polynomial of order n (for review of Legendre polynomials see Appendix 2). These functions satisfy the differential equations

$$\begin{aligned} u'' + \gamma^2\pi^2 u &= 0, \\ [(1 - x^2)v']' + n(n + 1)v &= 0 \end{aligned} \qquad 0 < x < 1$$

and clearly $(1 - x^2) \leq 1$ on the interval $[0,1]$. If $\gamma > m$, $u(x)$ has at least m zeros in $(0,1)$. It follows then from Theorem 4 that $v(x)$ has at least m zeros in $(0,1)$ whenever $n(n + 1) > m^2\pi^2$. If we didn't have any other information about Legendre polynomials, therefore, this simple comparison would have told us that, for example, P_3 has at least one zero in $(0,1)$ and P_6 has at least two. (P_6, of course, has three zeros in $(0,1)$.) □

There is a form of converse to Theorem 4 that will be helpful in our further discussion of the Sturm-Liouville problem (2.3–1) and that we now state. However, since the proof of this result also follows by an appropriate application of the Picone formula (2.3-4), we leave the demonstration to the interested reader (Problem 9.)

Theorem 5:

Let $u(x)$, $v(x)$ be real nontrivial solutions of (2.3–2), (2.3–3), respectively, where $P(x) \leq p(x)$ and $Q(x) \leq q(x)$ on $[a,b]$ and either $\beta = 0$ or $P(a)\, A/B \leq p(a)\, \alpha/\beta$ with $\beta B \neq 0$. Assume that neither u nor v vanishes at a fixed point c in $(a,b]$. Then if u and v both have the *same* number of zeros in the interval $(a,c]$, it follows that

$$P(c)\, v'(c)/v(c) \leq p(c)\, u'(c)/u(c),$$

with strict inequality unless $P \equiv p$, $Q \equiv q$.

Theorem 4 above showed that if $p(x)$ and/or $q(x)$ is decreased while α and β are held fixed, the solutions to the initial-value problem (2.3–2) exhibit increasing oscillatory behavior. Since the boundary condition at the left-hand endpoint a is specified, any new zeros that appear in the interval $[a,b]$ must enter from the right, that is, at $x = b$, and migrate (as p, q decrease) toward $x = a$. Theorem 5 then implies that at any point c in $(a,b]$ that is not a zero, changes in the quotient $p(c)\ u'(c)/u(c)$ are in the same direction as changes in p and q. As p, q decrease, the quotient decreases, becoming unboundedly large and negative as a zero approaches c (from the right); immediately after the zero has passed, the quotient is unboundedly large and positive.

Armed with these results, we are now in a position to complete the analysis of the eigenvalues and eigenfunctions associated with the given Sturm-Liouville problem.

2.4 Further eigenvalue/eigenfunction behavior

The eigenvalue problem under consideration is the Sturm-Liouville problem (2.3–1), which we restate here for convenience:

$$\begin{aligned}(py')' + (\lambda r - q)y &= 0 \qquad a < x < b \\ Ay(a) - By'(a) &= 0 \\ Gy(b) + Hy'(b) &= 0.\end{aligned} \tag{2.4–1}$$

Let $y \equiv Y(x,\lambda)$ designate *the* solution of the above differential equation that satisfies the first of the boundary conditions and for which $Y(a,\lambda) = c$, a specified constant. (If $B = 0$, then $c = 0$; otherwise, for definiteness we let $c = 1$). As noted earlier, the eigenvalues of (2.4–1) are then those λ for which $Y(x,\lambda)$ also satisfies the second boundary condition, that is, the eigenvalues are given by the implicit relation

$$GY(b,\lambda) + HY'(b,\lambda) = 0. \tag{2.4–2}$$

In Problem 12 the reader is asked to make some comparisons between the solutions of (2.3–1) and pure sinusoids. From those results it follows that if, for some fixed λ,

$$\min(\lambda r - q) > \left(\frac{m\pi}{b - a}\right)^2 \max p$$

(where the min and the max are taken with respect to x in $[a,b]$), then $Y(x,\lambda)$ has at least m zeros in (a,b). For continuous p, q, and r, therefore, we can choose λ large enough so that $Y(x,\lambda)$ has any prescribed number of zeros in the fundamental interval. At the other extreme, it turns out that we can choose λ so small (that is, large and negative) that $Y(x,\lambda)$ is zero-free in (a,b). Now we want to allow λ to vary between these extremes and to investigate $Y(x,\lambda)$, especially as regards the number and nature of its zeros as a function of λ.

The key ingredient in our analysis will be an application of the various comparison results of the previous section. Although in Theorems 3, 4, and 5 of Section 2.3 the real parameter λ was assumed fixed, since $r(x)$ is positive in $[a,b]$, changes in λ can be interpreted as if they were changes in q. In other words, for the purposes of utilizing Theorems 3–5, an increase in λ of the amount $\Delta\lambda$ can be viewed as if we have replaced q by $q - r\Delta\lambda$, leaving λ unchanged.

Keeping these observations in mind, we now proceed to the main result and its proof:

Theorem:

The Sturm-Liouville problem (2.4–1) has a countable infinity of (real, discrete) eigenvalues λ_m, $m = 0, 1, 2, \ldots$, with $\lambda_m \to \infty$ as $m \to \infty$. If these eigenvalues are arranged in order of increasing size, that is,

$$-\infty < \lambda_0 < \lambda_1 < \cdots < \lambda_m < \cdots,$$

then for each m, the eigenfunction associated with λ_m has precisely m zeros in (a,b).

Proof (1): We begin by showing that there is a least eigenvalue λ_0 of (2.4–1) with a corresponding eigenfunction that is zero-free. Let $P \equiv \min p(x)$ and $Q(\lambda) \equiv \min(q - \lambda r)$, where these minima are taken over $[a,b]$. Since r, p, and hence P are all positive under our original hypotheses, we may denote the quotient $Q(\lambda)/P$ by λ^2 and agree that $\lambda(\lambda)$ is positive if λ is sufficiently large and negative. Indeed, $\gamma(\lambda) \to +\infty$ as $\lambda \to -\infty$.

We shall use the equation

$$Py'' - Q(\lambda)y = 0 \tag{*}$$

as a comparison equation for (2.4–1) when $\lambda \to -\infty$. In this regard we observe that the solution $\tilde{y}(x,\lambda)$ of this constant-coefficient differential equation that satisfies the first of the boundary conditions in (2.4–1) can be expressed as

$$\tilde{y}(x,\lambda) = \text{const.}[(A/\gamma)\sinh\gamma(x - a) + B\cosh\gamma(x - a)];$$

and clearly it is zero-free in $(a,b]$ for λ sufficiently large and negative. Thus, if we associate $\tilde{y}(x,\lambda)$ with $v(x)$ and the solution $Y(x,\lambda)$ of (2.4–1) defined earlier with $u(x)$, Theorem 4 of Section 2.3 implies that $Y(x,\lambda)$ also has no zeros in $(a,b]$ for such λ.

As a consequence of the above reasoning, there exists an interval $I \equiv (-\infty, \mu_1)$ such that if λ is in I, $Y(x,\lambda)$ is zero-free in $(a,b]$. The constant μ_1 is finite and is given by the least value of λ that satisfies $Y(b,\lambda) = 0$. (Recall that for

$$\min(\lambda r - q) > \left(\frac{\pi}{b - a}\right)^2 \max p\,,$$

$Y(x,\lambda)$ must have at least one zero in (a,b), and any new zeros that appear in the interval enter from the right.)

If $H = 0$ in the eigencondition (2.4–2), μ_1 is the desired least eigenvalue λ_0. Otherwise, we inspect the quotient

$$Y'(b,\lambda)/Y(b,\lambda).$$

By comparison with the solution $\tilde{y}(x,\lambda)$ of equation (*) above we can deduce from Theorem 5 of Section 2.3 that

$$p(b)\, Y'(b,\lambda)/Y(b,\lambda) \geq P\tilde{y}'(b,\lambda)/\tilde{y}(b,\lambda) \qquad \textbf{(2.4–3)}$$

for $-\infty < \lambda < \mu_1$. The right side of this inequality becomes unboundedly large and positive as λ decreases without limit (recall that $\lim_{\lambda \to -\infty} \gamma(\lambda) = +\infty$). It follows then that the left-hand side of (2.4–3), which is strictly decreasing with λ in $(-\infty, \mu_1)$ from Theorem 5, actually decreases from $+\infty$ to $-\infty$ as λ increases from $-\infty$ to μ_1.

By continuity, therefore, there must be a unique value of λ in the interval $(-\infty, \mu_1)$ for which

$$p(b)\, Y'(b,\lambda)/Y(b,\lambda) = -p(b)\, G/H.$$

Because of (2.4–2), this value of λ is the eigenvalue λ_0 in this case, and, as noted earlier, the associated eigenfunction $Y(x,\lambda_0)$ is zero-free in (a,b).

(2) Since we have settled the existence question for the least eigenvalue, we turn now to a consideration of other eigenvalues. Beginning with the μ_1 given above, we define the increasing sequence of real constants $\mu_1 < \mu_2 < \mu_3 < \ldots < \mu_m < \ldots$ as the successive values of λ that satisfy $Y(b,\lambda) = 0$. Note that if $\mu_m < \lambda \leq \mu_{m+1}$, $Y(x,\lambda)$ has precisely m zeros in (a,b). It then follows from Theorem 1 of Section 2.3 that $\mu_m \to \infty$ as $m \to \infty$.

In the same way that $\lambda_0 = \mu_1$ for vanishing H, the other μ_i $(i = 2,3, \ldots)$ are the higher-order eigenvalues of (2.4–1) whenever $H = 0$. The proof of the theorem in this case follows by making the association $\lambda_m = \mu_{m+1}$. For more general H we again inspect the quotient $Y'(b,\lambda)/Y(b,\lambda)$ as a function of λ. By virtue of Theorem 5 of Section 2.3 and interpreting changes in λ as corresponding changes in q, this quotient decreases strictly monotonically from $+\infty$ to $-\infty$ as λ increases from any arbitrary μ_m to the next succeeding μ_{m+1}. Since the functions involved are continuous, in each of these intervals there is a unique value of λ such that

$$p(b)\, Y'(b,\lambda)/Y(b,\lambda) = -p(b)\, G/H.$$

In view of (2.4–2), these are the eigenvalues we seek. If we index them so that $\mu_m < \lambda_m < \mu_{m+1}$ and associate with each the eigenfunction $Y(x,\lambda_m)$, the demonstration is finally complete. □

The importance of the theorem we have just proved cannot be overemphasized. As suggested earlier, it is the basic generalization of the classic problem

$$\begin{aligned} y'' + \lambda y &= 0 \qquad 0 < x < \pi \\ y(0) &= 0 \\ y(\pi) &= 0. \end{aligned}$$

The result is fundamental to differential equations theory and underlies the successful analyses of untold numbers of practical applications that have been modelled as eigenvalue problems.

It is also essential to mention that with slight modifications in the proofs, the above results can be generalized to encompass the situation wherein the coefficients and constants in (2.4–1) depend upon λ, so long as $p(x,\lambda)$, $q(x,\lambda)$, $p(a,\lambda)A(\lambda)/B(\lambda)$, and $p(b,\lambda)G(\lambda)/H(\lambda)$ are strictly monotonically decreasing functions of λ. Other extensions of the theory allow $r(x)$ to vanish at a and/or b, or even at other isolated points in (a,b). If r changes sign in (a,b), then, provided $q(x)$ and the products AB and GH are all nonnegative, the spectrum associated with (2.4–1) splits into two parts (λ positive and λ negative) and both positive and negative infinity serve as eigenvalue limit points (see Problem 10 for the effect of these conditions when r is of one sign). Lastly, the vastly more complex "singular" cases, wherein the fundamental interval (a,b) is infinite or semi-infinite, or $p(x)$ vanishes at one or both of the endpoints, or $r(x)$ and/or $q(x)$ becomes unbounded near an endpoint, can be treated, in a number of special situations, by natural boundedness assumptions and limiting procedures. (See Coddington and Levinson (1955), Birkhoff and Rota (1978), among others.)* The resulting eigenvalue problems and corresponding eigenfunctions are among the most important in mathematical physics. For example, such extensions are pertinent to Legendre and Chebyshev polynomials on $(-1,1)$, Bessel functions on $(0,b)$ and Hermite functions on $(-\infty,\infty)$. (In Appendix 2 we remind the reader of the more significant properties associated with the Legendre polynomials and with Bessel functions; also see Chapter 3.) Standard references such as Coddington and Levinson (1955), Ince (1953), or Swanson (1968) should be consulted for further applications of the theory and related results.

Before leaving this section let us look briefly at one nontrivial practical problem that should be familiar to most readers.

Example The vibrations of a circular membrane of radius b are governed by the wave equation

$$\nabla^2\phi - \frac{1}{c^2}\frac{\partial^2\phi}{\partial t^2} = 0$$

*See Weinberger (1965) for cases in which the desired theory does *not* carry over.

which is subject, say, to the boundary condition $\phi = 0$ on the edge $\rho = b$ (see Dettman (1969), pp. 210–211, for example). If we separate variables and introduce the normalized radial variable $x = \rho/b$, the essential underlying eigenvalue problem takes the form

$$(xy')' + (\lambda b^2 x - n^2/x)y = 0,$$
$$y(1) = 0$$
$$y(0) \text{ finite.}$$

Owing to our Sturmian theory and its extensions, we can make the following qualitative statements without ever solving explicitly for the well-known Bessel function solutions:

(i) For each value of n there exist an infinite number of eigenvalues λ_{nm}, all of which are positive;

(ii) As n increases, each λ_{nm} also increases;

(iii) If the radius b is increased, then all the λ_{nm} decrease.

(See Problems 10, 14.) Since the separation parameters n and λ couple the radial solutions to the angular and time components, respectively, these conclusions have immediate relevance to the behavior of the resonance modes associated with the given membrane. □

2.5 Eigenfunction expansions

In the previous two sections we have studied the eigenvalue problem (2.3–1).* We have shown that this Sturm-Liouville problem has only real eigenvalues and (by convention) real eigenfunctions; the former are all simple, discrete, and countably infinite in number, while each of the latter has a specified number of zeros in the fundamental domain. A number of additional results of importance are included in the problems at the end of this chapter. In particular, we discuss there the effect on the eigenvalues of changes in $q(x)$, $r(x)$, and the boundary conditions.

With this foundation, the reader should feel reasonably comfortable in talking, at least in general terms, about the behavior of the eigenvalues and eigenfunctions of Sturm-Liouville problems. On the other hand, the reader is undoubtedly aware that, the concluding example of Section 2.4 notwithstanding, the determination of *specific* eigenvalues and eigenfunctions associated with a given problem of the form (2.3–1) can be a nontrivial undertaking. For this reason, Chapter 3 is devoted largely to a careful consideration of the individual solutions of various second-order differential equations that turn up repeatedly in applied endeavors. The especially important cases of Bessel functions and Legendre polynomials are also reviewed in Appendix 2.

In the meantime, let us assume that we have at our disposal an infinite set of functions such as the real eigenfunctions y_i ($i = 0, 1, 2, \ldots$) of a particular

---

*Coddington and Levinson (1955), amongst others, treats the more general self-adjoint problem involving nth order operators and mixed boundary conditions.

Sturm-Liouville problem. From the discussion at the beginning of Section 2.3, we know that these functions form an orthogonal set, that is,

$$\int_a^b r\, y_i y_j\, dx = 0 \quad \text{if} \quad i \neq j.$$

For convenience, we also presume that each function has been normalized. Now we want to ask what can be said about representing some arbitrary continuous real-valued function by a series expansion of the y_i.

This is a question of considerable practical importance. In numerous applied problems, it is helpful to be able to use an underlying orthonormal set of y_i as the building blocks for functions. In this way, knowledge about "normal modes" and "resonant vibrations" can be used to infer information regarding other functions either appearing elsewhere in the problem at hand or of importance in some closely-related investigation.

Historically, Daniel Bernoulli (1700–1782) and Euler had encountered this expansion question in their investigations of vibrating strings, elastic rods, and the like. However, the French mathematician Joseph Fourier (1768–1830) made the first systematic study of the problem, at least for the case of pure sinusoids, in connection with his development of the mathematical theory of heat in solids. The analysis of these eigenfunction expansions is therefore commonly termed the theory of *generalized Fourier series*.

In what follows we shall denote the generalized real *inner product* and *norm*, with respect to the (nonnegative) weight function $r(x)$, by

$$(f,g) \equiv \int_a^b f(x)\, g(x)\, r(x)\, dx$$

and

$$\|f\| \equiv (f,f)^{1/2} = \left[\int_a^b f^2(x)\, r(x)\, dx\right]^{1/2},$$

respectively.

Given an arbitrary function f, how well a finite series of orthonormal functions approximates it is customarily measured by calculating

$$\|f - \sum_{i=0}^{n} \alpha_i\, y_i\|.^*$$

Since integration is an averaging process, this quantity provides an assessment of the approximation error in some "mean" sense. Readers should recall, or prove anew for themselves (see Problem 20), that the error is smallest when the constants α_i in the sum are chosen to be the (generalized) *Fourier coefficients* (f,y_i) of f with

*Utilization of this quantity allows the extension of the approximation concepts that follow to discontinuous functions, since for such functions it is inappropriate from a practical point of view to talk about the difference $f - \Sigma\alpha_i y_i$ *without* the averaging provided by the operation of taking the norm. As an example, consider the case when f has jump discontinuities (such as are present in step functions) but all of the y_i are continuous.

respect to the $\{y_i\}$. Moreover, the same argument shows that these Fourier coefficients are square-summable. The specific result is

Property 1
If $f(x)$ is real-valued and continuous on $[a,b]$ and $\{y_i\}$ is a sequence of real-valued functions, orthonormal on $[a,b]$, then the series $\sum_i (f,y_i)^2$ is convergent and satisfies

$$\sum_{i=0}^{\infty} (f,y_i)^2 \leq \|f\|^2 \qquad \text{(Bessel's inequality)}$$

(see Birkhoff and Rota (1978), p. 299, for example).

On the basis of this property it follows that the expansion $\Sigma\ (f,y_i)y_i$ converges, at least in the average sense, to some function. (This is a ready consequence of a famous theorem of Riesz and Fischer.) Whether the series converges to the original function f depends upon a further property of the sequence $\{y_i\}$, namely its *completeness*. Completeness amounts to the absence of nontrivial functions orthogonal to all of the y_i. In other words, for complete sets $\{y_i\}$, if $(f,y_i) = 0$ for all values of i and f is continuous, then f vanishes identically. For sets of eigenfunctions it can be proved that

Property 2
The eigenfunctions of the Sturm-Liouville problem (2.3–1) form a complete set (Coddington and Levinson (1955), pp. 197ff; also see the discussion and proof in Section 10.5 of this text).

It follows from this and the preceding comments that

Property 3
If $f(x)$ is continuous on $[a,b]$ and $\{y_i\}$ is the orthonormal sequence of eigenfunctions associated with a given eigenvalue problem of the form (2.3–1) then

$$f = \sum_{i=0}^{\infty} (f,y_i)\, y_i, \qquad \textbf{(2.5–1)}$$

where the equality is meant in the sense that

$$\lim_{n\to\infty} \|f - \sum_{i=0}^{n} (f,y_i)\, y_i\| = 0.$$

Moreover, *Parseval's equality* is valid:

$$\sum_{i=0}^{\infty} (f,y_i)^2 = \|f\|^2. \qquad \textbf{(2.5–2)*}$$

*Interested readers may want to consider the finite-dimensional analogues of (2.5–1) (2.5–2), which are valid when the y_i are the eigenvectors of a given hermitian matrix A of order n and f is an arbitrary n-vector.

If $f(x)$ has a continuous derivative and satisfies the same boundary conditions at $x = a,b$ as the $\{y_i\}$, then the convergence of the series representation (2.5–1) can be shown to be actually uniform and absolute (see Dettman (1969), pp. 180ff., for example).

2.6 Specific examples: channels of various cross-section

Now we return to the analysis of shallow water waves initiated in Section 2.1. For harmonic motion the equation of interest is (2.1–7), namely

$$(bhy')' + (b\omega^2/g)y = 0, \qquad \textbf{(2.6–1)}$$

where $b(x)$ and $h(x)$ are the channel breadth and depth, respectively, g is the constant of gravitational acceleration, and ω the oscillatory frequency. This equation has the form of the standardized Sturm-Liouville problem (2.3–1) when we make the identifications $p = bh$, $q = 0$, $r = b/g$ and $\lambda = \omega^2$. The various assumptions of the basic theory will accordingly apply for all reasonable channel configurations. To note but one consequence, the eigenvalues λ will generally be nonnegative (see Problem 10) and thus real "eigenfrequencies" ω will result.

Constant depth channels

If both h and b are constant in (2.6–1), all nontrivial solutions are pure sinusoids. Of more interest is the case where $h(x)$ = constant but $b(x) = \alpha x$. Here (2.6–1) becomes Bessel's equation of zero order

$$(xy')' + k^2xy = 0, \qquad \textbf{(2.6–2)}$$

with $k^2 = \omega^2/(hg)$. (Review material on Bessel functions appears in Appendix 2.) If this wedge-shaped channel of uniform depth extends from $x = 0$ to $x = a$, the associated Sturm-Liouville problem on the interval $[0,a]$ is singular since $p(0) = 0$. However, as mentioned earlier, the theory goes through with the sole change being perhaps the form of the boundary condition at $x = 0$. In this case, it may be shown that no change is needed and the appropriate condition is merely $y'(0) = 0$.*

A variant of practical interest is a channel closed at the end $x = 0$ but coupled to the open sea at the end $x = a$. If we assume that the sea exhibits tidal oscillations of its own with amplitude C and frequency ω, then the coupling into the channel can be accomplished by specifying that $y(a)$ takes on this same value C.

---

*In view of the relationships of Section 2.1 connecting $y(x)$ to the fluid velocity $u(x,t)$, this boundary condition is equivalent to the vanishing of the velocity at the closed end of the channel $x = 0$.

It should be noted that the boundary-value problem in this case is not of Sturm-Liouville type. The allowable value of the eigenparameter is determined a priori by physical considerations, and the boundary conditions are not of the form (2.3–1b). Nevertheless, much of our earlier analysis is relevant. We forego details, but refer the reader to Lamb (1945).

The solution of (2.6–2) subject to the conditions $y'(0) = 0$ and $y(a) = C$ has the form

$$y(x) = C\,\frac{J_0(kx)}{J_0(ka)}.$$

Due to the behavior of the Bessel function, the oscillations that are induced in this channel have maximum amplitude at the closed end $x = 0$. As k increases (recall $k^2 = \omega^2/hg$), additional wave nodes (zeros) appear, as predicted by the general theory. Thus more peaks and valleys (troughs) are present in shallower channels than in deeper channels. The distance between troughs, as measured from one function minimum to the next succeeding function minimum, is, however, nearly the same throughout the channel (see Appendix 2).

Channels of constant breadth

If $b(x) =$ constant but $h(x) = \alpha x$, then (2.6–1) becomes

$$(xy')' + k^2 y = 0,$$

where now $k^2 = \omega^2/(\alpha g)$. This equation is also of Bessel-function type. The change of variable $s = 2\sqrt{x}$ transforms it to

$$\frac{d}{ds}\left(s\,\frac{dy}{ds}\right) + k^2 s y = 0.$$

If we assume that the channel is extended somewhat in the negative x-direction, thus allowing waves to come up "on the beach" so to speak, we can analyze a physical situation analogous to the previous example of a channel communicating with the open sea. Insisting that $y'(x)$ remain finite (that is, the fluid velocity remain bounded) at $x = 0$ and again setting $y(a) = C$ gives rise to the solution

$$y(x) = C\,\frac{J_0\,(2k\sqrt{x})}{J_0(2k\sqrt{a})}.$$

For this configuration, a bottom of more gentle slope generates a greater number of troughs. Moreover, the distance between troughs, by no means constant because of the presence of the radical $\sqrt{x}$, decreases toward the closed end of the channel.

Problem 21 provides a noteworthy example of a channel in which both depth and breadth vary in linear fashion.

Closed channels

As in any resonant structure, free oscillations can occur in closed channels (just imagine resonant waves in a lake. In a small container, such as a fuel tank, we might call this *sloshing*.) To illustrate this phenomenon, we consider a channel of constant breadth with a depth that varies according to the formula

$$h(x) = h_o\left(1 - \frac{x^2}{a^2}\right) \qquad -a < x < a$$

where h_o is constant. Equation (2.6–1) now takes the form

$$\left[\left(1 - \frac{x^2}{a^2}\right)y'\right]' + \left(\frac{\omega^2}{h_o g}\right)y = 0. \qquad \textbf{(2.6–3)}$$

As in our previous examples, since $p(x)$ vanishes at the endpoints of the interval, the natural boundary conditions suggested by physical considerations may have to be appropriately modified. In this case, insisting on the boundedness of $y'(x)$, and hence $y(x)$, at the endpoints $x = \pm a$ again suffices.

Although the Sturm-Liouville problem consisting of (2.6–3) and the boundary conditions is singular, our earlier theory is still applicable (see Birkhoff and Rota (1969), for example; also see Appendix 2). We find that $a^2\omega^2/(h_o g) = n(n+1)$, where n is integral, and that the normal modes (eigenfunctions) are expressible as

$$y(x) = c\, P_n(x/a),$$

where P_n is the Legendre polynomial of order n.

If we compare the eigenfrequencies ω associated with this channel configuration to those that occur in a closed channel of the same length but of *uniform** depth $\frac{2}{3}\, h_o$ (so that the fluid volumes are identical), we find that they are in the ratio

$$\sqrt{n(n+1)} \Big/ (n\pi/\sqrt{6}).$$

The two fundamental (lowest frequency) modes correspond to the case $n = 1$. They are both asymmetrical about the channel midpoint and have frequencies in the ratio $2\sqrt{3}/\pi \doteq 1.103$.

Actually this last comparison is of more than passing interest. As Troesch (1965), (1967) has shown, amongst all symmetrical convex channels enclosing identical fluid volumes, flat and parabolic bottoms lead to the smallest and largest fundamental eigenfrequencies, respectively. (See Problem 22 for an instructive intermediate case.)

---

*The uniform depth channel can of course be viewed as a limiting case. Troesch (1967) does this in connection with an interesting analysis of symmetrical channels with trapezoidal depth functions. He also has treated comparable symmetric two-dimensional configurations.

Final remarks

The several examples treated above are typical of what can be deduced using shallow water theory in one dimension. Closed channels exhibit the full resonance behavior expected of such structures. Open channel problems, on the other hand, display the consequences of coupling with the outside ocean and suggest the marked effects that oceanic tides have upon shallow seas, bays, and estuaries. Perhaps surprisingly, two-dimensional shallow-water theory is grossly applicable even to the case of tidal motion in the oceans themselves since the important parameter—the ratio of water depth to wavelength, or equivalently, to the curvature of the tidal wave surface—is small. Stoker (1957) should be consulted for other applications of the theory to seiches and related oscillations in lakes and harbors. Lamb (1945), however, remains the classic reference for the subject.

References

Birkhoff, G. and G.-C. Rota (1978): *Ordinary Differential Equations*, 3rd ed., John Wiley, New York; Chapters 10, 11.

Boyd, T. J. M. and J. J. Sanderson (1969): *Plasma Dynamics*, Barnes and Noble, New York.

Coddington, E. A. and N. Levinson (1955): *Theory of Ordinary Differential Equations*, McGraw-Hill, New York; Chapters 7, 8, 9.

Cole, R. H. (1968): *Theory of Ordinary Differential Equations*, Appleton-Century-Crofts, New York; Chapters 8, 9.

Dettman, J. W. (1969): *Mathematical Methods in Physics and Engineering*, 2nd ed., McGraw-Hill, New York; Chapter 4.

Greenspan, D. (1973): *Discrete Models*, Addison-Wesley, Reading, Mass.; Chapters 1, 7, 8, 9.

Ince, E. L. (1953): *Ordinary Differential Equations*, Dover, New York; Longmans, London (1927); Chapters IX, X, XI.

Lamb, H. (1945): *Hydrodynamics*, 6th ed., Dover, New York; Cambridge Univ. Press (1932); Chapter VIII.

Lighthill, J. (1975): "Mathematical Biofluidynamics", *Regional Conference Series in Applied Math. #17*, Soc. for Indust. and Appl. Math., Philadelphia.

Stoker, J. J. (1957): *Water Waves*, Wiley Interscience, New York; Chapters 2, 10.

Swanson, C. A. (1968): *Comparison and Oscillation Theory of Linear Differential Equations*, Academic Press, New York; Chapter 1.

Szego, G. (1975): "Orthogonal Polynomials," *Colloquium Publications Vol. XXIII*, 4th ed., Amer. Math. Soc., Providence, R.I.

Taylor, G. I. (1921): "Tides in the Bristol Channel," *Proc. Cambridge Phil. Soc.* 20: 320–325.

Troesch, B. A. (1965): "An Isoperimetric Sloshing Problem," *Comm. Pure Appl. Math.* 18: 319–338.

——— (1967): "Fluid Motion in a Shallow Trapezoidal Container," *SIAM J. Appl. Math.* 15: 627–636.

Weinberger, H. F. (1965): *A First Course in Partial Differential Equations with Complex Variables and Transform Methods*, John Wiley, New York; Chapter VII.

Problems

Sections 2.1, 2.2

1. Let the two differential operators M and N be defined by

$$My \equiv \alpha y'' + \beta y' + \gamma y \qquad Ny \equiv (\alpha y)'' - (\beta y)' + \gamma y$$

where α, β, γ are functions of the independent variable x. The operator N is called the *adjoint* of the operator M. Verify that

(a) for any two functions $u(x)$ and $v(x)$,

$$uMv - vNu = [\alpha W(u,v) - (\alpha' - \beta)uv]';$$

(b) M is the adjoint of N;

(c) $M = N$ if and only if $\beta = \alpha'$.
[Note that this relation characterizes operators of the form (2.2–1) and justifies the name "self-adjoint."]

2. The results of Problem 1 can be extended to higher-order linear ordinary differential operators. In this more general setting a differential operator L such that $uLv - vLu$ can be written as a perfect derivative is termed *self-adjoint*. A homogeneous linear ordinary differential equation $Ly = 0$ with a self-adjoint operator L is said to be in *self-adjoint form*.

(a) Show that the equation

$$\alpha y'' + \beta y' + \gamma y = 0$$

with $\alpha(x) \neq 0$ can be transformed into the equivalent self-adjoint form

$$(py')' - qy = 0.$$

(b) Can third-order homogeneous linear ordinary differential equations be put into some analogous canonical self-adjoint form?

(c) What about fourth-order equations?

★3. Verify that the general mixed homogeneous boundary conditions given by Definition 1 of Section 2.2 are regular if and only if

$$p(a) \det \begin{bmatrix} C & D \\ G & H \end{bmatrix} = p(b) \det \begin{bmatrix} A & B \\ E & F \end{bmatrix}.$$

(This is a nontrivial problem, algebraically; see Ince (1953), pp. 216–7).

Section 2.3

4. (a) Find the eigenvalues and normalized eigenfunctions associated with the boundary value problem

$$y'' + \lambda y = 0 \qquad 0 < x < 1$$
$$y(0) = 0, \quad \alpha y(1) + y'(1) = 0 \qquad (\alpha \text{ real constant}).$$

[An explicit formula will not be possible for the eigenvalues.]

(b) Describe the approximate location of the eigenvalues of large index.

(c) Verify the orthogonality of the eigenfunctions by direct integration using trigonometric identities.

5. Demonstrate that the eigenvalues of the boundary value problem (2.3–1) are all real. [Hint: recall the analysis leading up to Property 1 of Section 2.3. Also, compare with the proof of Property 1 of Section 1.5.]

6. Let $y_m(x)$ be the eigenfunction of the Sturm-Liouville problem (2.3–1) associated with the eigenvalue λ_m, and assume $Y(x,\lambda)$ is some other solution of the equation (2.3–1a) that merely satisfies the *first* of the boundary conditions (2.3–1b). Define $F(\lambda) \equiv GY(b,\lambda) + HY'(b,\lambda)$.

(a) Establish the relation

$$(\lambda_m - \lambda) \int_a^b r\, Yy_m \, dx = p(b)\, F(\lambda) \begin{cases} y_m(b)/H \\ -y_m'(b)/G \end{cases}.$$

(b) Use the result of part (a) to provide an alternative proof of the simplicity (and discreteness) of the eigenvalues of the Sturm-Liouville problem (2.3–1).

7. Prove Theorem 2 of Section 2.3. [Hint: Recall $p(x)W(u,v) = \text{constant}$.]

8. Let $u(x)$ and $v(x)$ be solutions of the differential equations in (2.3–2) and (2.3–3), respectively. Show formally that whenever $v \neq 0$,

$$(q - Q)u^2 + (p - P)\,(u')^2 + P[W(u,v)/v]^2 = [u^2(pu'/u - Pv'/v)]',$$

and hence the Picone identity (2.3–4) is valid.

9. Use the relation (2.3–4) to establish Theorem 5 of Section 2.3.

Sections 2.4, 2.5

10. (a) If $q(x)$ as well as the products AB and GH are nonnegative, establish that all the eigenvalues of the Sturm-Liouville problem (2.3–1) are likewise nonnegative. [Hint: Look at (y, Ly).]

(b) Show that a similar conclusion is valid if $p(a) = p(b)$ and the boundary conditions (2.3–1b) are replaced by the periodic conditions of Property 2 of Section 2.2.

(c) Are there mixed homogeneous boundary conditions, other than periodic ones, for which the eigenvalues of the associated Sturm-Liouville problem are nonnegative?

★11. Take $p = 1$, $q = 0$ and $[a,b] = [0,1]$ in the Sturm-Liouville problem (2.3–1). Assume that the boundary conditions are of the form

$$y(0) = 0, \quad y'(1) = \alpha y(1) \qquad (\alpha \text{ arbitrary, but fixed}),$$

and that $r(x)$ is a nonnegative function of x. Demonstrate that there can be at most one eigenvalue that is nonpositive. [Hint: Determine the number of zeros a solution y can have if $\lambda \leq 0$.]

12. Take $\lambda = 0$ in the self-adjoint equation (2.3–1a) and assume that the constants p_1, p_2, q_1, q_2 are such that $0 < p_1 \leq p(x) \leq p_2$, $q_1 \leq q(x) \leq q_2$.

    (a) Using a comparison argument, show that if $-q_1 < \left(\dfrac{\pi}{b-a}\right)^2 p_1$ (which is satisfied, in particular, if $q_1 \geq 0$) , then all nontrivial solutions of this equation have no more than one zero in (a,b).

    (b) In like manner, demonstrate that the solutions of this equation have at least m zeros in $(a,b]$ whenever $-q_2 \geq \left(\dfrac{m\pi}{b-a}\right)^2 p_2$.

13. Reason as in Problem 12 to deduce that if the eigenvalues $\{\lambda_m\}$ of the full Sturm-Liouville problem (2.3–1) are enumerated according to size, then

$$\frac{p_1}{r_2} \leq \lim_{m \to \infty} \lambda_m \left(\frac{b-a}{m\pi}\right)^2 \leq \frac{p_2}{r_1},$$

where $0 < p_1 \leq p(x) \leq p_2$ and $0 < r_1 \leq r(x) \leq r_2$. [See Birkhoff and Rota (1978) and Dettman (1969) for improvements to this result.]

★14. (a) Show by means of a perturbation argument that if $q(x)$ is replaced by $Q(x)$ where $Q(x) \leq q(x)$ (that is, $q(x)$ "decreases"), then all of the eigenvalues of the Sturm-Liouville problem (2.3–1) also decrease.

    (b) On the other hand, verify that if $r(x)$ "decreases," any negative eigenvalue associated with (2.3–1) decreases, but all the positive eigenvalues increase.

15. Consider two Sturm-Liouville problems formed with the same equation (2.3–1a) but with differing boundary conditions, namely

$$\text{(I)}\quad \begin{aligned} Ay(a) - By'(a) &= 0 \\ Gy(b) + Hy'(b) &= 0 \end{aligned} \qquad \text{and} \qquad \text{(II)}\quad \begin{aligned} Cy(a) - Dy'(a) &= 0 \\ Ey(b) + Fy'(b) &= 0. \end{aligned}$$

    (a) If $\{\lambda_i\}$, $\{\mu_i\}$ denote the ordered sequence of eigenvalues for the problems I and II, respectively, show that

$$\lambda_i \leq \mu_i < \lambda_{i+1} \qquad (i = 0,1,2,\ldots)$$

    whenever $C/D \geq A/B$ (or $B/A \geq D/C$) and $E/F \geq G/H$ (or $H/G \geq F/E$).

    (b) Infer from the result of part (a) that the eigenvalues associated with vanishing function values on the boundary (so-called homogeneous Dirichlet conditions) always dominate the corresponding eigenvalues associated with vanishing derivatives on the boundary (homogeneous Neumann conditions).

★16. (a) For the equation $y'' + Q(x)y = 0$, show that the product yy' is an increasing function in every interval in which $Q(x) < 0$, and therefore nontrivial solutions have at most one zero in such intervals. On the other hand, verify that if $Q(x)$ is positive and increases without bound, nontrivial solutions possess a countable infinity of zeros.

(b) Apply the results of part (a) to establish that every nontrivial solution of the Airy equation $y'' = xy$ has at most one positive zero, but an infinity of negative zeros.

(c) What are the comparable conclusions for the solutions of the generalized Airy equation $y'' = x^n y$, where n is integral?

★17. If, in the equation $y'' + Q(x)y = 0$, $Q(x)$ is increasing, demonstrate that

$$x_i - x_{i-1} < x_{i+1} - x_i$$

where $x_1 < x_2 < \dots$ are the zeros of a nontrivial solution y.

18. Use a comparison argument to show that if ν is real, every real nontrivial solution of Bessel's differential equation

$$x^2 y'' + xy' + (x^2 - \nu^2)y = 0$$

has an infinite number of positive zeros. [Hint: first let $y = Y/\sqrt{x}$.]

19. Design a Sturm-Liouville problem for which the eigenvalues are precisely the zeros of the Bessel function cross-product

$$f_n(k,\lambda) \equiv J_n(\lambda)Y_n(k\lambda) - Y_n(\lambda)J_n(k\lambda)$$

where n is integral and $k > 1$.

20. Prove Property 1 of Section 2.5.

Section 2.6

21. Consider a shallow-water channel in which the breadth and depth are each proportional to x; that is, $b(x) = \alpha x$ and $h(x) = \beta x$ with constant α,β. Analyze and discuss the nature of the wave oscillations that occur if the channel communicates with the open sea as in the open-channel examples of Section 2.6.

[This configuration provides a satisfactory description of a number of existing channels including the renowned Bristol Channel; see Taylor (1921).]

★22. Let a closed shallow-water channel of uniform breadth have length $2a$. Assume that the channel bottom slopes uniformly from either end to the middle and thus

$$h(x) = \begin{cases} h^* x/a & 0 \le x \le a \\ h^*(2 - x/a) & a \le x \le 2a \end{cases}$$

for some constant h^*.

(a) Show that the normal modes (eigenfunctions) for the free wave oscillations that this resonant configuration can support fall naturally into two classes determined by their symmetry about the channel midpoint.

(b) Determine the eigenfrequencies ω associated with the two lowest-order nontrivial modes, and compare these with the corresponding quantities determined for the closed channels considered in Section 2.6. (For the purpose of this comparison let $h^* = 4h_o/3$ so that all channels enclose the same volume of fluid.)

3

The special functions of applied mathematics

3.1 Electromagnetic propagation in curved waveguides

In Chapter 2 we discussed a class of applications which could be modelled mathematically as eigenvalue problems of the Sturm-Liouville type. There, we assumed that the solutions of these problems (which turned out to be Bessel functions and Legendre polynomials in the special cases considered) were all well-known functions with which we were reasonably acquainted. In this chapter, we want to take additional steps toward ensuring that the reader is indeed familiar with these functions.

In the next several sections, then, we will carry out a careful analysis of the more important of those functions that might be termed the *special functions of applied mathematics*. These are the functions that continually recur as solutions of differential equations and related boundary- and initial-value problems in a vast array of applied contexts. Our particular interest is in those functions that satisfy second-order equations, and we will consider properties unique to individual functions, as well as qualitatively similar behavior shared by the members of certain large classes of such functions.

In order to motivate our discussion, we want to analyze in some detail an extremely interesting problem in microwave propagation. Since the pertinent geometry is cylindrical, the solution turns out to involve Bessel functions. However, for this particular problem we need to have a deeper understanding of the nature of Bessel functions of both the first and second kinds than was necessary for the applications we encountered in Chapter 2. Indeed, the complete resolution of our microwave application requires some fairly detailed facts about the behavior of various combinations of Bessel functions as functions of their order.

Consider then electromagnetic waves propagating along the continuously curved waveguide of rectangular cross-section depicted in Figure 3.1. These waves can be described in terms of electric and magnetic field vectors **E** and **H** in much the same way as we would characterize static configurations of electric and magnetic charges. When the charges are moving, however, a time dependence is introduced into these functions and, in view of Faraday's law of induction, the two fields are coupled together. The Scottish physicist James Clerk Maxwell (1831–

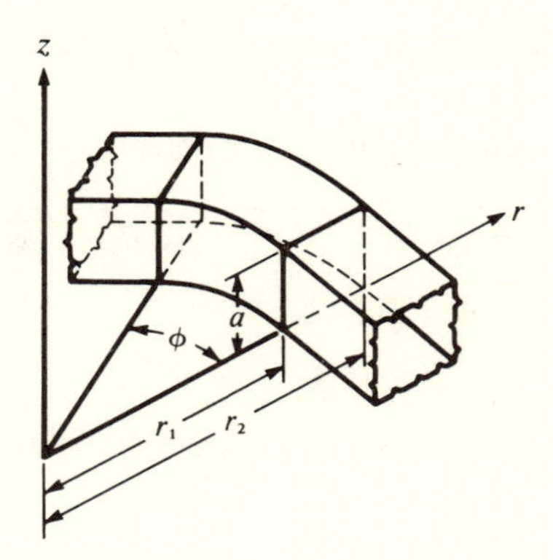

Figure 3.1 Curved waveguide of rectangular cross-section

1879) was the first to suggest how such an electrodynamical configuration could be modelled mathematically. In the most general case of a moving amorphous medium, his equations are rather complicated. Fortunately, for our problem it will suffice to assume that the waveguide of interest is stationary and is filled with a source-free nonconducting uniform material. In this situation Maxwell's equations take the form

$$\begin{aligned} \operatorname{curl} \mathbf{E} &= -\mu \frac{\partial \mathbf{H}}{\partial t} \qquad & \operatorname{div} \mathbf{E} &= 0 \\ \operatorname{curl} \mathbf{H} &= \epsilon \frac{\partial \mathbf{E}}{\partial t} \qquad & \operatorname{div} \mathbf{H} &= 0. \end{aligned} \tag{3.1–1}$$

[For further background see standard references such as Smythe (1968), Stratton (1941) or Panofsky and Phillips (1962).] Here the constants μ and ε are the medium permeability and permittivity, respectively. If a harmonic time dependence $e^{i\omega t}$ is assumed for the electromagnetic field, both **E** and **H** satisfy the reduced *vector* wave equation

$$\operatorname{curl} \operatorname{curl} \mathbf{F} - k^2\mathbf{F} = 0, \tag{3.1–2}$$

where $k^2 = \omega^2\mu\epsilon$ with ω the radian frequency.

As is often discussed in vector analysis courses, two noteworthy independent families of vector solutions of (3.1–2) may be expressed in terms of a constant vector **c** and the so-called Hertz potentials u as

$$\begin{aligned} \mathbf{M} &= \operatorname{curl} (\mathbf{c}u) \\ \mathbf{N} &= \frac{1}{k} \operatorname{curl} \operatorname{curl} (\mathbf{c}u). \end{aligned} \tag{3.1–3}$$

Classical vector identities suffice to show that if these special functions u satisfy the reduced *scalar* wave or Helmholtz equation

$$\nabla^2 u + k^2 u = 0,$$

then both **M** and **N** are solutions of the vector equation (3.1–2). By making the

Hertz potentials central to our considerations, therefore, we greatly simplify the analysis.

In rectangular coordinates, the Hertz potentials are given by

$$u = \frac{\sin}{\cos} \sqrt{h^2 - \nu^2}\, x \frac{\sin}{\cos} qz\, e^{\pm i\nu y}$$

with ν, q undetermined constants and

$$h \equiv (k^2 - q^2)^{1/2} \tag{3.1–4}$$

(see Problem 1). The choice of $\mathbf{c}$ as a unit vector parallel to the z-axis is convenient and leads to

$$\begin{aligned} \mathbf{M} &= \mathbf{e}_x \frac{\partial u}{\partial y} - \mathbf{e}_y \frac{\partial u}{\partial x} \\ &= \left\{ \pm i\nu \mathbf{e}_x \frac{\sin}{\cos} \sqrt{h^2 - \nu^2}\, x - \sqrt{h^2 - \nu^2}\, \mathbf{e}_y \right. \\ &\quad \left. \frac{\cos}{-\sin} \sqrt{h^2 - \nu^2}\, x \right\} \frac{\sin}{\cos} qz\, e^{\pm i\nu y} \end{aligned} \tag{3.1–5}$$

$$\begin{aligned} \text{and} \quad \mathbf{N} &= \frac{1}{k} \left[\mathbf{e}_x \frac{\partial^2 u}{\partial x \partial z} + \mathbf{e}_y \frac{\partial^2 u}{\partial y \partial z} + \mathbf{e}_y \left(\frac{\partial^2 u}{\partial z^2} + k^2 u \right) \right] \\ &= \left\{ \frac{q}{k} \left[\sqrt{h^2 - \nu^2}\, \mathbf{e}_x \frac{\cos}{-\sin} \sqrt{h^2 - \nu^2}\, x \right. \right. \\ &\quad \left. \pm i\nu\, \mathbf{e}_y \frac{\sin}{\cos} \sqrt{h^2 - \nu^2}\, x \right] \frac{\cos}{-\sin} qz \\ &\quad \left. + \frac{h^2}{k} \mathbf{e}_z \frac{\sin}{\cos} \sqrt{h^2 - \nu^2}\, x \frac{\sin}{\cos} qz \right\} e^{\pm i\nu y}, \end{aligned}$$

where $\mathbf{e}_x$, $\mathbf{e}_y$, and $\mathbf{e}_z$ designate unit vectors along the coordinate axes.

In a cylindrical coordinate system (r, φ, z) the Hertz potentials can be expressed as

$$u = C_\nu(hr) \frac{\sin}{\cos} qz\, e^{\pm i\nu\varphi}.$$

Here C_ν is a general Bessel function of order ν (some linear combination of J_ν and Y_ν), and h,q are again related by (3.1–4). If the same choice as before is made for $\mathbf{c}$ in (3.1–3), we find

$$\begin{aligned} \mathbf{M} &= \left\{ \pm \frac{i\nu}{r} C_\nu(hr)\mathbf{e}_r - hC_\nu'(hr)\mathbf{e}_\varphi \right\} \frac{\sin}{\cos} qz\, e^{\pm i\nu\varphi} \\ \mathbf{N} &= \left\{ \frac{q}{k} [hC_\nu'(hr)\mathbf{e}_r \pm \frac{i\nu}{r} C_\nu(hr)\mathbf{e}_\varphi] \frac{\cos}{-\sin} qz + \frac{h^2}{k} C_\nu(hr) \frac{\sin}{\cos} qz\, \mathbf{e}_z \right\} e^{\pm i\nu\varphi}, \end{aligned} \tag{3.1–6}$$

with $'$ denoting differentiation with respect to the argument.

The expressions (3.1–6) are relevant to our electromagnetic application; each such **M** or **N** is a conceivable field configuration. The waveguide bend of Figure 3.1, however, will support only those waves that also satisfy appropriate boundary conditions. The usual assumption is that the waveguide walls are infinitely conducting; and therefore, these boundary conditions express the vanishing at $r = r_1, r_2$ (with $0 < r_1 < r_2$) and at $z = 0, a$ of the components of electric field intensity tangent to the walls.

The admissible electromagnetic waves may be conveniently classified into two categories according to whether the electric field is represented by a vector function of type **M** or **N**. (Recall that div curl of any vector function vanishes.) If **M** is utilized, since the resulting electric field has no component normal to the plane of the bend, such waves may be descriptively termed *Longitudinal Electric* (LE) [see Buchholz (1939), Cochran and Pecina (1966)]. Analogously, vector functions of type **N** lead to *Longitudinal Magnetic* (LM) waves. It turns out that a complete solution to the problem of propagation in waveguide bends of constant curvature can be given in terms of LE and LM waves.

LE waves

The selection of the vector function **M** as the electric field gives rise to

$$E_r = \pm \frac{i\nu}{r} C_\nu(hr) \sin\left(\frac{n\pi z}{a}\right) e^{\pm i\nu\varphi}$$

$$E_\varphi = -h\, C_\nu'(hr) \sin\left(\frac{n\pi z}{a}\right) e^{\pm i\nu\varphi}$$

$$E_z = 0,$$

where $0 < r_1 \le r \le r_2$ and $0 \le z \le a$. The components of the associated magnetic field are found from Maxwell's equations (3.1–1):

$$H_r = \frac{ih}{\omega\mu}\left(\frac{n\pi}{a}\right) C_\nu'(hr) \cos\left(\frac{n\pi z}{a}\right) e^{\pm i\nu\varphi}$$

$$H_\varphi = \mp \frac{1}{\omega\mu}\left(\frac{n\pi}{a}\right)\left(\frac{\nu}{r}\right) C_\nu(hr) \cos\left(\frac{n\pi z}{a}\right) e^{\pm i\nu\varphi}$$

$$H_z = \frac{ih^2}{\omega\mu} C_\nu(hr) \sin\left(\frac{n\pi z}{a}\right) e^{\pm i\nu\varphi}$$

In these relations n is a positive integer, equation (3.1–4) for h has become

$$h = \left[k^2 - \left(\frac{n\pi}{a}\right)^2\right]^{1/2}, \tag{3.1–7}$$

and, apart from a multiplicative scale factor, C_ν is the following linear combination of Bessel functions of the first and second kind:

$$C_\nu(hr) = J_\nu'(hr_1)\, Y_\nu(hr) - Y_\nu'(hr_1) J_\nu(hr).$$

From the vanishing of E_φ at $r = r_1, r_2$ we conclude that the admissible values of the angular propagation constant ν are determined as the implicit solutions of the transcendental equation $C_\nu'(hr_2) = 0$, that is,

$$J_\nu'(hr_1) Y_\nu'(hr_2) - Y_\nu'(hr_1) J_\nu'(hr_2) = 0. \tag{3.1–8}$$

LM waves

Selecting **N** as the electric field, we obtain in like fashion

$$E_r = \frac{h}{k}\left(\frac{n\pi}{a}\right) C_\nu'(hr) \sin\left(\frac{n\pi z}{a}\right) e^{\pm i}{}_{\nu\varphi}$$

$$E_\varphi = \pm \frac{i\nu}{kr}\left(\frac{n\pi}{a}\right) C_\nu(hr) \sin\left(\frac{n\pi z}{a}\right) e^{\pm i\nu\varphi}$$

$$E_z = -\frac{h^2}{k} C_\nu(hr) \cos\left(\frac{n\pi z}{a}\right) e^{\pm i\nu\varphi}$$

and

$$H_r = \pm \frac{k}{\omega\mu}\left(\frac{\nu}{r}\right) C_\nu(hr) \cos\left(\frac{n\pi z}{a}\right) e^{\pm i\nu\varphi}$$

$$H_\varphi = \frac{ikh}{\omega\mu} C_\nu'(hr) \cos\left(\frac{n\pi z}{a}\right) e^{\pm i\nu\varphi}$$

$$H_z = 0$$

with $0 < r_1 \leq r \leq r_2$ and $0 \leq z \leq a$. Here h is again specified by equation (3.1–7), though now

$$C_\nu(hr) = J_\nu(hr_1) Y_\nu(hr) - Y_\nu(hr_1) J_\nu(hr),$$

and values $n = 0,1,2, \ldots$ are allowed. The analogue of (3.1–8) for the permissible values of ν associated with the waves in this case is $C_\nu(hr_2) = 0$; that is,

$$J_\nu(hr_1) Y_\nu(hr_2) - Y_\nu(hr_1) J_\nu(hr_2) = 0. \tag{3.1–9}$$

The eigenvalue problems

Our analysis of wave propagation in continuously curved waveguides of rectangular cross-section has led us to consideration of solutions of Bessel's differential equation, which we write in self-adjoint form

$$\frac{d}{dr}\left(r\frac{dy}{dr}\right) + \left(h^2 r - \frac{\nu^2}{r}\right)y = 0 \qquad r_1 \leq r \leq r_2, \tag{3.1–10}$$

with one of the two sets of boundary conditions

$$y(r_1) = 0, \qquad y(r_2) = 0$$

or

$$y'(r_1) = 0, \qquad y'(r_2) = 0.$$

Since $r_1 > 0$, these are obviously examples of the standard Sturm-Liouville problem considered earlier. The associations

$$x = r, \qquad p(x) = r, \qquad q(x) = -h^2 r, \qquad r(x) = \frac{1}{r} \qquad \text{and} \qquad \lambda = -\nu^2$$

prompt the following qualitative conclusions on the basis of the theory developed in Chapter 2:

Property 1
There exists a countable infinity of allowable LE and LM waves (modes). Waves of either type may be indexed by pairs (m,n) of integers that characterize the number of amplitude nulls (zeros) in the two transverse directions r and z. Different waves within each set are pairwise orthogonal (the orthogonality is with respect to the weight function $1/r$), and the sets themselves are complete.

Property 2
Only a finite number of the allowable propagation constants ν are real; the remaining ones are purely imaginary. If for a given value of n, we have

$$k^2 \leq (n\pi/a)^2$$

(that is, the frequency ω is below "cutoff" in the z-direction), then all admissible values of ν are purely imaginary (see Problem 10, Chapter 2). In view of the phase factor $\exp(\pm i\nu\phi)$, the resulting waves are evanescent (attenuated).

In order to deduce additional conclusions, particularly those of a more quantitative character, we need a deeper understanding of some properties peculiar to Bessel functions. This then is our rationale for further study of the special functions of applied mathematics in general, and Bessel functions in particular.

3.2 An overview of some important equations

Ordinary differential equations can be classified on the basis of their singularities (see Appendix 3 at the end of the book). The most elementary second-order differential equation having regular singular points is the well-known Cauchy-Euler equation

$$x^2y'' + axy' + by = 0 \qquad (a,b \text{ constant}). \qquad \textbf{(3.2–1)}$$

Perhaps surprisingly, other second-order equations that occur frequently in practical situations are only slightly more complicated in form; indeed, many of them can be obtained as special cases of the following equation with polynomial coefficients (see Hildebrand (1976)):

$$x^2(1 + rx^m)y'' + x(p_0 + p_1x^m)y' + (q_0 + q_1x^m)y = 0 \qquad \textbf{(3.2–2)}$$

where m is a positive integer. (The choice $m = 0$ leads to the Cauchy-Euler equation.) As important examples we mention

(I) $(1 - x^2)y'' - 2xy' + p(p + 1)y = 0$ (Legendre's equation)
where $m = 2, r = -1, p_0 = 0, p_1 = -2, q_0 = 0, q_1 = p(p + 1)$;

(II) $(1 - x^2)y'' - xy' + p^2y = 0$ (Chebyshev's equation)
where $m = 2, r = -1, p_0 = 0, p_1 = -1, q_0 = 0, q_1 = p^2$;

(III) $(1 - x^2)y'' - 2(\alpha + 1)xy' + p(p + 2\alpha + 1)y = 0$ (Jacobi's equation)
where $m = 2, r = -1, p_0 = 0, p_1 = -2(\alpha + 1), q_0 = 0,$
$q_1 = p(p + 2\alpha + 1)$;

and

(IV) $x(1 - x)y'' + [c - (a + b + 1)x]y' - aby = 0$
(the hypergeometric equation)
where $m = 1, r = -1, p_0 = c, p_1 = -(a + b + 1), q_0 = 0, q_1 = -ab$.

When p is an integer n, equation (I) has the Legendre polynomial $P_n(x)$ as a solution, while the Chebyshev polynomial $T_n(x)$ is a solution of (II). Equation (III) is a particular case of the more general equation

$$(1 - x^2)y'' + [\beta - \alpha - (\alpha + \beta + 2)x]y' + p(p + \alpha + \beta + 1)y = 0 \quad \textbf{(3.2–3)}$$

which, when p is an integer, has the Jacobi polynomial $P_n^{(\alpha,\beta)}(x)$ as a solution. Equations (I) and (II) are obtained from (III) with $\alpha = 0$ or $-\frac{1}{2}$, respectively, so Legendre and Chebyshev polynomials are special cases of Jacobi polynomials. The precise relationships are

$$P_n(x) = P_n^{(0,0)}(x)$$

and

$$T_n(x) = \frac{n!\sqrt{\pi}}{\Gamma(n + \frac{1}{2})} P_n^{(-\frac{1}{2},-\frac{1}{2})}(x)$$

[see Abramowitz and Stegun (1965)].

In this same vein, (3.2–3) can be obtained from the hypergeometric equation by writing

$$a = -p, \qquad b = p + \alpha + \beta + 1, \qquad c = \alpha + 1,$$

and changing the independent variable from x to $1 - 2x$. Equation (III), hence also (I) and (II), are accordingly subsumed in (IV). By convention then, notwithstanding the generality of Jacobi's equation (3.2–3), the hypergeometric equation (IV) is taken as the prototype for the other equations listed (and many additional ones). Because of the importance of these various equations in diverse practical applications, we will study this hypergeometric equation in some detail in what follows.

Before considering equation (IV), however, we record some additional important cases of (3.2–2):

(V) $x^2y'' + xy' + (x^2 - \nu^2)y = 0$ (Bessel's equation)
where $m = 2, r = 0, p_0 = 1, p_1 = 0, q_0 = -\nu^2, q_1 = 1$;

(VI) $y'' - 2xy' + 2py = 0$ (Hermite's equation)
where $m = 2, r = 0, p_0 = 0, p_1 = -2, q_0 = 0, q_1 = 2p$;

(VII) $y'' - xy = 0$ (Airy's equation)
where $m = 3, r = 0, p_0 = 0, p_1 = 0, q_0 = 0, q_1 = -1$;

and

(VIII) $xy'' + (c - x)y' - ay = 0$ (Kummer's equation)
where $m = 1, r = 0, p_0 = c, p_1 = -1, q_0 = 0, q_1 = -a$.

One solution to equation (VIII) is the *confluent hypergeometric function*, customarily designated as $M(a;c;x)$. In certain cases $M(a;c;x)$ is an elementary function; for example,

$$M(a;a;x) = e^x \qquad \text{and} \qquad M(1;2;2x) = \frac{e^x}{x}\sinh x.$$

More typically, the confluent hypergeometric function is rather complicated, as illustrated by the relation

$$M(\nu + \tfrac{1}{2};2\nu + 1;2ix) = \Gamma(1 + \nu)e^{ix}(x/2)^{-\nu}J_\nu(x), \tag{3.2–4}$$

where $J_\nu(x)$ is the Bessel function of the first kind of order ν. In fact, it turns out that each of the other equations (V), (VI), and (VII) mentioned above can be obtained from Kummer's equation by an appropriate choice of the parameters a and c and change of the independent and dependent variables (see Problems 9, 10). Like the hypergeometric equation, therefore, Kummer's equation is central to the study of special functions and will receive its due consideration later in this chapter.

3.3 Hypergeometric functions and the Gauss series

All of the second-order equations in which we are interested have polynomial coefficients, that is, the $f(x)$, $g(x)$ in the standard form (see equation (A3–2) of Appendix 3) are rational functions. This ensures (Property 1, Appendix 3) that each equation has at most a finite number of singularities. If there are just two singular points, both regular, then the equation is essentially of Cauchy-Euler type (Problem 6). Equations with more singularities are, of course, more complicated.

The most general equation with three regular singular points placed, say, at 0,1, and ∞ is the *Riemann Differential Equation*, which we can express in standard form as

$$y'' + \left(\frac{A}{x} + \frac{B}{1-x}\right)y' + \left(\frac{C}{x^2} + \frac{D}{(1-x)^2} + \frac{E}{x(1-x)}\right)y = 0 \qquad \textbf{(3.3–1)}$$

(see Problem 11). Its interest stems from

Theorem 1:

Every second-order linear differential equation with polynomial coefficients that has precisely three singular points, all regular, can be transformed into the Riemann differential equation (3.3–1) by a change of independent variable of the form $t = (ax + b)/(cx + d)$ with $ad \neq bc$.

Proof: Problem 12. □

More importantly, however,

Theorem 2:

Every Riemann differential equation can be transformed into the hypergeometric equation

$$x(1 - x)y'' + [c - (a + b + 1)x]y' - aby = 0 \qquad \textbf{(3.3–2)}$$

by a change of dependent variable of the form $y(x) = x^\lambda(1-x)^\mu Y(x)$.

Proof: Make the suggested change of variable in (3.3–1) to yield the equation

$$Y'' + \left[\frac{A + 2\lambda}{x} + \frac{B - 2\mu}{1 - x}\right] Y' + \left[\frac{C + \lambda(\lambda - 1 + A)}{x^2} + \frac{D + \mu(\mu - 1 - B)}{(1 - x)^2} + \frac{E + B\lambda - A\mu - 2\lambda\mu}{x(1 - x)}\right] Y = 0.$$

Choose λ, μ so that

$$\lambda(\lambda - 1 + A) + C = 0,$$
$$\mu(\mu - 1 - B) + D = 0.$$

The differential equation for $Y(x)$ then becomes

$$x(1 - x)Y'' + [A + 2\lambda - x(A + 2\lambda + 2\mu - B)]Y' - (2\lambda\mu + A\mu - B\lambda - E)Y = 0.$$

Set $c = A + 2\lambda$ and determine a, b from

$$\begin{aligned} a + b + 1 &= A + 2\lambda + 2\mu - B \\ ab &= 2\lambda\mu + A\mu - B\lambda - E. \end{aligned}$$

The result is (3.3–2). □

Taken together, the above two theorems imply that all other second-order equations with three regular singular points can be derived from the hypergeometric equation by choosing suitable values for the parameters a,b,c and perhaps making appropriate changes of the independent and dependent variables. In view of this, the solutions of (3.3–2) are certainly worthy of our attention.

The Gauss hypergeometric series

Associated with the singularity of (3.3–2) at $x = 0$ is the indicial equation

$$\alpha^2 + (c - 1)\alpha = 0,$$

which has the two roots $\alpha = 0$ and $\alpha = 1 - c$. If c is nonintegral, therefore, we know that two linearly independent solutions can be obtained by the classical method of Frobenius (see Appendix 3). Substitution of (A3–6), with $x_0 = 0$ and $\alpha = 0$, into (3.3–2) yields the recurrence relation

$$c_k = c_{k-1}\frac{(a + k - 1)(b + k - 1)}{(c + k - 1)k} \qquad k \geq 1 \tag{3.3–3}$$

for the coefficients of this solution. Repeated use of (3.3–3) gives

$$c_k = c_0\frac{a(a + 1)\cdots(a + k - 1)b(b + 1)\cdots(b + k - 1)}{c(c + 1)\cdots(c + k - 1)1\cdot 2\cdots k}$$

or, written more compactly in terms of the gamma function,

$$c_k = c_0\frac{\Gamma(a + k)}{\Gamma(a)}\frac{\Gamma(b + k)}{\Gamma(b)}\frac{\Gamma(c)}{\Gamma(c + k)}\frac{1}{k!}.$$

For $c_0 = 1$ we obtain the solution regular at the origin with power series expansion

$${}_2F_1(a,b;c;x) \equiv \frac{\Gamma(c)}{\Gamma(a)\Gamma(b)}\sum_{k=0}^{\infty}\frac{\Gamma(a + k)\Gamma(b + k)}{\Gamma(c + k)}\frac{x^k}{k!}. \tag{3.3–4}$$

(See Problem 13 for the second solution.)

The expression (3.3–4) introduces the customary notation*. The subscripts on F remind us that there are two factorial (gamma) functions in the numerator of

*This notation is particularly helpful when working with generalized hypergeometric functions; see, for example, Bailey (1935) or Askey (1975).

each coefficient and one in the denominator. (We do not count the $k!$). The parameters a,b,c within the parentheses specify which factorials are involved. Note that ${}_2F_1(a,b;c;0) = 1$ and ${}_2F_1(a,b;c;x) = {}_2F_1(b,a;c;x)$.

Unless c is a nonpositive integer, the series expansion for ${}_2F_1(a,b;c;x)$ converges uniformly and absolutely for $|x| \leq R < 1$, a conclusion that is equally valid for complex values of x. For complex x of unit modulus the series diverges when $\mathrm{Re}(c - a - b) \leq -1$, converges absolutely when $\mathrm{Re}(c - a - b) > 0$ and, excluding the point $x = 1$ (where it diverges), converges conditionally in between. If $a = 1$ and $b = c$, the expansion reduces to the geometric series $\sum_k x^k$, which provides the rationale for the name *hyper*geometric function. Other special cases include the elementary functions

$$\begin{aligned} {}_2F_1(a,c;c;x) &= (1 - x)^{-a}, \qquad\qquad \textbf{(3.3–5)} \\ {}_2F_1(1,1;2;x) &= -x^{-1}\ln(1 - x), \\ {}_2F_1(\tfrac{1}{2};1;\tfrac{3}{2};-x^2) &= x^{-1}\arctan x, \\ {}_2F_1(-a,a;\tfrac{1}{2};\sin^2 x) &= \cos 2ax. \end{aligned}$$

From the series expansion (3.3–4), we also observe that

$$\frac{d}{dx}\,{}_2F_1(a,b;c;x) = \frac{ab}{c}\,{}_2F_1(a+1,b+1;c+1;x).$$

Although the first known result that in essence concerns hypergeometric functions is to be found in an early 14th century Chinese manuscript [see Askey (1975)], the fundamental studies of the functions were carried out by Euler (some of whose inventions we encountered earlier)*, the German mathematician and astronomer Karl Friedrich Gauss (1777–1855), and his countryman Ernst Kummer (1810–1893). Many of their results, including differentiation and transformation formulas, relations for contiguous functions (differing by 1 in the parameter values), and various integral representations, are gathered together in standard references such as Erdélyi et al (1953a) or Abramowitz and Stegun (1965). Because it will be extremely useful later, we derive one such special result here. This particular integral representation was originally deduced by Euler more than 200 years ago.

Euler's formula

We express $(1 - xt)^{-b}$ by means of the binomial series expansion (3.3–5), and multiply the result by $t^{a-1}(1 - t)^{c-a-1}$ to obtain

$$(1 - xt)^{-b}\, t^{a-1}(1 - t)^{c-a-1} = \sum_{k=0}^{\infty} \frac{\Gamma(b + k)}{\Gamma(b)k!} x^k t^{a+k-1}(1 - t)^{c-a-1}.$$

---

*As Askey (1975) has pointed out, it was Euler, not Gauss, who originally provided the definition (3.3–4) and showed that this infinite series satisfied the hypergeometric equation (3.3–2).

If we integrate the right-hand side of this expression term-by-term from $t = 0$ to $t = 1$, we are called upon to evaluate integrals of the form

$$\int_0^1 t^{\alpha-1}(1 - t)^{\beta-1}dt.$$

These may be recognized as beta functions (see Appendix 5) so that when $\operatorname{Re} c > \operatorname{Re} a > 0$ and $|x| < 1$,

$$\int_0^1 (1 - xt)^{-b}t^{a-1}(1 - t)^{c-a-1}dt = \sum_{k=0}^{\infty} \frac{\Gamma(b + k)}{\Gamma(b)k!} x^k \frac{\Gamma(a + k)\Gamma(c - a)}{\Gamma(c + k)}.$$

It follows that

$$_2F_1(a,b;c;x) = \frac{\Gamma(c)}{\Gamma(a)\Gamma(c - a)} \int_0^1 (1 - xt)^{-b}\, t^{a-1}\, (1 - t)^{c-a-1}dt. \quad \textbf{(3.3–6)*}$$

In the same way that the familiar integral representation for the gamma function provides that function's analytic continuation, the Euler formula (3.3–6) extends the domain of the basic hypergeometric function $_2F_1$ to all complex values of x satisfying $|\arg(1 - x)| < \pi$.

3.4 Chebyshev functions

As we have seen earlier, several important second-order equations are special cases of the hypergeometric equation (3.3–2). Consequently, various well-known special functions are expressible in terms of the hypergeometric function $_2F_1$. Two notable examples are

$$_2F_1\left(-p, p+1; 1; \frac{1-x}{2}\right) = P_p(x) \qquad \text{(Legendre function)}, \qquad \textbf{(3.4–1)}$$

$$_2F_1\left(-p, p; ½; \frac{1-x}{2}\right) = T_p(x) \qquad \text{(Chebyshev function)}. \qquad \textbf{(3.4–2)}$$

Since Legendre functions, and in particular Legendre polynomials (when $p = n$), are treated in a variety of undergraduate courses (also see Appendix 2), we concentrate here on the Chebyshev functions.

From the representation (3.3–4) and the relation (3.4–2), we see that

$$T_p(x) = 1 + \frac{(-p)p}{\left(\frac{1}{2}\right)}\left(\frac{1 - x}{2}\right) + \frac{(-p)(1 - p)p(p + 1)}{\left(\frac{1}{2}\right)\left(\frac{3}{2}\right)2!}\left(\frac{1 - x}{2}\right)^2 + \cdots.$$

---

*As a special case of this relation we have

$$_2F_1(a,b;c;1) = \frac{\Gamma(c)\Gamma(c-a-b)}{\Gamma(c-a)\Gamma(c-b)},$$

which turns out to be valid for all $\operatorname{Re}(c-a-b) > 0$; see Problem 14.

If p is integral, the series terminates and defines the Chebyshev polynomials

$$T_0(x) = 1, \quad T_1(x) = x, \quad T_2(x) = 2x^2 - 1, \quad T_3(x) = 4x^3 - 3x, \cdots .$$

$T_n(x)$ is an odd or even function according as n is odd or even, as the recurrence relation

$$T_{p+1}(x) = 2xT_p(x) - T_{p-1}(x) \tag{3.4–3}$$

confirms. The several cases also suggest that

$$T_n(1) = 1, \qquad T_n(-1) = (-1)^n,$$

and, using (3.4–3) again, we can readily verify this fact. Indeed, from (3.4–2) we obtain for general p, after some simplification using the result given in the footnote on page 67,

$$T_p(1) = 1, \quad T_p(-1) = \cos \pi p.$$

Although it is seldom employed, the following elementary expression provides an explicit representation of the Chebyshev functions:

$$T_p(x) = \frac{1}{2}\left[(x + i\sqrt{1 - x^2})^p + (x - i\sqrt{1 - x^2})^p\right]$$

(see Abramowitz and Stegun (1965), p. 556). If we set $x = \cos \theta$ we derive

$$T_p(\cos \theta) = \cos (p\theta), \tag{3.4–4}$$

an important relation that not only verifies the aforementioned values of T_p at $x = \pm 1$ but also shows that for p real and x in $[-1,1]$

$$|T_p(x)| \leq 1.$$

Other valuable properties of the Chebyshev functions that can be readily inferred from the identity (3.4–4) are the locations of the zeros of $T_p(x)$ in the fundamental interval

$$x_m(p) = \cos \frac{2m - 1}{2p} \pi, \qquad m = 1,2, \cdots ; m \leq |p| + \tfrac{1}{2}$$

at which the derivative satisfies

$$T_p{}'(x_m(p)) = (-1)^{m+1}p \Big/ \sin \frac{2m - 1}{2p} \pi,$$

as well as the extrema of T_p

$$x_m{}'(p) = \cos \frac{m\pi}{p}, \qquad m = 0,1, \cdots ; m \leq |p|$$

at which we find

$$T_p(x_m{}'(p)) = (-1)^m.$$

The Chebyshev functions play important roles in the approximation of functions, numerical quadrature, and microwave filter design where the uniform (in θ) spacing of the zeros and equiripple behavior noted above are essential. (See Fox and Parker (1968), Stroud (1974), and Humpherys (1970) for detailed consideration of these matters.)

Before closing our discussion of Chebyshev functions we should observe that Chebyshev's equation, rewritten in Sturm-Liouville form, is

$$\left(\sqrt{1 - x^2}\, y'\right)' + \frac{p^2}{\sqrt{1 - x^2}} y = 0 \qquad -1 < x < 1.$$

As we observed earlier in Section 2.4, even though a boundary-value problem posed for this equation would technically be singular, much of the theory of that section is still applicable in this case. In particular, solutions corresponding to different values of p^2 are orthogonal with respect to the weight function $(1 - x^2)^{-1/2}$. Moreover, if we single out the Chebyshev functions T_p as the solutions of interest,* we see that for positive p, $T_p(x)$ has $[p + 1/2]$ zeros in $[-1,1)$. (Note that the usual roles of the endpoints are reversed.) As p increases, additional zeros appear at $x = -1$ (whenever p is half an odd integer) and migrate across the interval toward $x = 1$.

3.5 Kummer's equation and confluent hypergeometric functions

Now we turn to a brief analysis of Kummer's equation

$$xy'' + (c - x)y' - ay = 0, \tag{3.5–1}$$

which, we recall, possesses generality akin to the hypergeometric equation considered above. In actuality, the generality of (3.5–1) is *inherited* from that of the hypergeometric equation. To see this we rewrite (3.3–2), replacing x with x/b. This has the effect of repositioning to $x = b$ the regular singular point normally at $x = 1$ and leads to the equation

$$x(1 - x/b)y'' + [c - x - (a + 1)x/b]y' - ay = 0.$$

In the limit $b \to \infty$, this equation becomes (3.5–1).

Viewed in this light, Kummer's equation and its solutions are often termed the *confluent* hypergeometric equation and *confluent* hypergeometric functions. The point at ∞ has become an irregular singular point formed out of the coalescence of two of the regular singularities (at $x = b, \infty$) associated with transformed versions of the original hypergeometric equation. However, the origin is still a regular singular point of (3.5–1), and the associated indicial equation is identical to that of (3.3–2), namely

$$\alpha^2 + (c - 1)\,\alpha = 0. \tag{3.5–2}$$

*This can be done by requiring differentiability in $(-1,1]$. For differentiability at $x = -1$ also, p must be integral (see Problem 19).

As before, therefore, whenever c is nonintegral the Frobenius method leads to two linearly independent solutions. Corresponding to the root $\alpha = 0$ of (3.5–2) is the solution

$$M(a;c;x) = \frac{\Gamma(c)}{\Gamma(a)} \sum_{k=0}^{\infty} \frac{\Gamma(a + k)}{\Gamma(c + k)} \frac{x^k}{k!} \tag{3.5–3}$$

(see Problem 20 for the second solution).

In keeping with our earlier remarks, an alternative notation for the confluent hypergeometric function (3.5–3) is ${}_1F_1(a;c;x)$. Unless $c = 0, -1, -2, \ldots$, the power series is obviously convergent for all finite x. Furthermore, the series terminates whenever $a = -n$, whereupon $M(a;c;x)$ becomes a polynomial.

Kummer's equation and the confluent hypergeometric functions have been well-studied through the years. The basic results, including recurrence and differential properties, integral representations, and the like, may be found in Erdélyi et al.(1953a) or Abramowitz and Stegun (1965). Additional discussion appears in Buchholz (1969) and Slater (1960). We will be satisfied to note the integral representation that results when we replace x by x/b in the expression (3.3–6) and pass to the limit $b \to \infty$:

$$M(a;c;x) = \frac{\Gamma(c)}{\Gamma(a)\Gamma(c - a)} \int_0^1 e^{xt}\, t^{a-1}(1 - t)^{c-a-1} dt \qquad (\mathrm{Re}\, c > \mathrm{Re}\, a > 0). \tag{3.5–4}$$

3.6 More on Bessel functions

Basic definitions

As suggested in Section 3.3, a number of important special functions can be represented in terms of confluent hypergeometric functions with particular parameter values. These include Hermite functions, Laguerre functions, Coulomb wave functions, Weber or parabolic cylinder functions, Airy functions, and, of course, Bessel functions. In view of the practical application with which we introduced this chapter, we will concentrate on the last of these. However, rather than using (3.2–4) to deduce all of our results from the comparable relations for the confluent hypergeometric functions, we will derive some of them directly. Our discussion builds upon the review material in Appendix 2.

For Bessel's equation

$$x^2y'' + xy' + (x^2 - \nu^2)y = 0 \tag{3.6–1}$$

the origin is a regular singularity associated with the characteristic exponents $\alpha = \nu$ and $\alpha = -\nu$. Corresponding to the former, Frobenius' method leads to the Bessel function of the first kind of argument x and order ν

$$J_\nu(x) \equiv \sum_{k=0}^{\infty} \frac{(-1)^k (x/2)^{2k+\nu}}{\Gamma(k + \nu + 1)k!}. \tag{3.6–2}$$

When ν is nonintegral, $J_{-\nu}(x)$ obtained from (3.6–2) by replacing ν with $-\nu$ provides an acceptable second solution of Bessel's equation. However, since their linear independence breaks down for integer ν, it is customary to use as the second solution a particular linear combination of J_ν and $J_{-\nu}$, which avoids this difficulty. The Bessel function of the second kind

$$Y_\nu(x) \equiv \frac{\cos \nu\pi \, J_\nu(x) - J_{-\nu}(x)}{\sin \nu\pi} \tag{3.6–3}$$

has become the standard in this regard. Applying L'Hôpital's rule to (3.6–3) leads to

$$Y_0(x) = \frac{2}{\pi}\left[\ln(x/2)J_0(x) - \sum_{k=0}^{\infty} \frac{\psi(k + 1)\,(-1)^k (x/2)^{2k}}{(k!)^2}\right] \tag{3.6–4}$$

and

$$Y_n(x) = \frac{1}{\pi}\left[2\ln(x/2)J_n(x) - \sum_{k=0}^{n-1} \frac{(n - k - 1)!\,(x/2)^{2k-n}}{k!} - \sum_{k=0}^{\infty}\left\{\psi(k + 1) + \psi(n + k + 1)\right\}\frac{(-1)^k (x/2)^{2k+n}}{k!(k + n)!}\right] \quad n \geq 1 \tag{3.6–5}$$

in the case of integer ν (see Appendix 5 for remarks on the psi function; also see Problem 21 for an alternative derivation of these expressions).

One of the more useful Bessel function relations is the integral representation for $J_\nu(x)$ that can be derived from (3.5–4) by making use of (3.2–4). We obtain (for Re $\nu > -\tfrac{1}{2}$)

$$J_\nu(x) = \frac{\Gamma(2\nu + 1)e^{-ix}(x/2)^\nu}{\Gamma^2(\nu + \tfrac{1}{2})\Gamma(\nu + 1)} \int_0^1 e^{2ix\tau}\, \tau^{\nu-1/2}(1 - \tau)^{\nu-1/2}\, d\tau.$$

This expression can be simplified considerably by performing the change of variable $\tau = \tfrac{1}{2}(1 + t)$ in the integral and applying Legendre's duplication formula (see Appendix 5) for $\Gamma(2\nu)$. The end result is

$$J_\nu(x) = \frac{2(x/2)^\nu}{\sqrt{\pi}\,\Gamma(\nu + \tfrac{1}{2})} \int_0^1 \cos xt \quad (1 - t^2)^{\nu-1/2}\, dt \qquad (\text{Re}\ \nu > -\tfrac{1}{2}). \tag{3.6–6}$$

Comparable representations for the modified Bessel functions can be developed:

$$I_\nu(x) = \frac{2(x/2)^\nu}{\sqrt{\pi}\,\Gamma(\nu + \frac{1}{2})} \int_0^1 \cosh xt \quad (1 - t^2)^{\nu - \frac{1}{2}}\, dt \qquad (\text{Re } \nu > -\tfrac{1}{2})$$

and

$$K_\nu(x) = \frac{\sqrt{\pi}\,(x/2)^\nu}{\Gamma(\nu + \frac{1}{2})} \int_1^\infty e^{-xt}(t^2 - 1)^{\nu - \frac{1}{2}}\, dt \qquad (\text{Re } \nu > -\tfrac{1}{2},\ \text{Re } x > 0).$$

The first of these expressions follows easily from (3.6–6). The second is a consequence of a generalization of the representation (3.6–6) due to Hermann Hankel (1839–1873). Additional formulas may be found in the classic treatise of Watson (1958) or in Abramowitz and Stegun (1965) and Erdélyi et al (1953b).

The representation (3.6–6) is a rather useful identity that we will study in the limit of large $|x|$ later in Chapter 11. For the present we merely note that when $\nu = 0$, it implies that

$$\begin{aligned} J_0(x) &= \frac{2}{\pi} \int_0^1 \cos xt\, (1 - t^2)^{-\frac{1}{2}}\, dt \\ &= \frac{2}{\pi} \int_0^{\pi/2} \cos (x \sin \theta) d\theta \\ &= \frac{1}{\pi} \int_0^{\pi} \cos (x \sin \theta) d\theta. \end{aligned}$$

In similar fashion,

$$\begin{aligned} J_1(x) &= \frac{x}{\pi} \int_0^{\pi} \cos (x \sin \theta) \cos^2\theta\, d\theta \\ &= \frac{1}{\pi} \int_0^{\pi} \sin (x \sin \theta) \sin \theta\, d\theta \\ &= \frac{1}{\pi} \int_0^{\pi} \cos (x \sin \theta - \theta) d\theta, \end{aligned}$$

and

$$\begin{aligned} J_2(x) &= \frac{x^2}{3\pi} \int_0^{\pi} \cos (x \sin \theta) \cos^4\theta\, d\theta \\ &= \frac{x}{\pi} \int_0^{\pi} \sin (x \sin \theta) \cos^2\theta \sin \theta\, d\theta \\ &= \frac{1}{\pi} \int_0^{\pi} \cos (x \sin \theta) \cos 2\theta\, d\theta \\ &= \frac{1}{\pi} \int_0^{\pi} \cos (x \sin \theta - 2\theta) d\theta. \end{aligned}$$

These results suggest that for integral orders the very simple alternative representation

$$J_n(x) = \frac{1}{\pi}\int_0^\pi \cos(x\sin\theta - n\theta)d\theta \qquad \textbf{(3.6–7)}$$

is valid, and this can indeed be verified inductively using a classical differential recursion relation (see Appendix 2).

If ν is half an odd integer, the integration in (3.6–6) can be carried out explicitly. Such Bessel functions turn out to be combinations of half-integer powers of x multiplied by $\sin x$ and/or $\cos x$. For example,

$$J_{1/2}(x) = \sqrt{\frac{2}{\pi x}}\sin x \quad \text{and} \quad J_{3/2}(x) = \sqrt{\frac{2}{\pi x}}\left[\frac{\sin x}{x} - \cos x\right].$$

Using the differential relations and the definition (3.6–3), we then find

$$Y_{-1/2}(x) = J_{1/2}(x) \quad \text{while} \quad Y_{1/2}(x) = -J_{-1/2}(x) = -\sqrt{\frac{2}{\pi x}}\cos x.$$

It follows that

$$H^{(1)}_{-1/2}(x) = \sqrt{\frac{2}{\pi x}}e^{ix},\ I_{-1/2}(x) = \sqrt{\frac{2}{\pi x}}\cosh x, \quad \text{and} \quad K_{-1/2}(x) = \sqrt{\frac{\pi}{2x}}e^{-x}.$$

For small x and fixed ν, we can derive limiting forms of $J_\nu(x)$ and $Y_\nu(x)$ from the series (3.6–2) and the relation (3.6–3):

$$\begin{aligned}
&J_\nu(x) \sim (x/2)^\nu/\Gamma(\nu+1) \qquad (\nu \neq -1, -2, \ldots),\\
&Y_0(x) \sim (2/\pi)\ln x,\\
&Y_\nu(x) \sim -(2/x)^\nu\,\Gamma(\nu)/\pi \qquad (\operatorname{Re}\nu > 0),\\
&Y_\nu(x) \sim -(x/2)^\nu \cos\nu\pi\,\Gamma(-\nu)/\pi \qquad (\operatorname{Re}\nu < 0).
\end{aligned}$$

The asymptotic behavior of the Bessel functions for large x and fixed ν can be obtained in formal fashion as suggested in Problem 22. Alternatively, as we shall see in Chapter 11, these results can be developed via integration by parts applied to the representation (3.6–6) and use of (3.6–3). By either technique, we find that the expansions have as leading terms

$$\begin{aligned}
&J_\nu(x) \sim \sqrt{\frac{2}{\pi x}}\cos(x - \nu\pi/2 - \pi/4),\\
&Y_\nu(x) \sim \sqrt{\frac{2}{\pi x}}\sin(x - \nu\pi/2 - \pi/4).
\end{aligned} \qquad \textbf{(3.6–8)}$$

The corresponding expressions for the modified Bessel functions are

$$I_\nu(x) \sim \frac{e^x}{\sqrt{2\pi x}} \qquad \text{and} \qquad K_\nu(x) \sim \sqrt{\frac{\pi}{2x}}\, e^{-x}.$$

From (3.6–8) we can easily deduce that for fixed real ν, $J_\nu(x)$ and $Y_\nu(x)$ have an infinite number of real zeros. (This can also be established using Sturmian theory as in Problem 18 of Chapter 2). If we index the zeros with the integer s, by reversion of the asymptotic expansions of which (3.6–8) provide the lead terms, we find

$$\begin{aligned} x^J_{\nu,s} &\sim s\pi + \nu\pi/2 - \pi/4 \\ x^Y_{\nu,s} &\sim s\pi + \nu\pi/2 - 3\pi/4 \end{aligned} \qquad \textbf{(3.6–9)}$$

as $s \to \infty$. To this level of accuracy, the right-hand sides of these expressions also represent the leading terms in the asymptotic development, for large s, of the "turning points" of $Y_\nu(x)$ and $J_\nu(x)$, respectively, that is, of the zeros of $Y_\nu'(x)$ and $J_\nu'(x)$. (See p. 371 of Abramowitz and Stegun (1965) for additional terms in these expansions; also see Appendix 2.)

Dependence upon the order ν

The results on Bessel functions presented so far are appropriate for the case of variable argument, but fixed order. The practical problem of Section 3.1, however, necessitates our understanding the behavior of these functions for ν varying also. Although most readers are familiar with the usual plots of $J_\nu(x)$, $Y_\nu(x)$ versus x (see Figure 3.2 for a typical sketch), probably few have seen these Bessel functions

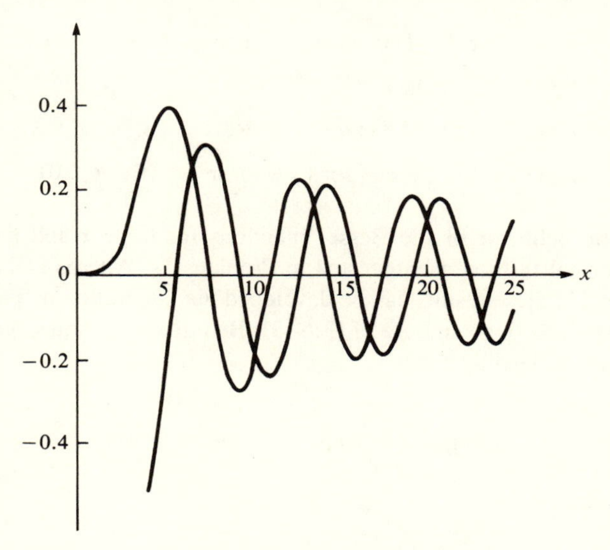

Figure 3.2 Sketch of $J_\nu(x)$, $Y_\nu(x)$ as functions of x for $\nu = 4$

graphed for fixed x as functions of ν. In Figure 3.3 we give such graphs for a representative case. These plots show clearly the finite numbers of positive ν-zeros that occur when x is fixed and positive.* Moreover, the figures suggest exponential growth of $Y_\nu(x)$ and decay of $J_\nu(x)$ for large ν.

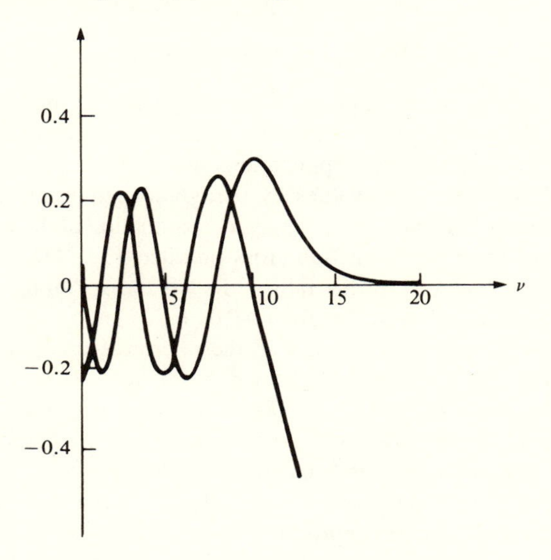

Figure 3.2 Sketch of $J_\nu(x)$, $Y_\nu(x)$ as functions of ν for $x = 12$

This asymptotic behavior of $J_\nu(x)$ for fixed $x > 0$ and $\nu \to \infty$ can be determined from the series (3.6–2). When ν is considerably greater than $x^2/4$, the first term in that series dominates. Hence

$$J_\nu(x) \sim (x/2)^\nu/\Gamma(\nu + 1),$$

or, using Stirling's formula for the gamma function (see Appendix 5),

$$J_\nu(x) \sim \frac{1}{\sqrt{2\pi\nu}}\left(\frac{ex}{2\nu}\right)^\nu .$$

The corresponding result for $Y_\nu(x)$, which is substantially more difficult to derive, has the form

$$Y_\nu(x) \sim -\sqrt{\frac{2}{\pi\nu}}\left(\frac{ex}{2\nu}\right)^{-\nu} .$$

A plethora of additional facts about Bessel functions is catalogued in Erdélyi et al (1953b) and in Abramowitz and Stegun (1965). Rather than repeat these we choose to return to the analysis of electromagnetic propagation in curved waveguides begun in Section 3.1, and to determine more precisely the appropriate direction for our final inquiries.

*Neither $J_\nu(x)$ nor $Y_\nu(x)$ can vanish for $\nu > x > 0$.

3.7 The practical problem revisited

The transcendental equations (3.1–8) and (3.1–9), namely

$$J_\nu(hr_1)Y_\nu(hr_2) - Y_\nu(hr_1)J_\nu(hr_2) = 0 \tag{3.7–1}$$

$$J'_\nu(hr_1)Y'_\nu(hr_2) - Y'_\nu(hr_1)J'_\nu(hr_2) = 0, \tag{3.7–2}$$

are not novel in applied investigations. Over the years they have arisen in a wide variety of physical problems with circular or cylindrical geometry and they continue to recur. Typically, however, the applications are such that the parameter ν is *a priori restricted* to certain *fixed* values by periodicity requirements on an angular variable φ. Interest then usually centers upon determination of the zeros with respect to the argument of the Bessel function cross-products (3.7–1), (3.7–2). (Cochran and Pecina (1966) references several tables of such zeros; Abramowitz and Stegun (1965), p. 415 provides a few values for $\nu = 0, 1$.)

Our viewpoint in connection with the microwave propagation problem is somewhat different. In this case r_1, r_2 and $h = [\omega^2\mu\epsilon - (n\pi/a)^2]^{1/2}$ are known in advance; it is the zeros of (3.7–1), (3.7–2), *for fixed argument,* which must be determined. These are the so-called *ν-zeros*, and it is the dependence of these ν-zeros on ω, r_1 and r_2 that demands our further investigation.

Extensive plots of the ν-zeros of (3.7–1), (3.7–2) as functions of h, r_1, and r_2 for values of these parameters appropriate to propagation in waveguides of typical microwave dimensions can be found in Cochran and Pecina (1966). Although discussion of computational routines necessary to perform such calculations with precision might be interesting to many readers, we merely mention here that the programs utilize the series representation (3.6–2) and various recursion relations from Appendix 2 combined with alternative forms and extensions of the asymptotic results (3.6–8) and the analogous expressions for large ν.

Approximations

The nature of the ν-zeros of the cross-products (3.7–1), (3.7–2) in two limiting regimes, namely when h is large (high frequency) and when r_1, r_2 are large but $r_2 - r_1$ is constant (gradual bends), is of distinct practical significance. It turns out that we can systematically analyze both these cases using asymptotic expressions due to Olver and obtain simple formulas that are remarkably accurate for a wide range of parameter values. (Since the analysis is rather difficult, we shall only give the highlights here.) The leading terms of the appropriate large-order Bessel function expansions have the form (see Olver (1974) or Abramowitz and Stegun (1965))

$$\begin{aligned} J_\nu(\nu z) &\sim \Phi(\zeta)\,\frac{Ai(\nu^{2/3}\zeta)}{\nu^{1/3}}, \\ Y_\nu(\nu z) &\sim -\,\Phi(\zeta)\,\frac{Bi(\nu^{2/3}\zeta)}{\nu^{1/3}}, \end{aligned} \tag{3.7–3}$$

where

$$\zeta \equiv -\left[\frac{3}{2}\int_1^z \frac{(t^2-1)^{1/2}}{t}\,dt\right]^{2/3}$$

$$= \begin{cases} \left[\frac{3}{2}\ln\frac{1+(1-z^2)^{1/2}}{z} - \frac{3}{2}(1-z^2)^{1/2}\right]^{2/3} & (0 < z \le 1) \\ -\left[\frac{3}{2}(z^2-1)^{1/2} - \frac{3}{2}\sec^{-1}z\right]^{2/3} & (z \ge 1) \end{cases} \tag{3.7–4}$$

and

$$\Phi(\zeta) = \left(\frac{4\zeta}{1-z^2}\right)^{1/4}.$$

The Ai and Bi appearing in (3.7–3) are the well-studied Airy functions of the first and second kind. They are appropriately normalized real solutions of the Airy equation mentioned earlier as Example VII of Section 3.2. Each exhibits sinusoidal behavior for negative argument values and decays (grows) exponentially for positive arguments. [We will encounter these functions again in Chapter 11; see also Abramowitz and Stegun (1965) and Problem 16 of Chapter 2.]

Thanks to (3.7–3), it is possible to show (see Cochran and Pecina (1966) and the references cited there for details) that for large h the ν-zeros of (3.7–1), (3.7–2) satisfy

$$\begin{aligned} \nu_s &\sim hr_2 + a_s(hr_2/2)^{1/3}, \\ \nu_s' &\sim hr_2 + a_s'(hr_2/2)^{1/3}. \end{aligned} \tag{3.7–5}$$

Here a_s, $s = 1,2, \ldots$, and a_s', $s = 0,1, \ldots$, are the zeros and turning points of the Airy function of the first kind, ordered according to decreasing size (recall that they are all negative). The case of a gradual bend, in which $r_2 - r_1 << r_1$, can be analyzed in related fashion. In this situation we find for both sets of zeros

$$\begin{Bmatrix} \nu_s \\ \nu_s' \end{Bmatrix} \sim \frac{r_1 + r_2}{2(r_2 - r_1)}[h^2(r_2 - r_1)^2 - s^2\pi^2]^{1/2}. \tag{3.7–6}$$

(In deriving this last result one makes use of an asymptotic expression, similar to (3.6–9), for the zeros of an Airy function cross-product.)

A typical application

We conclude our study of this electromagnetic propagation problem with some representative numerical results from which the accuracy of the above approximations can be estimated. Suppose $r_2 - r_1 = a = 1.872$ in (4.755 cm), a standard waveguide dimension, and the operating frequency is 7.0 GHz (that is, $\omega = 14\pi \times 10^9$ rad/sec). These parameter values are appropriate for microwave propagation in what is termed the C-band, which is one of the frequency ranges used in various intracontinental line-of-sight microwave communication links. Unless r_1 is small, such a guide will support only three LE modes [LE(0,1), LE(0,2), and LE(1,1)] and three LM modes [LM(1,0), LM(2,0), and LM(1,1)] (recall the nomenclature suggested by Property 1, Section 3.1).

The precise values of the lowest-order ν-zeros (actually the largest real ν-zeros) of (3.7–1) and (3.7–2), rounded to one decimal place, are shown in the accompanying tables for various values of hr_1. These are the angular propagation constants associated with the fundamental LM(1,0) and LE(0,1) modes. Also displayed there are the approximate values of the propagation constants provided by the "high-frequency" and "gradual bend" expressions (3.7–5) and (3.7–6), respectively.

Table 1: Exact and approximate values of the angular propagation constant; fundamental LM(1,0) mode in curved guide, $r_2 - r_1 = a = 4.755$ cm, $f = 7.0$ GHz.

hr_1	hr_2	ν	ν_{hf}	ν_{gb}
3.0	9.971	6.0	6.0	5.8
6.0	12.971	8.6	8.6	8.5
9.0	15.971	11.4	11.3	11.1
12.0	18.971	14.2	14.0	13.8
60.0	66.971	56.7	59.4	56.7

Table 2: Exact and approximate values of the angular propagation constant; fundamental LE(0,1) mode in curved guide, $r_2 - r_1 = a = 4.755$ cm, $f = 7.0$ GHz.

hr_1	hr_2	ν'	ν'_{hf}	ν'_{gb}
3.0	9.224	7.6	7.6	6.1
6.0	12.224	10.4	10.4	9.1
9.0	15.224	13.3	13.3	12.1
12.0	18.224	16.1	16.1	15.1
60.0	66.224	63.5	63.0	63.1

The approximate calculations for the LM(1,0) mode are made with $n = 0$ so that h as given by (3.1–7) takes on the value $h = 14\pi/30\ \text{cm}^{-1} \doteq 1.466\ \text{cm}^{-1}$, and $s = m = 1$. Analogously, for the LE(0,1) mode, $n = 1$, $h \doteq 1.309\ \text{cm}^{-1}$, and $s = m = 0$. The necessary values for the lead zero and turning point of the Airy function of the first kind are $a_1 \doteq -2.3381$ and $a_0' \doteq -1.0188$.

Although the precision of the high-frequency approximation is impaired when $(r_2 - r_1)/r_1$ is small (precisely where the gradual bend formula is best), both approximate expressions yield rather accurate values for the propagation constants in most of the other cases listed. This suggests that the simple formulas (3.7–5),

(3.7–6) may be put to good use in the initial stages of microwave design work or in other applications where the time or expense of determining an exact solution is unwarranted. Such is precisely the situation that frequently occurs in practice.

References

Abramowitz, M. and I. A. Stegun (1965): *Handbook of Mathematical Functions*, Dover, New York; NBS Applied Math Series 55, U.S. Dept. of Commerce, Washington, D.C. (1964); Chapters 8, 9, 10, 13, 15, 22.

Askey, R. (1975): "Orthogonal Polynomials and Special Functions," *Regional Conference Series in Applied Math. #21*, Soc. for Indust. and Appl. Math., Philadelphia.

Bailey, W. N. (1972): *Generalized Hypergeometric Series*, Cambridge Univ. Press, Cambridge.

Birkhoff, G. and G.-C. Rota (1978): *Ordinary Differential Equations*, 3rd ed. Wiley, New York; Chapter 9.

Buchholz, H. (1939): "Der Einfluss der Krümmung von rechteckigen Hohlleitern auf das Phasenmass ultrakurzer Wellen," *Elek. Nachrichtentech.* 16(No. 3): 73-85.

——— (1969): *The Confluent Hypergeometric Function*, Springer-Verlag, New York; Transl. of German edition, Springer-Verlag, Berlin (1953).

Cochran, J. A. and R. G. Pecina (1966): "Mode Propagation in Continuously Curved Waveguides," *Radio Sci.* 1(N.S.): 679–696.

Coddington, E. A. and N. Levinson (1955): *Theory of Ordinary Differential Equations*, McGraw-Hill, New York; Chapters 4, 5.

Dettman, J. W. (1974): *Introduction to Linear Algebra and Differential Equations*, McGraw-Hill, New York; Chapter 8.

Erdélyi, A., et al.(1953a): *Higher Transcendental Functions, Vol. 1*, McGraw-Hill, New York; Chapters II, III, VI.

——— (1953b): *Higher Transcendental Functions, Vol. 2*, McGraw-Hill, New York; Chapters VII, X.

Fox, L. and I. B. Parker (1968): *Chebyshev Polynomials in Numerical Analysis*, Oxford Univ. Press, London.

Hildebrand, F. B. (1976): *Advanced Calculus for Applications*, 2nd ed. Prentice-Hall, Englewood Cliffs, N.J.; Chapter 4.

Humpherys, D. S. (1970): *The Analysis, Design and Synthesis of Electrical Filters*, Prentice-Hall, Englewood Cliffs, N.J.

Ince, E. L. (1953): *Ordinary Differential Equations*, Dover, New York; Longmans, London (1927); Chapters XVII, XX.

Kaplan, W. (1958): *Ordinary Differential Equations*, Addison-Wesley, Reading, Mass.; Chapters 9, 12.

Olver, F. W. J. (1974): *Asymptotics and Special Functions*, Academic Press, New York; Chapters 2, 6, 7, 11.

Panofsky, W. K. H. and M. Phillips (1962): *Classical Electricity and Magnetism*, 2nd ed., Addison-Wesley, Reading, Mass.

Slater, L. J. (1960): *Confluent Hypergeometric Functions*, Cambridge Univ. Press, Cambridge.

Smythe, W. R. (1968): *Static and Dynamic Electricity*, 3rd ed., McGraw-Hill, New York.

Stratton, J. A. (1941): *Electromagnetic Theory*, McGraw-Hill, New York.

Stroud, A. H. (1974): "Numerical Quadrature and Solution of Ordinary Differential Equations," *Applied Math. Sciences Vol. 10*, Springer-Verlag, New York.

Watson, G. N. (1958): *A Treatise on the Theory of Bessel Functions*, 2nd ed., Cambridge Univ. Press, Cambridge.

Problems

Sections 3.1, 3.2

1. (a) Separate variables for the Helmholtz equation

$$\frac{\partial^2 u}{\partial x^2} + \frac{\partial^2 u}{\partial y^2} + \frac{\partial^2 u}{\partial z^2} + k^2 u = 0$$

in rectangular coordinates (x,y,z), and determine the form of the separate solutions.

   (b) Use the definitions (3.1–3), (3.1–4) and the result of part (a) to verify the expressions (3.1–5).

2. The differential equation

$$x^2 y'' + (x^2 + x)y' - y = 0$$

has only one solution in the form of a power series convergent in the neighborhood of the origin. Find this solution and express it in terms of well-known functions.

★3. Consider an ordinary differential equation of the form (A3–2, Appendix 3) with a regular singular point at $x = 0$. Assume that the roots of the indicial equation (A3–7) are real and differ by an integer. If the Frobenius method leads to only one solution of the form (A3–6), show that an independent second solution can be represented as in (A3–8). [Hildebrand (1976), pp. 135–138, has the details.]

4. Use the method of Frobenius to obtain the *general solution* of

$$x^2 y'' + (x^2 - 4x)y' + (4 - x)y = 0,$$

valid near the origin $x = 0$. *Express your result in closed form*. [This is an example wherein every solution of the equation has a power series representation convergent about the origin even though $x = 0$ is a singularity of the differential equation.]

Sections 3.2, 3.3

5. Assume that a second-order ordinary differential equation is in the standard form, namely
$$y'' + f(x)y' + g(x)y = 0.$$
State criteria, in terms of $f(x)$ and $g(x)$, for $x = \infty$ to be an ordinary point; a regular singular point. [Hint: Let $t = 1/x$ and consider the behavior of the resulting equation at the point $t = 0$.]

6. (a) Show that the unique second-order linear differential equation with polynomial coefficients that has precisely two singular points, both regular, placed at $x = 0$ and $x = \infty$ is the Cauchy-Euler equation (3.2–1).
   (b) Characterize the form of the Cauchy-Euler equation that corresponds to the point at ∞ being an ordinary point.

7. Find the general solution of the equation
$$x^2y'' + x^2y' - 2y = 0.$$
Express your answer in terms of well-known functions.

8. Determine representations in powers of x for the two linearly independent solutions of
$$xy'' + y' - y = 0.$$

9. Consider Kummer's equation
$$xy'' + (c - x)y' - ay = 0.$$
   (a) Demonstrate that if $a = \frac{1}{2}(1 - p)$ and $c = \frac{3}{2}$, the change of variables
$$x = t^2 \quad \text{and} \quad y(x) = t^{-1}u(t)$$
transforms this equation into Hermite's equation
$$\frac{d^2u}{dt^2} - 2t\frac{du}{dt} + 2pu = 0.$$
   (b) In similar fashion, show that when $a = \nu + \frac{1}{2}$, $c = 2\nu + 1$, $x = 2it$ and $y(x) = t^{-\nu}e^{it}u(t)$, Kummer's equation is equivalent to Bessel's equation
$$t^2\frac{d^2u}{dt^2} + t\frac{du}{dt} + (t^2 - \nu^2)u = 0.$$

10. In a manner analogous to the one used used in Problem 9, employ the substitutions $a = \frac{5}{6}$, $c = \frac{5}{3}$, $x = \dfrac{4}{3}t^{3/2}$, and $y(x) = t^{-1}e^{\frac{2}{3}t^{3/2}}u(t)$ to convert Kummer's equation into Airy's equation
$$\frac{d^2u}{dt^2} - tu = 0.$$

11. Consider a second-order linear differential equation with polynomial coefficients. Assume that all but three of the values of the independent variable

are ordinary points and the three singular points are regular. If these regular singularities are $x = 0, 1$, and ∞, show that the resulting differential equation is Riemann's (3.3–1).

★12. Establish Theorem 1 of Section 3.3 (p. 64).

13. Verify that if c is nonintegral, the solution, in the vicinity of the origin, of the hypergeometric equation (3.3–2) associated with the exponent $\alpha = 1 - c$ can be expressed as

$$y(x) = x^{1-c}{}_2F_1(a - c + 1, b - c + 1; 2 - c; x).$$

14. Satisfy yourself that as a special case of Euler's integral representation,

$${}_2F_1(a,b;c;1) = \frac{1}{\Gamma(c)}\sum_{k=0}^{\infty}\frac{(a)_k\,(b)_k}{(c)_k\,k!} = \frac{\Gamma(c-a-b)}{\Gamma(c-a)\Gamma(c-b)} \qquad \mathrm{Re}(c-a-b) > 0,$$

and hence, when $b = -n$, with n integral,

$${}_2F_1(a,-n;c;1) = \sum_{k=0}^{n}\frac{(a)_k(-n)_k}{(c)_k\,k!} = \frac{(c-a)_n}{(c)_n}. \qquad (*)$$

Here $(a)_k \equiv \Gamma(a + k)/\Gamma(a)$ is Pochhammer's symbol, that is, $(a)_0 = 1$ and $(a)_k = a(a+1)(a+2)\cdots(a+k-1)$ for $k = 1,2,\ldots$. [The result (*) is usually attributed to Vandermonde. Askey (1975), pp. 59f, observes that it was actually first published by a Chinese mathematician Chu in 1303.]

15. The familiar binomial series (see (3.3–5)) has the form

$$\sum_{k=0}^{\infty}\frac{(a)_k x^k}{k!} = (1-x)^{-a} \qquad |x| < 1.$$

(a) Multiply two such series together and match coefficients of like terms to obtain the identity

$$\sum_{k=0}^{n}\frac{(a)_k(b)_{n-k}}{k!\,(n-k)!} = \frac{(a+b)_n}{n!}.$$

(b) Provide an alternative derivation of the relation (*) of Problem 14 by transforming the result of part (a).

★16. For Legendre's equation

$$(1 - x^2)y'' - 2xy' + p(p + 1)y = 0$$

with real p:

(a) Derive a series representation for the one solution that is finite at $x = 1$. [Hint: Expand about the point $x = 1$.]

(b) Demonstrate that this solution is meaningful at $x = -1$ if and only if p is integral.

17. Introduce the change of variables $t = (1 - \tau)/(1 - x\tau)$ into the Euler representation (3.3–6) and thereby deduce that

$${}_2F_1(a,b;c;x) = (1-x)^{c-b-a}\,{}_2F_1(c-a,c-b;c;x)$$

for $|x| < 1$.

18. Employ Euler's integral representation (3.3–6) to demonstrate that

$$\text{(a)}\quad {}_2F_1(a,b;\, a-b+1;\, -1) = \frac{\Gamma(\tfrac{1}{2}a)\,\Gamma(1+a-b)}{2\Gamma(a)\Gamma(1+\tfrac{1}{2}a-b)}$$

$$= \frac{\sqrt{\pi}\,2^{-a}\Gamma(1+a-b)}{\Gamma(\tfrac{1}{2}+\tfrac{1}{2}a)\,\Gamma(1+\tfrac{1}{2}a-b)},$$

and

$$\text{(b)}\quad {}_2F_1(a,1-a;\, b;\, \tfrac{1}{2}) = 2^{-a}\frac{\Gamma(b)\Gamma(\tfrac{1}{2}b-\tfrac{1}{2}a)}{\Gamma(b-a)\Gamma(\tfrac{1}{2}b+\tfrac{1}{2}a)}$$

$$= \frac{\sqrt{\pi}\,2^{1-b}\Gamma(b)}{\Gamma(\tfrac{1}{2}b-\tfrac{1}{2}a+\tfrac{1}{2})\Gamma(\tfrac{1}{2}b+\tfrac{1}{2}a)}.$$

[These summation formulas complement that given in the footnote on page 67.]

Sections 3.4, 3.5

19. Assume that the Chebyshev function $T_p(x)$ is defined by (3.4–2) and let $V_p(x)$ be given by

$$V_p(x) = (1-x)^{\frac{1}{2}}\,{}_2F_1\left(\tfrac{1}{2}-p,\, \tfrac{1}{2}+p;\, \tfrac{3}{2};\, \frac{1-x}{2}\right).$$

(a) Verify that $V_p(x)$ is a second solution of Chebyshev's equation

$$(1-x^2)y'' - xy' + p^2y = 0,$$

which is linearly independent of $T_p(x)$.

(b) Show that for general real p

(i) $V_p(1) = 0$ and $V_p(-1) = \sin p\pi/\sqrt{2}p$

(ii) $V_p'(x)$ is undefined at $x = 1$; it is also undefined at $x = -1$ unless p is half an odd integer;

(iii) $T_p'(1) = p^2$;

(iv) $T_p'(-1)$ is undefined unless p is integral, in which case $T_n'(-1) = n^2(-1)^{n+1}$.

20. Verify that

$$U(a;c;x) \equiv \frac{\pi}{\sin \pi c}\left[\frac{M(a;c;x)}{\Gamma(1+a-c)\Gamma(c)} - x^{1-c}\frac{M(1+a-c;\, 2-c;\, x)}{\Gamma(a)\Gamma(2-c)}\right]$$

gives a second solution of Kummer's equation

$$xy'' + (c-x)y' - ay = 0,$$

which is linearly independent of $M(a;c;x)$.

Sections 3.6, 3.7

21. Use the procedure suggested in connection with formula (A3–8) of Appendix 3 in order to derive the expression (3.6–5) for the Bessel function of the second kind $Y_n(x)$ with positive integer order.

★22. Even though $x = \infty$ is an irregular singular point of Bessel's equation

$$x^2y'' + xy' + (x^2-\nu^2)y = 0,$$

a modified Frobenius technique can be used to derive asymptotic representations valid for large x. Explore this approach by determining values of α and β so that

$$e^{\beta x} \sum_{k=0}^{\infty} c_k x^{-k-\alpha}$$

formally satisfies the above differential equation. On the basis of other known results about Bessel functions, choose appropriate combinations of the resulting expressions in order to obtain at least the first several terms of the classical large-argument asymptotic expansion of $J_\nu(x)$ and $Y_\nu(x)$. [The leading terms in these expansions are displayed in (3.6–8).]

23. (a) Show that near $z = 1$, $\zeta(z)$ given by (3.7–4) satisfies

$$\zeta \sim 2^{1/3}(1 - z).$$

(b) Use this result and the asymptotic representations (3.7–3) of Olver to verify that, within the order of accuracy of these expressions, the ν-zeros of the Bessel function cross-products (3.7–1), (3.7–2) are given approximately by (3.7–5). Here $h \to \infty$ and r_1, r_2 are constants.

4

Optimization and the calculus of variations

4.1 An optimal racing strategy

The mathematical descriptions of the principal applications considered in the first three chapters of this book are unified by their nature as eigenvalue problems. In addition, however, the models share the common characteristic of linearity. In the next several sections we want to widen our horizons and include physical situations for which the natural model may be nonlinear. We have in mind the broad class of "optimization" problems in which one is interested in maximizing or minimizing some given quantity while, perhaps, keeping other quantities fixed. When the mathematical decription contains only algebraic relations, the model becomes what is commonly termed a programming problem, and matrix-related principles and procedures are often applicable (see Hillier and Lieberman (1980) or Wagner (1975), for example). However, when continuous functions, integrals, and differential equations are involved, rather different techniques are needed. It is this latter variety of problem that will occupy our attention in this chapter.

The type of optimization problems of interest to us have a long and rich history. Virgil's Aeneid, Book I suggests that the legendary Queen Dido of Carthage "solved" a variant of the *isoperimetric problem* (which we shall discuss in Section 4.5) around 850 B.C. Somewhat later Galileo Galilei (1564–1642), Christian Huygens (1629–1695), Isaac Newton (1642–1729), the Bernoulli brothers (John, 1667–1748, and James, 1654–1705) and others posed and/or solved a variety of "chrone" (from the Greek, meaning time) problems including the celebrated *brachistochrone problem*. In much more recent times, *optimal control problems* arising, for example, in connection with process management, rocket design, and space exploration have occupied center stage.

An intriguingly different physical problem of this last genre was examined in the early 1970's by Joseph Keller (see Keller (1973, 1974)). In view of the appealing uniqueness of the Keller application, we will use this problem as the point-of-departure for our further considerations.

Imagine that you are a world-class runner interested, naturally, in doing the best you possibly can at your distance. A biomathematician friend of yours suggests that she can help you achieve an optimum performance by developing and then solving an appropriate minimization problem. Her reasoning is as follows: In a typical race your velocity (speed) at any time t is $v(t)$. If you run a race of distance D in total time T, then D, T, and $v(t)$ are related by

$$D = \int_0^T v(t)dt. \tag{4.1–1}$$

Your racing speed, however, is not entirely unrestricted. Initially, assuming the usual stationary start,

$$v(0) = 0. \tag{4.1–2}$$

Moreover, as a consequence of momentum conservation (remember (2.1–1)), $v(t)$ must satisfy an equation such as

$$\frac{dv}{dt} + \alpha v = f(t). \tag{4.1–3}$$

Here $f(t)$ is the propulsive force per unit mass which you are able to generate during the race, and it must be balanced off against *both* your resulting acceleration *and* a natural resistive force per unit mass. We have assumed that this latter force is proportional to your velocity with proportionality constant α.

Unfortunately, the three equations (4.1–1)–(4.1–3) are not quite the entire story, your friend continues. Your propulsive force $f(t)$ also enters into two additional relationships which influence your racing speed. Firstly, it is natural to assume that $f(t)$ is bounded, that is,

$$f(t) \leq F, \tag{4.1–4}$$

where F is some appropriate maximum value. Secondly, throughout your entire race you are "burning" energy. However, this energy consumption can be modelled by analyzing your oxygen utilization since it is this essential ingredient that makes possible the exo-energetic reactions necessary for your running. If the amount $E(t)$ of oxygen that is available in your muscles per unit mass, by virtue of breathing and circulation, is measured in units of the amount of energy it is converted into during running, a possible energy-balance equation is

$$\frac{dE}{dt} = \beta - fv. \tag{4.1–5}$$

In this relation your energy (oxygen) build-up is equated to the difference between your energy supply rate β and your demand rate fv. Here β is taken to be the (assumed) constant rate at which you supply energy (oxygen) during running, owing in part to faster breathing, *in excess* of that supplied during the non-running state. The simple product demand function is a plausible way to incorporate both the acceleration and the "coasting" (constant v) phases of the race into the model.

Two qualifying conditions that belong with the differential equation (4.1–5) are needed to complete the description of energy conservation. These are the initial condition

$$E(0) = E_0, \tag{4.1–6}$$

which gives the amount of normal energy (oxygen) stored, and hence available, at the start of the race and the nonnegativity assumption

$$E(t) \geq 0. \qquad \textbf{(4.1–7)}$$

These various equations and inequalities constitute the mathematical model of a typical race as envisioned by your friend. If you know your physiological parameters α, β, F, and E_0, she advises, your best racing strategy is to vary your racing speed $v(t)$, within the constraints implied by the relations (4.1–2) to (4.1–7), in such a way that your total time T for the race, defined by (4.1–1), is minimized.

The problem evolved by the biomathematical friend is of the constrained optimization type: an extremum is sought subject to various modifying conditions. Indeed, the problem appears to be (and is) not entirely trivial; this particular formulation involves three interrelated unknowns ($v, f,$ and E), two coupled differential equations, and several other constraints. Its resolution thus must await the development of some more basic tools, and therefore we turn our attention first to that task.

4.2 The simplest problem

In the differential calculus, one of the more fundamental problems is the determination of maxima and minima of functions. In this new subject area, classically termed the "calculus of variations," the emphasis shifts to finding extrema of functions of functions or, to be more precise, of *functionals*. Functionals arise in a vast array of applications and appear in any number of different forms. Here are some examples: the kinetic energy of a moving mass as a function of its velocity, the manufacturing cost of a particular product as a function of labor rates, rocket fuel consumption as a function of aerodynamic drag, and the area enclosed by a smooth convex curve in terms of the parametric representation of the curve. In each case there is a rule that assigns to every function in a given set of functions (usually a subspace of some normed vector space) a unique number.

One particular functional form that plays a key role in our discussion in this chapter is the definite integral

$$I(y) = \int_a^b f(x,y,y')dx, \qquad \textbf{(4.2–1)}$$

where f is some specified function defined for all points in some open set of three-space and a,b are fixed. In this case the functional $I(y)$ can be envisioned as a path-dependent line integral. For definiteness $y = y(x)$ is required to be continuously differentiable on the interval $[a,b]$.

Variations

We take our cue from the ordinary calculus. Since we are interested in extrema of functionals, we investigate what happens to a general functional, written $F(y)$ for short, as we vary the function y by an amount Δy. If we describe the variation of y in the form $\Delta y = \epsilon\eta$ where ϵ is assumed to be a "small" scalar and η is a function having the same smoothness as y, then

Definition 1 The functional $F(y)$ is said to have a **variation δF at y** given by

$$\delta F(y) = \lim_{\epsilon \to 0} \frac{F(y + \epsilon\eta) - F(y)}{\epsilon}$$

whenever this limit exists for all η in the subspace of interest.

The variation given in this definition has the form of a "directional derivative." However, in view of the manner in which ϵ has been introduced, there is an obvious alternative interpretation in terms of ordinary differentiation:

Property 1

$$\delta F(y) = \frac{d}{d\epsilon} F(y + \epsilon\eta)\bigg|_{\epsilon = 0.} \tag{4.2–2}$$ *

We shall commonly employ this latter rendition of the definition, since it relates the fundamental problem in the calculus of variations to the more familiar max-min problem for functions of a single real-variable.

For the particular functional $I(y)$ given by (4.2–1) we calculate, using the usual rules for differentiation under the integral sign,

$$\frac{d}{d\epsilon} I(y + \epsilon\eta) = \int_a^b \frac{d}{d\epsilon} f(x, y + \epsilon\eta, y' + \epsilon\eta')dx.$$

Thus, assuming the various derivatives are meaningful,

$$\delta I(y) = \int_a^b \left(\frac{\partial f}{\partial y}\eta + \frac{\partial f}{\partial y'}\eta'\right)dx. \tag{4.2–3}$$

The Euler-Lagrange equation

If $\bar{y}$ provides a local maximum for a given functional $F(y)$ within some function subspace of interest, then the functional values for all functions y "close" to $\bar{y}$ must satisfy $F(y) \leq F(\bar{y})$. This relation can obviously be rewritten in terms of the variation $\delta y = \epsilon\eta$ as

$$F(\bar{y} + \epsilon\eta) \leq F(\bar{y}) \qquad (\epsilon \text{ sufficiently small}).$$

---

*It should be understood that the variation $\delta F(y)$ is also a function of η, even though we have suppressed this in our notation (see Problems 1, 2).

Since the parameter ϵ can take on real values of both signs, it then follows that

$$\frac{F(\tilde{y} + \epsilon\eta) - F(\tilde{y})}{\epsilon} \leq 0$$

for all small $\epsilon > 0$ while

$$\frac{F(\tilde{y} + \epsilon\eta) - F(\tilde{y})}{\epsilon} \geq 0$$

for all small $\epsilon < 0$. Thus, if the limit inherent in Definition 1 exists, we must have $\delta F(\tilde{y}) = 0$. Analogous reasoning holds in case $\tilde{y}$ is a local minimum for $F(y)$ and hence

Theorem 1

If a functional $F(y)$ defined on a given function subspace has a local extremum at $\tilde{y}$ and if $\delta F(\tilde{y})$ exists, then

$$\delta F(\tilde{y}) = 0. \qquad \textbf{(4.2–4)}$$

This not unexpected result leads to a classical differential equation for the case wherein $F(y)$ has the form (4.2–1). In this situation we have from (4.2–3)

$$0 = \int_a^b \left(\frac{\partial f}{\partial y}\eta + \frac{\partial f}{\partial y'}\eta'\right)dx.$$

Integrating the second term by parts and assuming that $\eta(a) = \eta(b) = 0$ (which is equivalent to assuming that all the functions of interest have the same function value at $x = a$ as well as at $x = b$), we then deduce that

$$0 = \int_a^b \left[\frac{\partial f}{\partial y} - \frac{d}{dx}\left(\frac{\partial f}{\partial y'}\right)\right]\eta\, dx.$$

This relation must hold for all, say, continuously differentiable functions $\eta(x)$ that vanish at the endpoints of the interval $[a,b]$. In view of the arbitrariness of η, if the bracketed term in the integrand of this integral is continuous, it must actually vanish*, and we obtain the following.

Corollary:

If the functional $I(y) = \int_a^b f(x,y,y')dx$ is well-defined for all, say, continuously differentiable functions $y(x)$ satisfying $y(a) = y_i$, $y(b) = y_f$ for given y_i, y_f, and if $\tilde{y}(x)$ provides a local extremum for $I(y)$, then, assuming

*The argument can be traced back to the German mathematical analyst Paul du Bois-Reymond (1831–1889). This helpful result is often singled out and given a special name such as "A Basic Lemma" (Weinstock, 1952), "the Euler-Lagrange Lemma" (Young, 1969), or "the du Bois-Reymond Lemma" (Smith, 1974).

all the indicated derivatives exist and are continuous, the extremum function $\tilde{y}(x)$ must satisfy the *Euler-Lagrange equation*

$$\frac{\partial f(x,y,y')}{\partial y} - \frac{d}{dx}\frac{\partial f(x,y,y')}{\partial y'} = 0. \qquad \textbf{(4.2–5)}$$

As suggested by its name, the equation (4.2–5) was first derived by Euler and later rederived in a much simpler manner by Lagrange. For functionals $I(y)$ of the form (4.2–1), the Euler-Lagrange equation provides a *necessary condition* that potential extrema must satisfy. However, contrary to the way (4.2–5) has been used in practice for decades, solutions of this celebrated differential equation need not either maximize or minimize I.* The fact that in numerous inquiries the Euler-Lagrange equation has provided a convenient recipe for calculating functions that turn out subsequently to be the desired extrema is, logically, a fortuitous coincidence. (See Young (1969) for extensive comments in this regard presented in truly delightful fashion).

The Brachistochrone problem

Example The paramount "chrone" application of the late 17th century (which Newton amazingly solved in an evening's time upon first learning of it) was the *brachistochrone problem* of Euler's teacher John Bernoulli. In this practical example the smooth curve is sought along which a point mass moves, under the sole action of gravity, from one point to a nearby lower point *in the shortest time*.

If the x- and y-axes are taken in their conventional directions and the initial and final points are designated (a,y_i) and (b,y_f), respectively, then we can mathematically model this application very simply as follows. The total time of descent is given by

$$T = \int_0^T dt = \int_0^{s(T)} \frac{dt}{ds}\,ds$$
$$= \int_0^{s(T)} \frac{ds}{v} = \int_a^b \frac{\sqrt{1 + y'^2}}{v}\,dx$$

where $v = ds/dt$ is the speed of the descending mass along the curve $y = y(x)$ whose arclength is s. By virtue of conservation of energy we can compute this speed v as a function of the coordinates. Indeed, neglecting friction, the increase in kinetic energy is precisely balanced by the decrease in potential energy so that, assuming a start from rest,

$$\tfrac{1}{2}\, mv^2 = mg(y_i - y)$$

and hence

$$v = \sqrt{2g(y_i - y)}\,,$$

*These solutions are merely guaranteed to render $I(y)$ *stationary*.

where m is the mass of the particle and g is the usual constant acceleration due to gravity. The brachistochrone problem, therefore, is to minimize the functional

$$T(y) = \int_a^b \frac{\sqrt{1 + y'^2}}{\sqrt{2g(y_i - y)}}\,dx \tag{4.2–6}$$

amongst, say, continuously differentiable functions $y = y(x)$ satisfying $y(a) = y_i$, $y(b) = y_f$.

The integrand of (4.2–6) is in the standard form (4.2–1) to which the Euler-Lagrange equation is applicable. Performing the indicated differentiations we find, after simplification, that potential extremal curves must be solutions of the second-order equation

$$1 + y'^2 - 2y''(y_i - y) = 0.$$

This particular equation has a first integral

$$(1 + y'^2)\,(y_i - y) = 2c,$$

where c is a positive constant. Hence, choosing the physically meaningful negative sign for the square root, we obtain

$$y' = -\sqrt{\frac{2c}{y_i - y} - 1}\,. \tag{4.2–7}$$

There are a number of ways in which this latter relation can be integrated. One of the more direct methods is to use the substitution (see Weinstock (1952) or Smith (1974), for example)

$$y = y_i - c(1 - \cos\theta), \tag{4.2–8}$$

in which case (4.2–7) gives rise to

$$c(1 - \cos\theta)\theta' = 1.$$

It easily follows then that

$$x = a + c(\theta - \sin\theta). \tag{4.2–9}$$

Equations (4.2–8), (4.2–9) are to be recognized as the parametric representation for a *cycloid* generated by the motion of a fixed point on the circumference (rim) of a circle (wheel) of radius c, which rolls on the negative side (underside) of the line $y = y_i$ as shown in Figure 4.1 on p. 92.* The initial point (a,y_i) corresponds to $\theta = 0$ and the coordinates of the final point (b,y_f) determine the appropriate (unique!)† values of c and θ_f. Moreover, as calculated from (4.2–6) we have

$$T = \sqrt{\frac{c}{g}}\int_a^b \frac{dx}{(y_i - y)} = \sqrt{\frac{c}{g}}\int_0^{\theta_f} d\theta = \sqrt{\frac{c}{g}}\,\theta_f. \qquad \square$$

*Remarkably, the cycloid also crops up as the solution to another "chrone" problem; see the discussion of *Abel's integral equation* in Section 10.2.

†Note that $h(\theta) \equiv \dfrac{x - a}{y_i - y} = \dfrac{\theta - \sin\theta}{1 - \cos\theta}$ is a continuously differentiable function that increases monotonically, and unboundedly, as θ increases from 0 to 2π.

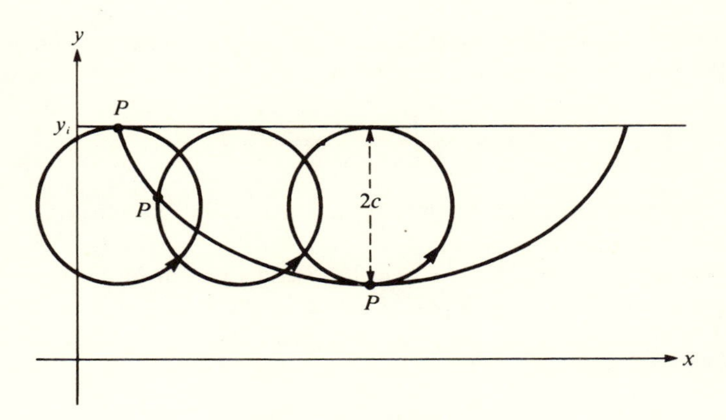

Figure 4.1 One arch of the cycloid generated by the motion of the point P on the circumference of the circle of radius c rolling along the underside of the line $y = y_i$.

The cycloid we have just derived is the putative minimizing extremal for the brachistochrone problem. Conveniently, it can be verified that this cycloid does indeed minimize the descent time T within the class of functions of interest. However, the demonstration even for this classic problem is anything but trivial (see Bliss (1925) or Young (1969) for instance), a fact that underscores the distinct difference between determining potential maximizing and minimizing functions and rigorously establishing their extremal character.

4.3 Practical extensions of the theory

Having laid out the rudimentary foundations of the theory, now we want to examine a number of special (and useful) topics in the calculus of variations that go beyond the basic discussion of Section 4.2. However, in order to keep our considerations within reasonable limits, we will have to be a bit more brief than usual in our explanations.

Some simplifications

If the function $f(x,y,y')$ appearing in the functional $I(y)$ of (4.2–1) has some additional structure, the Euler-Lagrange equation derived earlier can often be simplified. As examples we have:

$f(x,y,y')$ does not depend explicitly on y
In this case (4.2–5) becomes

$$\frac{d}{dx}\left(\frac{\partial f}{\partial y'}\right) = 0,$$

and hence

$$\frac{\partial f}{\partial y'} = C \tag{4.3–1}$$

where C is some constant of integration.

$f(x,y,y')$ does not depend explicitly on x
Here we use the readily verifiable identity

$$\frac{d}{dx}\left(f - y'\frac{\partial f}{\partial y'}\right) = \frac{\partial f}{\partial x} + \frac{\partial f}{\partial y}y' + \frac{\partial f}{\partial y'}y'' - y''\frac{\partial f}{\partial y'} - y'\frac{d}{dx}\left(\frac{\partial f}{\partial y'}\right)$$
$$= y'\left[\frac{\partial f}{\partial y} - \frac{d}{dx}\left(\frac{\partial f}{\partial y'}\right)\right].$$

If the right-hand side of this expression vanishes by virtue of (4.2–5), the left-hand side may be immediately integrated to give

$$f - y'\frac{\partial f}{\partial y'} = C \tag{4.3–2}$$

where C is again some arbitrary constant (not necessarily the same as above). In this special situation, therefore, we may replace the more complicated Euler-Lagrange second-order differential equation by the simpler first-order equation (4.3–2). (This is why we so easily obtained a first integral in the earlier brachistochrone example.)

The case wherein $f(x,y,y')$ does not depend explicitly on y' is relatively uninteresting (see Problem 6).

Variable endpoints

In the derivation of the Euler-Lagrange equation (4.2–5) for functionals of the form (4.2–1), we assumed that the endpoints of the functions of interest were fixed. Since this is too rigid a restriction for many applications, we want to extend our earlier analysis to the so-called *variable endpoint problem* in which the admissible functions are merely required to begin and end on certain prescribed curves. For simplicity, however, we study only the case of a variable right-hand endpoint; the corresponding result for a variable left-hand endpoint should be clear by analogy. If both endpoints are variable the partial results are combined in the obvious manner.

Our consideration begins as before with the application of Theorem 1 of Section 4.2 to the functional $I(y)$ given by

$$I(y) = \int_a^{x_f} f(x,y,y')dx. \tag{4.3–3}$$

The extremal function is sought amongst, say, the continuously differentiable functions $y(x)$ satisfying the usual condition $y(a) = y_i$ at the fixed left-hand endpoint, but now the new condition

$$g(x_f, y(x_f)) = 0 \tag{4.3–4}$$

at the variable right-hand endpoint. The pertinent variations are

$$\begin{aligned} x_f &= b + \epsilon\beta \\ y(x) &= \bar{y}(x) + \epsilon\eta(x) \end{aligned} \tag{4.3–5}$$

where b is the optimum endpoint value for x_f that will be determined as part of the problem, ϵ and β are scalars, and the function η (aside from differentiability restrictions) satisfies only the single endpoint condition $\eta(a) = 0$.

By an obvious generalization of Property 1 of Section 4.2, we calculate

$$\begin{aligned} 0 = \left.\frac{dI}{d\epsilon}\right|_{\epsilon=0} &= \beta f\Big|_{x=b} + \int_a^{x_f} \left(\frac{\partial f}{\partial y}\eta + \frac{\partial f}{\partial y'}\eta'\right) dx \Bigg|_{\epsilon=0} \\ &= \beta f\Big|_{x=b} + \eta \frac{\partial f}{\partial y'}\Big|_{x=b} + \int_a^b \left[\frac{\partial f}{\partial y} - \frac{d}{dx}\left(\frac{\partial f}{\partial y'}\right)\right] \eta\, dx. \end{aligned} \tag{4.3–6}$$

The first term on the right-hand side of (4.3–6) arises from the differentiation of $I(y)$ with respect to its variable upper limit; the second term is a consequence of integrating by parts. By virtue of (4.3–4) and (4.3–5), however, these two terms are not entirely independent. Indeed, if we differentiate (4.3–4) with respect to ϵ and use (4.3–5), we find

$$\frac{\partial g}{\partial x}\beta + \frac{\partial g}{\partial y}\left(\bar{y}'(x_f)\beta + \eta(x_f) + \epsilon\eta'(x_f)\beta\right) = 0.$$

Then, setting $\epsilon = 0$ and solving for β yields

$$\beta = -\frac{\eta(b)\, \partial g/\partial y_f}{\partial g/\partial b + y_f'\, \partial g/\partial y_f}. \tag{4.3–7}$$

Here we have used the shorthand notation $y_f \equiv \bar{y}(b)$, $y_f' \equiv \bar{y}'(b)$.

The implication of this last observation is that β and $\eta(b)$ are proportional. Equation (4.3–6) can be rewritten, therefore, so that *both* of the first two terms on the right-hand side have $\eta(b)$ as a common factor. Of course $\eta(b)$ can be zero, in which case (reasoning as previously) we deduce that potential extrema of (4.3–4) must still be solutions of the Euler-Lagrange equation (4.2–5). On the other hand, $\eta(b)$ need not vanish. It then follows, since the integral in (4.3–6) must now be zero, that the variable endpoint extrema must also satisfy

$$\beta f\Big|_{x=b} + \eta(b)\frac{\partial f}{\partial y'}\Big|_{x=b} = 0.$$

This is a *natural boundary* or *transversality condition* that can be rewritten, using (4.3–7) and simplifying, in the more familiar form

$$\frac{\partial g(b,y_f)}{\partial x}\frac{\partial f(b,y_f,y_f')}{\partial y'} + \frac{\partial g(b,y_f)}{\partial y}\left[y_f'\frac{\partial f(b,y_f,y_f')}{\partial y'} - f(b,y_f,y_f')\right] = 0. \quad \textbf{(4.3–8)}$$

We summarize the above results for the variable endpoint problem as

Theorem 1

If the functional

$$I(y) = \int_a^{x_f} f(x,y,y')dx$$

is well defined for all, say, continuously differentiable functions $y(x)$ satisfying

$$y(a) = y_i,$$
$$g(x_f,y(x_f)) = 0$$

for given constant y_i and prescribed function $g(x,y)$, and if $\tilde{y}(x)$ provides a local extremum for $I(y)$, then, assuming all the indicated derivatives exist and are continuous, the extremal function $\tilde{y}(x)$ must satisfy both the Euler-Lagrange equation (4.2–5) and the natural right-hand endpoint condition (4.3–8).

Two special cases and a useful extension of Theorem 1 merit individual mention:

Variable endpoint with fixed x-value

In this case the terminating curve is the vertical line $x = b$ and the function $g(x,y)$ has the form

$$g(x,y) = x - b.$$

The natural boundary condition (4.3–8) reduces to the elementary expression

$$\frac{\partial f(b,y_f,y_f')}{\partial y'} = 0. \quad \textbf{(4.3–9)}$$

Integrands of the form $f(x,y,y') = h(x,y)\sqrt{1 + y'^2}$

For such integrands (4.3–8) simplifies to

$$h(b,y_f)\left[\frac{\partial g(b,y_f)}{\partial x}y_f' - \frac{\partial g(b,y_f)}{\partial y}\right] = 0. \quad \textbf{(4.3–10)}$$

Since along the curve $g(x,y) = 0$ we must have

$$\frac{\partial g}{\partial x} + \frac{\partial g}{\partial y}y' = 0,$$

(4.3–10) merely requires that, if $h(b,y_f)$ is nonvanishing, then the extremal function $\tilde{y}(x)$ intersects the terminating curve *at right angles*. (Note that the brachistochrone integrand of (4.2–6) is of this form.)

Functionals of the form $F(y) = \int_a^{x_f} f\,dx + h(x_f)$

In order to handle this modification of functional form we need only to replace the $f(b, y_f, y_f')$ appearing in the bracketed term within the natural boundary condition (4.3–8) by the quantity $f + dh/dx_f$. The presence of the additional function h has no effect on the nature of the Euler-Lagrange equation and the rest of Theorem 1 remains as before (see Problem 11).

More than one dependent variable

The extension of our considerations to functionals depending upon several unknown functions, say, $y_1(x), y_2(x), \ldots, y_n(x)$ can be made formally by introducing vector notation and combining the scalar functions into a single vector function $\mathbf{y}(x) \equiv (y_1(x), y_2(x), \ldots, y_n(x))$. Then the analogue of (4.2–1) is

$$I(\mathbf{y}) = \int_a^b f(x, \mathbf{y}, \mathbf{y}')dx, \qquad \textbf{(4.3–11)}$$

where f is some specified function defined for all points in some open set of $(2n + 1)$-space and a,b, are fixed. If we are interested in characterizing the extrema of I among vectors $\mathbf{y}$ whose components are all continuously differentiable on $[a,b]$ and satisfy the fixed endpoint conditions $\mathbf{y}(a) = \mathbf{y}_i$, $\mathbf{y}(b) = \mathbf{y}_f$, then an approach like that used in Section 4.2 is applicable. The only change is the vectorial nature of the analysis.

As comparison function we now employ

$$\mathbf{y}(x) = \tilde{\mathbf{y}}(x) + \epsilon\,\boldsymbol{\eta}(x)$$

with $\boldsymbol{\eta}(a) = 0 = \boldsymbol{\eta}(b)$, where $\tilde{\mathbf{y}}(x)$ is the (assumed to exist) actual extremal vector function and $\boldsymbol{\eta}(x) \equiv (\eta_1(x), \eta_2(x), \ldots, \eta_n(x))$. Introducing this representation into (4.3–11) again renders I a function of the single variable ϵ with an extremum at $\epsilon = 0$. Just as before, therefore, we investigate the implications of $I'(0) = 0$. This time we find

$$0 = \sum_{j=1}^{n} \int_a^b \left[\frac{\partial f}{\partial y_j} \eta_j + \frac{\partial f}{\partial y_j'} \eta_j' \right] dx$$

$$= \sum_{j=1}^{n} \int_a^b \left[\frac{\partial f}{\partial y_j} - \frac{d}{dx} \frac{\partial f}{\partial y_j'} \right] \eta_j \, dx.$$

Since this relation must hold for all continuously differentiable η_j which vanish at the two endpoints, by *successively* choosing all but one of the $\eta_j(x)$ $(j = 1, 2, \ldots, n)$

to be *identically zero* and then using the same "basic lemma" as before on the one remaining integral, we are led to n separate Euler-Lagrange equations that must be satisfied by the n extremal functions $y_1(x), y_2(x), \ldots, y_n(x)$:

$$\frac{\partial f(x,\mathbf{y},\mathbf{y}')}{\partial y_j} - \frac{d}{dx}\frac{\partial f(x,\mathbf{y},\mathbf{y}')}{\partial y_j'} = 0 \qquad (j = 1,2, \ldots, n). \qquad \textbf{(4.3–12)}$$

Functionals with higher-order derivatives

Our now-familiar procedure can also be applied successfully to functionals of the form

$$I(y) = \int_a^b f(x,y,y',y'', \ldots, y^{(n)})\, dx. \qquad \textbf{(4.3–13)}$$

In the fixed endpoint problem we seek to maximize or minimize I among all functions $y = y(x)$ that, along with their derivatives up to the nth order, are continuous on $[a,b]$ and that satisfy the $2n$ prescribed boundary conditions

$$\begin{array}{ll} y(a) = y_i & y(b) = y_f \\ y'(a) = c_1 & y'(b) = d_1 \\ \vdots & \vdots \\ y^{(n-1)}(a) = c_{n-1} & y^{(n-1)}(b) = d_{n-1} \end{array}$$

for given constants y_i, y_f, and c_j, d_j $(j = 1,2, \ldots, n - 1)$. When we employ our usual function variation $\Delta y = \epsilon\eta$ and carry out the obvious steps, we are led to

$$0 = \int_a^b \left[\frac{\partial f}{\partial y}\eta + \frac{\partial f}{\partial y'}\eta' + \frac{\partial f}{\partial y''}\eta'' + \cdots + \frac{\partial f}{\partial y^{(n)}}\eta^{(n)} \right] dx.$$

If η and all its derivatives up to the $(n - 1)$st order are assumed to vanish at both of the endpoints a and b, integration by parts (as many times as needed) then can be performed to yield

$$0 = \int_a^b \left[\frac{\partial f}{\partial y} - \frac{d}{dx}\left(\frac{\partial f}{\partial y'}\right) + \frac{d^2}{dx^2}\left(\frac{\partial f}{\partial y''}\right) + \cdots + (-1)^n \frac{d^n}{dx^n}\left(\frac{\partial f}{\partial y^{(n)}}\right) \right] \eta dx.$$

From the arbitrariness of η it now follows, by virtue of the "basic lemma," that the extremal function $\tilde{y}(x)$ must satisfy the *generalized Euler-Lagrange equation*

$$\frac{\partial f}{\partial y} + \sum_{j=1}^{n} (-1)^j \frac{d^j}{dx^j}\left(\frac{\partial f}{\partial y^{(j)}}\right) = 0. \qquad \textbf{(4.3–14)}$$

We note in passing that equation (4.3–14) is in general a differential equation of $(2n)$th order, and that $2n$ boundary conditions are needed to uniquely specify a potential extremal solution.

Several independent variables

Numerous applications lead to functionals that depend on multidimensional functions. For our purposes, it will suffice to restrict our attention to the case of three independent variables and consider only functionals of the form

$$I(u) = \iiint_V f(x,y,z,u,u_x,u_y,u_z)dV. \qquad \textbf{(4.3–15)}$$

Here V will be taken to be some given fixed region in 3-space bounded by a smooth* surface S, dV is the infinitesimal volume element, f is a well-defined function of its arguments and is continuously twice differentiable with respect to any combination of them, u is a function of the three variables $x,y,z,$ and subscripts denote partial derivatives (that is, $u_x \equiv \dfrac{\partial u}{\partial x}$, etc.). The outward unit normal to the surface will be assumed to be well-defined at each point on S and will be designated by $\mathbf{n}$.

The analogue of the simplest problem in the calculus of variations for the functional given by (4.3–15) is the determination of the function $u = u(x,y,z)$ which maximizes or minimizes $I(u)$ amongst all functions that are continuously differentiable on the closed region $V + S$ and that satisfy the fixed boundary condition

$$u(x,y,z) = g(x,y,z) \qquad \text{on } S$$

with g a given function. Not surprisingly, to analyze this problem we can mimic our earlier approach: we first introduce the one-parameter family of comparison functions $u = \tilde{u} + \epsilon\eta$, where $\tilde{u} = \tilde{u}(x,y,z)$ is assumed to be the actual extremal function and $\eta = \eta(x,y,z)$ vanishes on S. We then calculate $dI/d\epsilon$, set $\epsilon = 0$ and find, since $I(0)$ is the extremum sought, that

$$0 = \iiint_V \left[\frac{\partial f}{\partial u}\eta + \frac{\partial f}{\partial u_x}\eta_x + \frac{\partial f}{\partial u_y}\eta_y + \frac{\partial f}{\partial u_z}\eta_z\right] dV. \qquad \textbf{(4.3–16)}$$

To effect the next step we need a multidimensional version of integration by parts. Fortunately, the well-known *Divergence Theorem* will do very nicely in this regard. We use it in the following form

---

*Smooth surfaces are those with the property that at each of their points there exists a tangent plane that varies continuously as the point moves continuously on the surface.

$$\iint_S \eta \mathbf{F} \cdot \mathbf{n}\, dS = \iiint_V \operatorname{div}(\eta \mathbf{F}) dV$$

$$= \iiint_V \operatorname{grad} \eta \cdot \mathbf{F}\, dV + \iiint_V \eta \operatorname{div} \mathbf{F}\, dV. \qquad \textbf{(4.3–17)}$$

If we make the association $\mathbf{F} = \mathbf{e}_x \dfrac{\partial f}{\partial u_x} + \mathbf{e}_y \dfrac{\partial f}{\partial u_y} + \mathbf{e}_z \dfrac{\partial f}{\partial u_z}$, the last three terms within the bracket of (4.3–16) are precisely the integrand of the first integral on the right-hand side of (4.3–17). Since $\eta = 0$ on S, we can thus use the negative of the second integral on the right-hand side of (4.3–17) to rewrite (4.3–16) as

$$0 = \iiint_V \left[\frac{\partial f}{\partial u} - \frac{\partial}{\partial x}\left(\frac{\partial f}{\partial u_x}\right) - \frac{\partial}{\partial y}\left(\frac{\partial f}{\partial u_y}\right) - \frac{\partial}{\partial z}\left(\frac{\partial f}{\partial u_z}\right) \right] \eta\, dV.$$

From this expression it is a simple matter to conclude (using the arbitrariness of η and employing an obvious generalization of the "basic lemma" p. 89) that for this multidimensional situation any local extrema $\hat{u}(x,y,z)$ must satisfy the partial differential equation

$$\frac{\partial f}{\partial u} - \frac{\partial}{\partial x}\left(\frac{\partial f}{\partial u_x}\right) - \frac{\partial}{\partial y}\left(\frac{\partial f}{\partial u_y}\right) - \frac{\partial}{\partial z}\left(\frac{\partial f}{\partial u_z}\right) = 0 \qquad \textbf{(4.3–18)}$$

within the region V in addition to the fixed boundary condition $u = g$ on S.

4.4 A potpourri of mini-applications

Like the differential and integral calculus, the scope of the calculus of variations can truly be sensed only as it is applied. In this section, therefore, we sample a variety of applications that suggest the breadth of utility of the theory as we have presented it so far.

Shortest distance

The most elementary minimum-distance problem asks, "What planar curve connecting two given points in 3-space has the smallest arclength?" By a suitable choice of coordinate system, the problem can be formulated in terms of minimizing the functional

$$I(y) = \int_a^b \sqrt{1 + y'^2}\, dx \qquad \textbf{(4.4–1)}$$

where $y(a) = y_i$, $y(b) = y_f$. As expected, the extremal solution turns out to be a straight line (see Problem 3).

A somewhat less trivial (and more practical) minimum distance problem can be posed by requiring the curve connecting the two fixed points to lie on a given smooth surface. The extremal arcs are then called *geodesic curves* for the surface and play important roles in fields as diverse as manufacturing and navigation. In order to formulate this more general problem, we assume that (i) the given surface can be expressed in terms of two parameters α and β as

$$x = x(\alpha,\beta), \qquad y = y(\alpha,\beta), \qquad z = z(\alpha,\beta);$$

(ii) the given fixed points have the parametric representation (α_i,β_i) and (α_f,β_f) with $\alpha_f > \alpha_i$; and (iii) the arcs joining the fixed points can be expressed in the form $\beta = \beta(\alpha)$. It follows that the appropriate functional is

$$I(\beta) = \int_{\alpha_i}^{\alpha_f} \sqrt{A(\alpha,\beta) + 2B(\alpha,\beta)\beta' + C(\alpha,\beta)\beta'^2}\, d\alpha \qquad \textbf{(4.4–2)}$$

where

$$A = \left(\frac{\partial x}{\partial \alpha}\right)^2 + \left(\frac{\partial y}{\partial \alpha}\right)^2 + \left(\frac{\partial z}{\partial \alpha}\right)^2,$$

$$B = \frac{\partial x}{\partial \alpha}\frac{\partial x}{\partial \beta} + \frac{\partial y}{\partial \alpha}\frac{\partial y}{\partial \beta} + \frac{\partial z}{\partial \alpha}\frac{\partial z}{\partial \beta},$$

$$C = \left(\frac{\partial x}{\partial \beta}\right)^2 + \left(\frac{\partial y}{\partial \beta}\right)^2 + \left(\frac{\partial z}{\partial \beta}\right)^2$$

(see Problem 15).

Example For the special case of the right circular cylinder of radius R we have

$$x = R\cos\phi, \qquad y = R\sin\phi, \qquad \text{and} \quad z = z.$$

If we let ϕ and z play the role of α and β, respectively, (4.4–2) then becomes, after some simplification,

$$I(z) = \int_{\phi_i}^{\phi_f} \sqrt{R^2 + z'^2}\, d\phi$$

where $z' \equiv dz/d\phi$ and $0 \le \phi_f - \phi_i \le \pi$.* Since the integrand of I is independent of both ϕ and z, either of the simplified forms (4.3–1) or (4.3–2) of the Euler-Lagrange equations can be employed. We find that the extremal function must satisfy

$$z'(\phi) = \text{const},$$

which finally leads to the linear relation

$$z = c\phi + d.$$

*We obviously want to choose the short way around the cylinder.

This last equation, coupled with the earlier expressions for x and y in terms of ϕ, provides the parametric representation of the so-called *circular helix*. When the constants c and d are selected so that the helix passes through the specified fixed points, we have determined the appropriate geodesic curve for the given right-circular cylinder. Moreover, if those specified fixed points have coordinates (x_i, y_i, z_i) and (x_f, y_f, z_f), then the resulting minimum distance can be shown to be

$$D = \sqrt{R^2\gamma^2 + (z_f - z_i)^2}\,,$$

where $\cos\gamma \equiv (x_i x_f + y_i y_f)/R^2$ with $0 \leq \gamma \leq \pi$. □

Another important special case, namely that of determining the geodesic curves for the sphere of radius R, is considered in Problem 16.

Least time

Minimum distance and minimum time problems are equivalent whenever the velocity function that relates the two quantities remains constant. However, if velocity varies as a function of position, as it does for example in the brachistochrone problem studied earlier, then minimum time problems must be treated separately. In the case of light propagation in inhomogeneous media, such problems have special importance owing to a principle first enunciated by the French mathematician Pierre de Fermat (1601–1665): *light rays travel along the quickest paths.*

The optimization problem inherent in Fermat's principle of least time can be easily modelled. We only need describe the possible optical paths joining two fixed points in terms of a suitable parameter α as

$$x = x(\alpha), \qquad y = y(\alpha), \qquad z = z(\alpha) \qquad \alpha_i \leq \alpha \leq \alpha_f.$$

Then the light signal travels along such a path between the two points in a time T given by

$$\begin{aligned} T &= \int_0^T dt = \int_0^{s(T)} \frac{dt}{ds}\,ds \\ &= \int_0^{s(T)} \frac{ds}{v} = \int_{\alpha_i}^{\alpha_f} \frac{(ds/d\alpha)}{v}\,d\alpha \\ &= \int_{\alpha_i}^{\alpha_f} \frac{\sqrt{x'^2 + y'^2 + z'^2}}{v(x,y,z)}\,d\alpha, \end{aligned} \qquad \textbf{(4.4–3)}$$

and this is recognizable as an extremal problem with more than one dependent variable, to which the Euler-Lagrange equations (4.3–12) are applicable.

Example An interesting application occurs when the light velocity v is a function of only a single coordinate variable, say y. In this case, the problem becomes essentially two-dimensional, the parameter α can be given up, say, in favor of the variable x, and (4.4–3) may be rewritten as

$$T = \int_{x_i}^{x_f} \frac{\sqrt{1 + y'^2}}{v(y)}\, dx. \qquad \textbf{(4.4–4)}$$

Since the integrand of this functional representation does not depend explicitly upon x, any function that minimizes the time functional of (4.4–4) must satisfy the integrated expression (4.3–2). When we apply this fact to our integrand, therefore, we find

$$v(y)\sqrt{1 + y'^2} = \text{const.} \qquad \textbf{(4.4–5)}$$

Not surprisingly, this relation gives rise to Snell's well-known law of refraction for the case of two different (but individually homogeneous) media separated by a planar interface (see Problem 17). □

Equation (4.4–5) is also useful in the analysis of sound propagation (which obeys similar physical laws) in layered media (see Brekhovskikh (1980), for example). Other applications are considered by Smith (1974), including some in which *maxima* are sought for (4.4–4) and/or (4.4–3).

Least action and Hamilton's principle in mechanics

In the analysis of the motion of elementary mechanical systems, the standard description involves the normal variables of distance and velocity. For more complicated applications, however, it is often advantageous to introduce *generalized coordinates* (angles, areas, and the like, as well as distances) and *generalized velocities* (the time derivatives of the generalized coordinates) into the problem formulation (recall the primary example of Chapter 1). For purposes of this discussion we will follow the latter approach. We denote the generalized coordinates by $q_j (j = 1, 2, \ldots, n)$, the generalized velocities by $\dot{q}_j (j = 1, 2, \ldots, n)$, and the potential and kinetic energies of a given dynamical system* by V and T, respectively. We assume that V is a function only of the coordinates q_j (that is, the system is *conservative*) and that T, while possibly dependent on both the q_j and the $\dot{q}_j$, is a *homogeneous function, of degree two,* of the second set of variables. The implication of this last requirement is that T satisfies the differential equation

$$\sum_{j=1}^{n} \dot{q}_j \frac{\partial T}{\partial \dot{q}_j} = 2T. \qquad \textbf{(4.4–6)}^{\dagger}$$

*We use the term dynamical system to refer to any mechanical system, composed of perhaps many components and thus needing perhaps a large number of generalized coordinates and velocities to specify its state, undergoing individual and/or collective motion.

†A function $f(x_1, \ldots, x_n, v_1, \ldots, v_m)$ is said to be *homogeneous of degree two* in the variables $v_1, \ldots, v_m$ if, for arbitrary h,

$$f(x_1, \ldots, x_n, hv_1, \ldots, hv_m) = h^2 f(x_1, \ldots, x_n, v_1, \ldots, v_m).$$

The quantities introduced above can be blended with the calculus of variations in a variety of ways so as to lead to a description of the motion of dynamical systems, which is alternative (but equivalent) to that based solely on Newton's second law. For example, already in the mid-eighteenth century Euler and the French astronomer Pierre de Maupertuis (1698–1759) had separately proposed what amounts to the following:

Principle of Least Action:

Consider a given conservative system characterized by a potential energy $V = V(\mathbf{q})$ and a kinetic energy $T = T(\mathbf{q},\dot{\mathbf{q}})$. Assume that the system moves from a configuration at time $t = t_i$ to a configuration at a later time $t = t_f$ in such a way that the total system energy $E = T + V$ remains constant. Then the motion of the system must be such that the *action integral*

$$I_M = 2\int_{t_i}^{t_f} T\,dt \qquad \textbf{(4.4–7)}$$

is minimized.

Almost a century later, the Irish mathematician William Hamilton (1805–1865) enunciated his version of a least-action principle:

Hamilton's Principle:

Let a given conservative system with potential energy $V = V(\mathbf{q})$ and kinetic energy $T = T(\mathbf{q},\dot{\mathbf{q}})$ be characterized by *a Lagrangian function*

$$L \equiv T - V.$$

Then the motion connecting two known configurations of the system, say at times $t = t_i$ and $t = t_f$, is the one that minimizes the *Hamilton action integral*

$$I_H = \int_{t_i}^{t_f} L\,dt. \qquad \textbf{(4.4–8)}$$

As might be expected, these two least-action principles are equivalent. The Euler-Lagrange equations for (4.4–8) are

$$\frac{\partial L}{\partial q_j} - \frac{d}{dt}\left(\frac{\partial L}{\partial \dot{q}_j}\right) = 0 \qquad (j = 1,2,\ldots,n),$$

which may be integrated to give

Differentiating both sides of this relation with respect to h, and then setting $h = 1$, it follows that

$$\sum_{j=1}^{m} v_j \frac{\partial f}{\partial v_j} = 2f.$$

$$\sum_{j=1}^{n} \dot{q}_j \frac{\partial L}{\partial \dot{q}_j} - L = E$$

where E is constant, since L does not depend explicitly on t (recall (4.3–2)). But $V = V(\mathbf{q})$ combined with (4.4–6) and the definition of L then implies

$$\begin{aligned} E &= 2T - (T - V) \\ &= T + V, \end{aligned}$$

so that the total system energy is automatically conserved during an extremal motion. Moreover, reading this relation in the opposite direction, we see that $2T$ and L differ only by the constant E so that the Euler-Lagrange equations for (4.4–7) and for (4.4–8) must be identical, and the minimum values of I_M and I_H are trivially related.*

The vibrating string

Example The analysis of the small vibrations of a flexible elastic string provides an instructive application for the use of Hamilton's principle. We assume that the string, at rest, is stretched under constant tension τ along the x-axis from the origin $x = 0$ to $x = l$. If vibrations are restricted to a plane (containing the x-axis), then at time t we can describe the displacement of the string by a suitably smooth function $u(x,t)$, $0 \le x \le l$. We assume that these vibrations are *small*, in the sense that $\left|\frac{\partial u}{\partial x}\right| << 1$, and that they are *transverse* so that each infinitesimal portion of the string moves only in a straight line perpendicular to the x-axis.

In order to apply Hamilton's principle to this dynamical system, we need formulas for both the associated potential and kinetic energies. For the former we find

$$\begin{aligned} V &= \tau\left[\int_0^{s(l)} ds - l\right] \\ &= \tau\int_0^l \left(\sqrt{1 + u_x^2} - 1\right)dx, \end{aligned}$$

since the only work being done on the string, owing to its flexibility, is that required to lengthen it.* If the linear mass density of the string is $\sigma(x)$ $(0 \le x \le l)$, then in view of the transverse nature of the vibrations, the comparable relation for the kinetic energy is

---

*This "demonstration" glosses over some subtle details; see Winestock (1952). This classic text is also a good source for applications. In addition, serious students of mechanics may want to consult Lanczos (1970), a valuable storehouse of information on these and other dynamical principles.

*Recall *energy* = *work* = *force* × *distance*, and the stretching force must balance the tension τ.

$$T = \int_0^l \tfrac{1}{2}\, \sigma\, u_t^2 \, dx.$$

Combining these two expressions we thus obtain

$$I_H = \int_{t_i}^{t_f} \int_0^l \left[\tfrac{1}{2}\, \sigma u_t^2 - \tau \left(\sqrt{1 + u_x^2} - 1 \right) \right] dx\, dt \tag{4.4–9}$$

for the Hamilton action integral (4.4–8).

The functional given by (4.4–9) is multi-dimensional. Therefore, extremals of I_H must satisfy an appropriate partial differential Euler-Lagrange equation. If we associate t with y and restrict equation (4.3–18) to two dimensions, after some simplification the result is

$$\frac{u_{xx}}{(1 + u_x^2)^{3/2}} = \frac{\sigma(x)}{\tau}\, u_{tt}\, . \tag{4.4–10}$$

According to Hamilton's principle, then, this equation characterizes the motion of our vibrating string. In the limit of small vibrations, where u_x^2 can be neglected relative to unity in the coefficient of u_{xx}, (4.4–10) becomes the well-known one-dimensional wave equation

$$u_{xx} - \frac{\sigma}{\tau}\, u_{tt} = 0, \tag{4.4–11}$$

already encountered in Section 2.1 and which we will meet again in Chapter 8. □

Almost-planar minimal surfaces

Example Another classic multidimensional optimization problem concerns the functional

$$I(u) = \iint_R \sqrt{1 + u_x^2(x,y) + u_y^2(x,y)}\, dx\, dy \tag{4.4–12}$$

where R is a given fixed bounded open region in the (x,y) plane with a smooth boundary curve and where $u = u(x,y)$ is a suitably smooth function defined for all points (x,y) in R. Since the value of $I(u)$ is precisely the *surface area* of the graph of $u(x,y)$ in three-space, minimizing $I(u)$ over all appropriate functions u that have the *same* three-dimensional perimeter contour gives rise to the so-called *minimal surface*.

The general minimal surface problem dates back to Lagrange and was studied at length experimentally by the Belgian physicist Joseph Plateau (1801–1883) using soap films. (A good historical sketch of Plateau's work and its modern ramifications is to be found in Nitsche (1974).) It turns out that these soap films, when allowed to span the given closed perimeter contour, generally assume the minimal configuration by virtue of the surface tension present within the film.

For the functional defined by (4.4–12), the Euler-Lagrange equation (4.3–18) assumes the form

$$\frac{\partial}{\partial x}\left(\frac{u_x}{\sqrt{1 + u_x^2 + u_y^2}}\right) + \frac{\partial}{\partial y}\left(\frac{u_y}{\sqrt{1 + u_x^2 + u_y^2}}\right) = 0. \qquad \textbf{(4.4–13)}$$

The resulting *minimal surface equation* proves to be rather difficult to analyze in detail. However, if we can neglect u_x^2 and u_y^2 relative to unity within the square roots in (4.4–13), as is the situation in the case when the minimal surface perimeter contour is nearly a horizontally plane curve and the minimal surface is thus itself almost planar, then the general equation reduces to

$$u_{xx} + u_{yy} = 0. \qquad \textbf{(4.4–14)}$$

We recognize (4.4–14) as the familiar Laplace's equation for two-dimensions, another fundamental equation that will be central to our inquiries in several subsequent chapters. □

Sturm-Liouville problems

In Chapter 2 we discussed at length the nature of eigenvalue problems associated with the differential equation

$$(py')' + (\lambda r - q)y = 0 \qquad a < x < b \qquad \textbf{(4.4–15)}$$

where $p'(x)$ and $q(x)$ are real-valued and continuous functions on the bounded interval $[a,b]$ and $p(x)$ and $r(x)$ are both positive. Readers are undoubtedly familiar with such Sturm-Liouville problems from numerous applications, for example, the eigenvalue problems that arise when separation of variables is applied to partial differential equations such as the wave equation (4.4–11) or Laplace's equation (4.4–14). Now we want to explore briefly the role that (4.4–15) can be shown to play in the calculus of variations.

For our purposes we need the two functionals

$$\begin{aligned} I_1(y) &\equiv \int_a^b (py'^2 + qy^2)dx, \\ I_2(y) &\equiv \int_a^b r\,y^2\,dx \end{aligned} \qquad \textbf{(4.4–16)}$$

and their quotient I_1/I_2, which we designate by λ, that is,

$$\lambda \equiv \frac{I_1(y)}{I_2(y)}. \qquad \textbf{(4.4–17)}$$

We note that $I_2(y)$ is the square of the norm of the function $y(x)$ with respect to the (nonnegative) weight function $r(x)$ and thus is positive for all nontrivial y. The quotient (4.4–17) is one form of the so-called *Rayleigh quotient*, which we consider in more detail in Chapter 10.

Let us suppose that we seek to minimize the Rayleigh quotient amongst all continuously differentiable functions $y = y(x)$ that satisfy, say, the fixed endpoint conditions $y(a) = 0 = y(b)$. Aside from the quotient character of the functional, this is merely a problem of the basic type we considered in Section 4.2. It follows then, designating the minimizing function by $\tilde{y}$ and the corresponding minimum value of λ by $\tilde{\lambda}$, that

$$0 = \delta\lambda = \frac{I_2(\tilde{y})\delta I_1(\tilde{y}) - I_1(\tilde{y})\delta I_2(\tilde{y})}{[I_2(\tilde{y})]^2}$$

$$= \frac{\delta[I_1 - \tilde{\lambda} I_2](\tilde{y})}{I_2(\tilde{y})}, \qquad \textbf{(4.4–18)}$$

where $[I_1 - \tilde{\lambda} I_2]$ is a composite integral functional of the standard form (4.2–1) with integrand

$$f_c \equiv py'^2 + (q - \tilde{\lambda}\, r)y^2.$$

But (4.4–18) implies that $\tilde{y}$ provides a local extremum for this composite functional as well, so that as a consequence of (4.2–5), $\tilde{y}$ must be a solution of the Euler-Lagrange equation

$$\frac{d}{dx}\left(\frac{\partial f_c}{\partial y'}\right) - \frac{\partial f_c}{\partial y} = 0.$$

When the indicated differentiations are performed we see that we have come full circle; the Euler-Lagrange equation is none other than the Sturm-Liouville equation (4.4–15) but with $\tilde{\lambda}$ in place of λ.

We can summarize our findings as follows: The optimization problem for the Rayleigh quotient functional (4.4–17) and the eigenvalue problem for the differential equation (4.4–15) are intimately intertwined. The Euler-Lagrange equation associated with (4.4–17) is precisely (4.4–15). The minimizing solution for the Rayleigh quotient is the eigenfunction for the corresponding Sturm-Liouville problem (with boundary conditions $y(a) = 0 = y(b)$)*, which is associated with the *least* eigenvalue (recall our discussion in Section 2.4), and this smallest eigenvalue is the minimum value of the quotient. Indeed, since nowhere in our variational considerations do we make use of the minimum character of $\tilde{y}$, *every* appropriate eigenfunction y_j ($j = 1,2, \ldots$) of (4.4–15) must render (4.4–17) stationary and give rise to a value for the Rayleigh quotient equal to the corresponding eigenvalue λ_j. These higher-order eigenfunctions can be singled out as minimizing solutions in their own right by adding suitable constraints to the basic problem (see Smith (1974), Weinstock (1952), or Sagan (1969), for example). More comprehensive discussions of variational approaches to general eigenvalue problems are to be found in Gould (1966), in Denn (1978), and in Weinstein and Stenger (1972).

---

*Actually, in view of the nature of the composite functional f_c and the form of the natural boundary condition (4.3–9), whenever $p(x)$ is strictly of one sign the boundary conditions may involve $y'(a) = 0$ and/or $y'(b) = 0$ as well.

Approximations

In some problems it is impractical (if not impossible) to obtain an exact solution. In such situations, techniques for obtaining an approximate solution then become the order of the day. The premier approximation approach for eigenvalue problems of Sturmian type and indeed for general extremal problems, at least in one independent variable, is the Rayleigh-Ritz procedure. However, since the relevance of this method to comparable problems in integral equations is equally as great, we delay consideration of this important method until Chapter 10. Here we content ourselves with some simpler, yet remarkably accurate, one-parameter approximation techniques that we illustrate with several examples:

Example Consider the familiar elementary eigenvalue problem

$$\begin{aligned} y'' + \lambda y &= 0 \qquad 0 < x < 1 \\ y(0) &= 0 \\ y(1) &= 0. \end{aligned} \tag{4.4–19}$$

In view of the form of (4.4–16), the corresponding extremal problem is to minimize the quotient

$$\lambda = \frac{\int_0^1 y'^2 dx}{\int_0^1 y^2 dx}.$$

As an approximate solution, we try the one-parameter family of functions $\phi_\alpha(x) = [x(1 - x)]^\alpha$ where α is a free parameter (greater than ½) which we will select in such a way as to make $\lambda(\phi_\alpha)$ as small as possible. (Note that $\phi_\alpha(x)$ has been chosen to satisfy the boundary conditions of (4.4–19).) Since $\phi_\alpha{}' = \alpha[x(1 - x)]^{\alpha-1}(1 - 2x)$, we calculate

$$\begin{aligned} \lambda(\phi_\alpha) &= \left[\frac{\alpha\Gamma(2\alpha + 1)\Gamma(2\alpha - 1)}{\Gamma(4\alpha)}\right] \Big/ \left[\frac{\Gamma^2(2\alpha + 1)}{\Gamma(4\alpha + 2)}\right] \\ &= \frac{2\alpha(4\alpha + 1)}{2\alpha - 1}. \end{aligned}$$

(Recall the integral representation of the beta function; see Appendix 5). It follows by classical minimization techniques that the best value of α satisfies

$$8\alpha^2 - 8\alpha - 1 = 0.$$

This leads to

$$\begin{aligned} \alpha &= \frac{2 + \sqrt{6}}{4} \\ &\doteq 1.11 \end{aligned}$$

and thence

$$\begin{aligned}\lambda(\phi_\alpha) &= 5 + 2\sqrt{6} \\ &\doteq 9.90.\end{aligned}$$

Since the least eigenvalue of the Sturm-Liouville problem (4.4–19) is $\lambda_0 = \pi^2 \doteq 9.87$, the eigenvalue approximation we have obtained is accurate to within 3 units in the third significant figure. □

Example As a second example we take

$$\begin{aligned} x^2y'' + xy' + (\lambda x^2 - 1)y &= 0 \qquad 0 < x < 1 \\ y(0) &= 0 \\ y(1) &= 0. \end{aligned} \tag{4.4–20}$$

This is Bessel's differential equation of order 1 with parameter $\sqrt{\lambda}$ (see Appendix 2). The eigenfunctions are $y_j = J_1(\sqrt{\lambda_j}x)$ with eigenvalues given by the solutions of

$$J_1(\sqrt{\lambda}) = 0.$$

The least eigenvalue of (4.4–20), therefore, is approximately

$$\lambda_0 \doteq 14.68.$$

Once the above Bessel equation has been put into the self-adjoint form of (4.4–15), we see that the extremal problem equivalent to (4.4–20) is to minimize the quotient

$$\lambda = \frac{\displaystyle\int_0^1 (xy'^2 + y^2/x)dx}{\displaystyle\int_0^1 xy^2dx}.$$

This time we consider as an approximate solution the function family

$$\phi_\alpha(x) = x - x^\alpha \qquad \alpha > 1.$$

Again the free parameter α is to be chosen so as to minimize $\lambda(\phi_\alpha)$ within this family. By elementary integrations, we find

$$\begin{aligned}\lambda(\phi_\alpha) &= \left[\frac{(\alpha - 1)^2}{2\alpha}\right] \Bigg/ \left[\frac{(\alpha - 1)^2}{4(\alpha + 3)(\alpha + 1)}\right] \\ &= \frac{2(\alpha + 3)(\alpha + 1)}{\alpha}.\end{aligned}$$

It follows then that the best value of α is such that

$$0 = 2\,\frac{\alpha(2\alpha + 4) - (\alpha + 3)(\alpha + 1)}{\alpha^2},$$

which yields

$$\alpha = \sqrt{3}.$$

The corresponding value of λ is

$$\begin{aligned}\lambda(\phi_\alpha) &= 8 + 4\sqrt{3} \\ &\doteq 14.93,\end{aligned}$$

an approximation to the least eigenvalue of the boundary-value problem (4.4–20) with a relative accuracy of approximately 1.7%. □

Example Our final example concerns the nonlinear equation

$$\sqrt{x}\, y'' - y^{3/2} = 0 \qquad x > 0. \tag{4.4–21}$$

One solution is easily determined to be $y_1 = 144/x^3$. There is another solution, however, which is finite at the origin, and it is this solution we seek to approximate. Following Irving and Mullineaux (1959), we set $y(0) = 1$, $\lim_{x\to\infty} y(x) = 0$ and investigate the functional

$$I(y) = \int_0^\infty \left[y'^2 + 4y^{5/2}/(5\sqrt{x}) \right] dx.$$

Since (4.4–21) is the Euler-Lagrange equation associated with $I(y)$, our desired solution should render $I(y)$ stationary.

As a reasonable one-parameter family of approximate solutions, we try

$$\phi_\alpha(x) = \frac{\alpha^2}{(x+\alpha)^2} \qquad \alpha > 0.$$

Performing the necessary integrations, we then find

$$I(\phi_\alpha) = \frac{4}{5\alpha} + \frac{7\pi\sqrt{\alpha}}{32},$$

which suggests that the best value of α is given by

$$\begin{aligned}\alpha &= 16\left(\frac{4}{35\pi}\right)^{2/3} \\ &\doteq 1.757.\end{aligned}$$

For this value of α, we obtain

$$\begin{aligned}I(\phi_\alpha) &= \frac{3}{20}\left(\frac{35\pi}{4}\right)^{2/3} \\ &\doteq 1.366,\end{aligned}$$

a result that agrees very favorably with the minimum value $I_{\min} \doteq 1.363$ calculated numerically for the true solution of (4.4–21). □

We note in passing that the last two examples are "singular" in nature. The eigenvalue problem (4.4–20) has a $p(x)$ that vanishes and a $q(x)$ that is unbounded at one end of the interval of interest, while the fundamental interval for the nonlinear equation (4.4–21) is infinite in extent. However, just as in the case of certain nonstandard problems in Sturm-Liouville theory, the essential details of the method are still applicable.

4.5 Constraints and controls

Up to now we have skirted around the question of constraints with the exception of endpoint or other boundary conditions. We need to rectify this situation, however, since constrained optimization problems are arising with increasing frequency in modern applications. In this final section of the chapter, then, we discuss three types of common constraints and suggest the modifications of our earlier theoretical results that are necessary to accommodate these new restrictions. We also return to the original racing strategy problem of Section 4.1 and complete its solution.

Isoperimetric constraints

Optimization problems in which there are additional restrictions having the form of prescribed values of certain auxiliary definite integrals are commonly termed *isoperimetric problems*. The simplest problem of this type would be to find the extreme value of the functional

$$I(y) = \int_a^b f(x,y,y')dx$$

amongst, say, continuously differentiable functions $y(x)$ that satisfy $y(a) = y_i$, $y(b) = y_f$ for given y_i, y_f and also for which the second functional

$$J(y) = \int_a^b h(x,y,y')dx \tag{4.5–1}$$

possesses a fixed prescribed value γ. (An alternative approach to Sturm-Liouville problems can be made using these ideas; see Problem 23.) As Euler (and later Lagrange) observed, this constrained problem can be treated by introducing the combined functional

$$I(y) = \mu_0 I(y) + \mu_1 J(y) \tag{4.5–2)*}$$

where μ_0, μ_1 are constants (Lagrange multipliers) to be determined as part of the analysis. If the set of functions y that satisfy the isoperimetric constraint $J(y) = \gamma$ is nonempty, the natural analogue of the Corollary of Theorem 1 of Section 4.2 then

*Often this is written with a single parameter by letting $\mu_0 = 1$, $\mu_1 = \lambda$. However, in certain applications it is helpful to have this two-parameter flexibility.

affirms that there exist certain suitable values for μ_0 and μ_1 (not both zero) such that every putative extremal function $\tilde{y}(x)$ for $I(y)$ taking on the prescribed value for $J(y)$ satisfies the Euler-Lagrange equation (4.2–5), but with f replaced by

$$\tilde{f} = \mu_0 f + \mu_1 h. \tag{4.5–3}$$

(See Problem 22; also see Smith (1974), pp. 62ff).* The most interesting cases are, of course, those for which μ_0 turns out to be different from zero.

In view of the form of the modified integrand $\tilde{f}$ of (4.5–3), the solution of the Euler-Lagrange equation (4.2–5) will generally lead to three undetermined constants: two from integration and the third, say, the ratio μ_1/μ_0. These quantities are then determined, whenever a specific solution of the given problem is known to exist, by imposing the fixed endpoint conditions and the prescribed value for the isoperimetric constraint (4.5–1).

As might be anticipated, the various extensions of the theory discussed in Section 4.3 all carry over to this new class of problems.

Example Consider the problem of maximizing the area under the smooth curve $y = y(x)$ of constant length l, which begins and ends on the x-axis and lies above this axis everywhere in between. Stated mathematically, we want to maximize

$$I(y) = \int_{x_i}^{x_f} y\, dx$$

amongst all nonnegative continuously differentiable functions $y(x)$ that satisfy

$$y(x_i) = 0 = y(x_f),$$

and

$$J(y) = \int_{x_i}^{x_f} \sqrt{1 + y'^2}dx = l,$$

where x_i, x_f are arbitrary except for the obvious restriction $x_i < x_f < x_i + l$.

This is a classic problem of the variable endpoint type to which Theorem 1 of Section 4.3 is applicable. From the Euler-Lagrange equation for the modified integrand $\tilde{f} = \mu_0 y + \mu_1 \sqrt{1 + y'^2}$ we obtain

$$\begin{aligned} C &= \tilde{f} - y' \frac{\partial \tilde{f}}{\partial y'} \\ &= \mu_0 y + \frac{\mu_1}{\sqrt{1 + y'^2}}, \end{aligned} \tag{4.5–4}$$

*In the case of multiple constraints $J_j(y) \equiv \int_a^b h_j(x,y,y')dx = \gamma_j$ $(j = 1,2, \ldots, n)$, the modified integrand assumes the form

$$\tilde{f} = \mu_0 f + \sum_{j=1}^{n} \mu_j h_j.$$

where C is constant (recall equation (4.3–2)). Solving for y' and integrating then leads to

$$(\mu_0 x - D)^2 + (\mu_0 y - C)^2 = \mu_1{}^2 \qquad \text{with } y \geq 0, \tag{4.5–5}$$

where D is also constant. The extremal curve is thus part of a circle of radius $|\mu_1/\mu_0|$.

Application of the natural boundary condition (4.3–8) is particularly easy since the g function of (4.3–4) is merely $g = y$. It follows immediately that $C = 0$ (compare (4.5–4) and (4.3.8)), and the circle given by (4.5–5) is centered on the x-axis. (The variable endpoint condition at $x = x_i$ would lead to the same conclusion.) Completing the analysis we find $|\mu_1/\mu_0| = l/\pi$, $x_f = x_i + 2l/\pi$, and

$$(x - x_i - l/\pi)^2 + y^2 = (l/\pi)^2 \qquad y \geq 0;$$

the desired curve turns out to be nothing more than the expected semicircle of radius l/π. □

Inequality constraints

Inequality constraints occur almost as a matter of course in discrete optimization or programming problems. Such constraints are also common in applications to which variational models are appropriate—just imagine a specified minimum production level, or a maximum cost, or a minimum area to be enclosed, or a maximum amount of fuel available. However, since the treatment of optimization problems with inequality constraints within the context of the calculus of variations easily leads to considerable complexity, we restrict our attention to one special (yet rather important) case.

Consider the problem of finding the extreme value of the functional

$$I(y) = \int_a^b f(x,y,y')\,dx$$

amongst functions $y(x)$ which, in addition to fixed or variable endpoint conditions, satisfy a *global pointwise inequality constraint* of the form

$$y(x) \geq \phi(x) \qquad \text{for } a \leq x \leq b \tag{4.5–6a}$$

$$y(x) \leq \psi(x) \qquad \text{for } a \leq x \leq b. \tag{4.5–6b}$$

Here it is assumed that $\phi(x)$ or $\psi(x)$, whichever is the applicable constraint, is a given known continuously differentiable function of x. As examples, the reader can imagine the brachistochrone problem of Section 4.2 with the restriction that $y(x) \geq y_0$ for some given constant y_0 or a shortest distance problem with the constraint that the extremal path must go "around" some intermediate "function obstacle." After suitable reflection, it appears reasonable to expect that the extremal function might well prove to be *composite*, agreeing with the constraint (4.5–6) along various subintervals of $[a,b]$ and having other appropriate character elsewhere in the fundamental interval. And this is what actually occurs.

As Smith (1974), pp. 236ff. nicely demonstrates, the problem of maximizing or minimizing $I(y)$ under, say, the inequality constraint (4.5–6a) can be reduced to a related *unconstrained* optimization problem by introduction of a *slack function* $S(x)$ (similar to the slack variable of linear programming theory) defined by

$$y(x) = \phi(x) + S^2(x).$$

Solutions to this related problem (in S) are then sought within the class of *piecewise* smooth functions, thus allowing for "corners" and a composite character for y. The analysis proceeds as follows with a comparable conclusion resulting in case the constraint has the form (4.5–6b). We rewrite the original functional $I(y)$ in terms of S as

$$\begin{aligned} I(y) &= \int_a^b f(x,y,y')dx \\ &= \int_a^b f(x,\phi + S^2, \phi' + 2SS')dx \\ &\equiv \int_a^b g(x,S,S')dx \\ &\equiv J(S). \end{aligned}$$

Therefore, the appropriate Euler-Lagrange equation for this related problem is

$$\frac{\partial g(x,S,S')}{\partial S} - \frac{d}{dx}\left[\frac{\partial g(x,S,S')}{\partial S'}\right] = 0.$$

However,

$$\frac{\partial g}{\partial S} = 2S\frac{\partial f}{\partial y} + 2S'\frac{\partial f}{\partial y'}$$

and

$$\frac{d}{dx}\left[\frac{\partial g}{\partial S'}\right] = \frac{d}{dx}\left[2S\frac{\partial f}{\partial y'}\right] = 2S'\frac{\partial f}{\partial y'} + 2S\frac{d}{dx}\left[\frac{\partial f}{\partial y'}\right].$$

When referred back to the original problem, the end result is the combined equation

$$S(x)\left\{\frac{\partial f(x,y,y')}{\partial y} - \frac{d}{dx}\left[\frac{\partial f(x,y,y')}{\partial y'}\right]\right\} = 0 \tag{4.5–7}$$

for the extremal function $\tilde{y}(x)$. This *complementary slackness* relation (4.5–7) shows quite clearly that for each point in $[a,b]$ *either $S(x)$ vanishes and $\tilde{y}(x) = \phi(x)$ or the usual Euler-Lagrange equations must hold*. Therefore, if $\tilde{y}$ is piecewise smooth, it generally has the composite nature suggested above with the precise position of the joins ("corners") determinable from other considerations. The solution of our racing strategy problem will be an example of this type of extremal function.

Differential equations as constraints

In a variety of modern applications the mathematical model involves optimizing a functional that depends upon several dependent variables (recall (4.3–11)) but has *differential equations as constraints*. Often these differential constraints, as well as the functional itself, depend upon a number of additional parameters (or parametric functions) that may themselves be constrained. Time t is generally the independent and $\mathbf{x}$ the dependent variable in these applications. Consequently, the problem of finding the maximum or minimum of

$$I(\mathbf{x}) = \int_{t_i}^{t_f} f(t,\mathbf{x},\mathbf{u})dt \qquad \textbf{(4.5–8)}$$

subject to

$$\dot{\mathbf{x}}(t) = \mathbf{G}(t,\mathbf{x},\mathbf{u}) \qquad \textbf{(4.5–9)}$$

represents a typical example. Here it is assumed, say, that the dependent variable vector $\mathbf{x} = \mathbf{x}(t)$ has n components, the parameter function vector $\mathbf{u} = \mathbf{u}(t)$ has m components (with m and n not necessarily the same), and $f,\mathbf{G}$ are given sufficiently smooth real-valued functions/vectors of the designated variables.

By convention $\mathbf{x}$ is usually called the *state* vector and $\mathbf{u}$ the *control* vector. On the one hand, the state vector describes the status or "state" of the "system" under investigation as a function of time, for example the position, attitude, and vector velocity of a rocket or the quality level and availability of a given consumer product. The control vector, for example the thrust of the rocket engine or the production rate of and sales demand for the product, on the other hand, serves to "guide" the evolution of the "system" by means of the differential equation(s) (4.5–9). Therefore, finding the extremum of (4.5–8) under these (and other constraints) may naturally be termed a problem in *optimal control*.

On occasion, the equations (4.5–9) may be solved explicitly for the control parameters in terms of the components of the state vector and their derivatives. In such a situation, substitution back into (4.5–8) then reduces the problem to standard form. More generally, however, we must resort to the reverse procedure; we (in effect) solve the differential equations (4.5–9), subject to given initial condition(s) (and smoothness assumptions), for $\mathbf{x}$ in terms of t and $\mathbf{u}$ and then (again, in effect) substitute the resulting solution $\mathbf{x}(t,\mathbf{u})$ into the functional (4.5–8). In the process $I(\mathbf{x})$ becomes a function *only* of the control vector $\mathbf{u}(t)$, namely

$$I(\mathbf{x}) = \int_{t_i}^{t_f} f(t,\mathbf{x}(t,\mathbf{u}),\, \mathbf{u}(t))dt \equiv J(\mathbf{u}), \qquad \textbf{(4.5–10)}$$

and the original problem devolves into optimizing the new functional $J(\mathbf{u})$ over all suitable $\mathbf{u}$.*

*We have used the phrase "in effect" in this discussion since, if we can explicitly carry out the indicated steps, the transformed problem is of standard form. Its analysis, therefore, would be relatively straightforward.

Smith (1974, pp. 288ff) has worked out the details for the case when both the state vector and the control vector are one-dimensional. Here the problem is to maximize or minimize the functional

$$I(x) = \int_{t_i}^{t_f} f(t,x,u)dt$$

subject to the single differential equation

$$\frac{dx}{dt} = g(t,x,u) \tag{4.5–11}$$

and an initial condition like $x(t_i) = x_i$. Due to the manner in which u enters into the transformed functional $J(u)$, the analogue of the Euler-Lagrange equation (4.2–5) turns out to be the expression

$$\frac{\partial f(t,x(t,u),u(t))}{\partial u} + B(t)\int_{t}^{t_f} \frac{\partial f(s,x(s,u),u(s))}{\partial x} e^{-\int_s^t A(\tau)d\tau} ds = 0 \tag{4.5–12*}$$

where

$$A(t) \equiv \left.\frac{\partial g(t,x,u)}{\partial x}\right|_{\substack{x=x(t,u)\\ u=u(t)}}$$

and

$$B(t) \equiv \left.\frac{\partial g(t,x,u)}{\partial u}\right|_{\substack{x=x(t,u)\\ u=u(t)}}$$

Simultaneous solution of the integro-differential equation (4.5–12) and the initial value problem for (4.5–11) leads to the (putative) extremal control u and the associated state function x.

The mathematical theory of dynamical systems optimization and control is large and growing. Interested readers may wish to consult references such as Young (1969), Hestenes (1980), or Lee and Markus (1967) where more advanced topics and techniques (such as those based on the maximum principle of Pontryagin) are discussed.† Our purposes are best served, however, by applying some of the above ideas to the solution of our racing strategy problem (also see Problem 25).

The road race revisited

We are now ready to attack the racing time optimization problem of Section 4.1. We begin by noting that the application as formulated there has given rise to a

*Note the "automatic" constraint that $\partial f/\partial u|_{t=t_f} = 0$; this plays the role of the natural boundary condition at the free endpoint.

†The relationship of *discrete* optimal control and some of the notions important to mathematical programming theory is explored in recent books such as Canon, et al (1970).

problem in optimal control with the propulsive force $f(t)$ playing the role of the control function. We also observe that f is limited explicitly by the inequality constraint (4.1–4) and, due to the relation (4.1–5), implicitly by (4.1–7).

If we were to follow the second procedure outlined in the immediately preceding subsection, we would eliminate the dependent variable v in favor of the control function f. Although this can be easily accomplished, subsequent calculations in the resolution of the transformed problem turn out to be inordinately complicated. Fortunately for us, then, in this problem the control enters in such a simple manner that we can easily remove it instead. (We can also eliminate the energy (oxygen) function E as well.) This approach has the added merit of keeping the statement of the problem in terms of more familiar physical quantities such as velocity, acceleration, and so on. Pursuing this latter course of action we find that our problem of interest becomes, for a given distance D, to *minimize* the time T, or *equivalently*, for a given time T to *maximize* the distance D given by

$$D = \int_0^T v(t)\, dt \qquad \textbf{(4.5–13)}$$

subject to the constraints

$$\frac{dv}{dt} + \alpha v \leq F, \qquad \textbf{(4.5–14)}$$

$$v(0) = 0, \qquad \textbf{(4.5–15)}$$

and

$$E_0 + \beta t - \tfrac{1}{2} v^2(t) - \alpha \int_0^t v^2(s)\, ds \geq 0. \qquad \textbf{(4.5–16)}$$

For definiteness we take as our values for the various physiological parameters

$$\begin{aligned} \alpha &= 1.12/\text{sec} \\ \beta &= 41.5 \text{ joules/kg sec} \\ F &= 12.2 \text{ m/sec}^2 \\ E_0 &= 2400 \text{ joules/kg.} \end{aligned} \qquad \textbf{(4.5–17)*}$$

We also assume that T is greater than 60 secs, that is, we are interested in somewhat longer races.

In view of the initial condition (4.5–15), the constraint (4.5–16) is automatically satisfied for small t. Moreover, in this regime the initial value problem consisting of (4.5–14) with (4.5–15) implies that

$$v(t) = \frac{F}{\alpha}(1 - e^{-\alpha t}). \qquad \textbf{(4.5–18)}$$

For small t, therefore, the optimization problem reduces to maximizing the functional $D(v)$ given by (4.5–13) subject to the single inequality constraint (4.5–18),

---

*These are the values developed by Keller (1973, 1974) using a least-squares fit of the predictions of the theory to world-record racing data.

and for this the earlier "complementary slackness" relation (4.5–7) is applicable. It follows that (4.5–18) must hold as an equality; and thus, for some initial interval $0 \le t \le t_1$, the optimum velocity should satisfy

$$v(t) = \frac{F}{\alpha}(1 - e^{-\alpha t}). \qquad \textbf{(4.5–19)}$$

We remark that t_1 must be less than T since, if $v(t)$ given by (4.5–19) is used in (4.5–16), the left-hand side of that inequality becomes negative long before $t = 60$ sec.

Now we shift our attention to the opposite end of the race and make the physically reasonable assumption that (4.5–16) holds as an equality when $t = T$, that is,

$$E_0 + \beta T - \tfrac{1}{2} v_T^2 - \alpha \int_0^T v^2(t)\, dt = 0 \qquad \textbf{(4.5–20)}$$

where $v_T \equiv v(T)$. Referring back to (4.1–5) we see that this equation says that $E(T) = 0$ and amounts to nothing more than the stipulation that the available energy (oxygen) should be completely used up at the finish of the race. It follows then from (4.5–20) and the nonnegativity assumption (4.1–7) that $dE/dt \le 0$ for t near T, and thus in this large t regime

$$v\frac{dv}{dt} + \alpha v^2 \ge \beta. \qquad \textbf{(4.5–21)}$$

Also observe that, owing to (4.5–14), $v_T F \ge \beta$.

In terms of the terminal velocity v_T the inequality (4.5–21), upon integration, leads to

$$v^2(t) \le \beta/\alpha + (v_T^2 - \beta/\alpha)\, e^{2\alpha(T-t)}, \qquad \textbf{(4.5–22)}$$

while (4.5–14) implies

$$v(t) \ge F/\alpha + (v_T - F/\alpha)e^{\alpha(T-t)}. \qquad \textbf{(4.5–23)}$$

These new inequalities are consistent for t near T.

Since we are interested in maximizing $D(v)$ given by (4.5–13), only the single inequality (4.5–22) needs to be considered for large t. Similar to the case of small t, therefore, the "complementary slackness" relation (4.5–7) is again applicable. As before, we find that the pertinent inequality holds as an equality, and thus

$$v(t) = [\beta/\alpha + (v_T^2 - \beta/\alpha)e^{2\alpha(T-t)}]^{1/2} \qquad \textbf{(4.5–24)}$$

for t in some terminal interval, say, $t_2 \le t \le T$. We assume in the remainder of the analysis that $v_T^2 \ne \beta/\alpha$ since otherwise we have the less-interesting situation wherein $v(t)$ remains constant throughout the large t regime.

At this juncture we have determined the putative extremal velocity function $v(t)$ for both small and large t. It remains to blend these pieces together in continuous fashion through the intermediate regime $t_1 \le t \le t_2$. Before doing this we use (4.5–19) and (4.5–24) to rewrite $D(v)$ and $E(T)$, which are given by (4.5–13) and

(4.5–20), respectively (remember the condition $E(T) = 0$ introduced in the analysis of the large t regime):

$$D(v) = \int_0^{t_1} \frac{F}{\alpha}(1 - e^{-\alpha t})dt + \int_{t_1}^{t_2} v(t)dt + \int_{t_2}^{T} [\beta/\alpha + (v_T^2 - \beta/\alpha)e^{2\alpha(T-t)}]^{1/2}dt, \quad \textbf{(4.5–25)}$$

$$E(T) = E_0 + \beta T - \tfrac{1}{2} v_T^2 - \alpha \int_0^{t_1} \left(\frac{F}{\alpha}\right)^2 (1 - e^{-\alpha t})^2 dt - \alpha \int_{t_1}^{t_2} v^2(t)dt - \alpha \int_{t_2}^{T} [\beta/\alpha + (v_T^2 - \beta/\alpha)e^{2\alpha(T-t)}]dt. \quad \textbf{(4.5–26)}$$

As these relations show, our original optimization problem, which has now become the problem of maximizing D subject to the constraint $E(T) = 0$, is nothing more than a variant of the basic isoperimetric problem with unrestricted variable endpoints.* In order to carry out the final solution, therefore, we introduce the combined functional $I(v) = \mu_0 D + \mu_1 E(T)$, where μ_0 and μ_1 are Lagrange multipliers, and calculate from the Euler-Lagrange equation (4.2–5)

$$\frac{d}{dv}[\mu_0 v - \mu_1 \alpha v^2] = 0.$$

Upon solving, we find

$$v(t) = \frac{\mu_0}{2\mu_1\alpha} \equiv C(\text{constant}) \quad \text{for} \quad t_1 \leq t \leq t_2. \quad \textbf{(4.5–27)}$$

Application of the natural boundary conditions at the *right-hand* endpoint $t = t_2$ gives rise to the relation†

$$\mu_0 v_2 - \mu_1 \alpha v_2^2 - \mu_0 u(t_2) + \mu_0 v_T v_T' \int_{t_2}^{T} e^{2\alpha(T-t)} \frac{dt}{u(t)} - \mu_1 v_T v_T' + \mu_1 \alpha u^2(t_2) - 2\mu_1 \alpha v_T v_T' \int_{t_2}^{T} e^{2\alpha(T-t)} dt = 0,$$

where $u(t) \equiv [\beta/\alpha + (v_T^2 - \beta/\alpha)e^{2\alpha(T-t)}]^{1/2}$, $v_2 \equiv v(t_2)$, and we have allowed for the (implicit) dependence of the terminal velocity v_T upon the variable endpoint t_2. This expression can be considerably simplified by explicitly performing the indicated integrations and recognizing that $u(t_2) = v_2 = C$ as given by (4.5–27). The end result is

$$v_T v_T'[v_2^2 - 2v_2 v_T + \beta/\alpha] = 0. \quad \textbf{(4.5–28)}$$

---

*The constraints (4.5–14), (4.5–16) turn out to be automatically satisified (with considerable slack) in this intermediate regime.

†Recall Theorem 1 of Section 4.3 and the discussion following; also see Problem 11. Here $g(t,v) = v(t) - v(t_2)$.

In an analogous manner, the natural boundary condition at the *left-hand* endpoint $t = t_1$ leads to

$$\mu_0 v_1 - \mu_1 \alpha v_1^2 - \mu_0 \frac{F}{\alpha}(1 - e^{-\alpha t_1}) + \mu_1 \alpha \left(\frac{F}{\alpha}\right)^2 (1 - e^{-\alpha t_1})^2 = 0.$$

But this last equation is an identity because of the form of $v(t)$ in the interval $0 \leq t \leq t_1$. The results for the intermediate interval $t_1 \leq t \leq t_2$, therefore, are contained entirely in (4.5–27) and (4.5–28).

Putting everything together, we see that the proposed optimal racing strategy calls for an initial period of rapid acceleration as prescribed by (4.5–19), an intermediate phase of "coasting" at the constant speed C, and then a final segment of modest deceleration (as opposed to a final "kick") given by (4.5–24). The constant C and the changeover times t_1 and t_2 are determined through the simultaneous application of continuity conditions and the constraints implied by (4.5–20) and (4.5–28). Since $v_T' = 0$ generally leads to an inconsistent solution, we have from this last equation that $v_T = \frac{1}{2}(v_2 + \beta/\alpha v_2)$, and these several relations can thus be rewritten simply as

$$v_2 = v_1 = C = \frac{F}{\alpha}(1 - e^{-\alpha t_1}), \qquad C^2[1 - 4e^{2\alpha(t_2 - T)}] = \beta/\alpha, \qquad \textbf{(4.5–29)}$$

and

$$E_0 + \beta t_2 - \alpha C^2(t_2 - t_1) - \frac{F}{\alpha}(Ft_1 - C) = 0.$$

The comparable expression for D is

$$D = \frac{F}{\alpha} t_1 + C(t_2 - t_1) - \frac{1}{2\alpha}\left(C + \frac{\beta}{\alpha C}\right) + \frac{1}{2\alpha}\sqrt{\beta/\alpha}\,\ln\frac{C + \sqrt{\beta/\alpha}}{C - \sqrt{\beta/\alpha}}. \qquad \textbf{(4.5–30)}$$

In his 1973 and 1974 papers, Keller compares the predictions of the theory as represented by (4.5–29) and (4.5–30) with various world records over a wide range of distances. As a single illustration of more than passing interest, we note that the solution of these equations for the case of $D = 1$ mile is approximately

$$C \doteq 6.80 \text{ m/sec}, \quad t_1 \doteq 0.87 \text{ sec}, \quad t_2 \doteq 235.94 \text{ sec}, \quad \text{and} \quad T \doteq 237.28 \text{ sec}.$$

(Note particularly the unusually short initial and final time intervals.) Since the calculations have led to a sub-four-minute total time for this distance, a runner following the velocity profile specified by the model we have developed in this chapter would most assuredly be of "world-class" stature.

References

Akhiezer, N. I. (1962): *The Calculus of Variations*, translated from the Russian by A. H. Frink, Blaisdell, New York.

Bliss, G. A. (1925): *Calculus of Variations*, Open Court Publishing, Chicago.

Bolza, O. (1973): *Lectures on the Calculus of Variations*, 3rd ed., Chelsea, New York.

Brekhovskikh, L. M. (1980): *Waves in Layered Media*, translated from the Russian by D. Lieberman, 2nd ed., Academic Press, New York.

Canon, M. D., C. D. Cullum, Jr., and E. Polak (1970): *Theory of Optimal Control and Mathematical Programming*, McGraw-Hill, New York.

Denn, M. M. (1978): *Optimization by Variational Methods*, reprint of 1969 ed., Krieger, New York.

Elsgolc, L. E. (1962): *Calculus of Variations*, translated from the Russian, Pergamon Press, London

Finlayson, B. A. (1972): *Method of Weighted Residuals and Variational Principles*, Academic Press, New York.

Gelfand, I. M., and S. V. Fomin (1963): *Calculus of Variations*, translated from the Russian by R. Silverman, Prentice Hall, Englewood Cliffs, New Jersey.

Gould, S. H. (1966): *Variational Methods for Eigenvalue Problems*, 2nd ed., University of Toronto Press, Toronto.

Hestenes, M. R. (1980): *Calculus of Variations and Optimal Control Theory*, reprint of 1966 ed., Krieger, New York.

Hillier, F. S., and G. J. Lieberman (1980): *Operations Research*, 3rd ed., Holden-Day, San Francisco.

Irving, J., and N. Mullineaux (1959): *Mathematics in Physics and Engineering*, Academic Press, New York; Chapter VII.

Keller, J. B. (1973): "A Theory of Competitive Running," *Physics Today, 26*, #9: 42–47.

——— (1974): "Optimal Velocity in a Race," *Amer. Math. Monthly, 81*: 474–480.

Lanczos, C. (1970): *The Variational Principles of Mechanics*, 4th ed., Univ. of Toronto Press, Toronto.

Lee, E. B., and L. Markus (1967): *Foundations of Optimal Control Theory*, John Wiley, New York.

Nitsche, J. C. C. (1974): "Plateau's Problems and Their Modern Ramifications," *Amer. Math. Monthly, 81*: 945–968.

Pars, L. A. (1962): *An Introduction to the Calculus of Variations*, Heinemann, London.

Sagan, H. (1969): *Introduction to the Calculus of Variations*, McGraw-Hill, New York.

Smith, D. R. (1974): *Variational Methods in Optimization*, Prentice-Hall, Englewood Cliffs, New Jersey.

Wagner, H. M. (1975): *Principles of Operations Research with Applications to Managerial Decisions*, 2nd ed., Prentice-Hall, Englewood Cliffs, New Jersey.

Weinstein, A., and W. Stenger (1972): *Methods of Intermediate Problems for Eigenvalues*, Academic Press, New York.

Weinstock, R. (1952): *Calculus of Variations*, McGraw-Hill, New York; also Dover, New York.

Young, L. C. (1969): *Lectures on the Calculus of Variations and Optimal Control Theory*, W. B. Saunders, Philadelphia.

Problems

Section 4.2

1. (a) Let $F(y)$ be a continuously differentiable function of a real variable y. Calculate the variation of $F(y)$ at y.

   (b) Generalize the result of part (a) to real-valued functions of vector arguments.

   (c) Discuss the implications of Theorem 1 of Section 4.2 for these two cases.

2. (a) Determine the variation of the functional

$$I(y) = \int_0^1 (y^2 + 2xy + y'^2)dx.$$

   (b) What is the relevant Euler-Lagrange equation for this functional?

★3. (a) Show that the only smooth solutions of the Euler-Lagrange equation associated with the functional

$$I(y) = \int_a^b \sqrt{1 + y'^2}\, dx$$

are linear functions. (This result is actually more generally valid whenever the integrand of the functional of interest depends only on y'.)

   (b) Prove explicitly that amongst all continuously differentiable functions $y = y(x)$ satisfying $y(a) = A$, $y(b) = B$, the linear function corresponding to the straight line joining the two points (a,A) and (b,B) gives the *absolute* minimum for I. (Hint: Think first of how to handle the case when $A = B$.)

★4. (a) Solve the Euler-Lagrange equation that is appropriate for the functional

$$I(y) = \int_a^b y\sqrt{1 + y'^2}\, dx.$$

(These are candidates for the curve joining the two points (a,A), (b,B) that, when revolved around the x-axis, generates the smallest surface area.)

   (b) Investigate the special case when $a = -b$ and $A = B$, and show that, depending upon the relative size of b, B, there may be none, one, or two candidate curves that satisfy the requisite endpoint conditions. (For further discussion of these *catenary* curves see Bliss (1925), pp. 85ff. or Weinstock (1952), pp. 30ff.)

Section 4.3

5. (a) Discuss the application of the simplified Euler-Lagrange equation (4.3–2) to the integrand of the *brachistochrone* functional (4.2–6).

   (b) Use (4.3–2) to help solve the Euler-Lagrange equation associated with the functional

$$I(y) = \int_a^b (y^2 - yy' + y'^2)dx.$$

6. Analyze in detail the simplifications that occur in the solution of the Euler-Lagrange equation (4.2–5) when the function $f(x,y,y')$ does *not* depend explicitly on y'.

7. Investigate the consequences of the Euler-Lagrange equation associated with the functional

$$I(y) = \int_a^b f(x,y,y')dx$$

in the case when f is explicitly the *total* derivative with respect to x of some function of x and y.

8. Show that if the fixed endpoint condition at $x = b$ in the *brachistochrone* problem of Section 4.2 is relaxed, that is, $y(b)$ is made arbitrary, then the natural boundary condition (4.3–8) implies $y'(b) = 0$ and hence

$$\theta_f = \pi, \quad C = \frac{b-a}{\pi}, \quad \text{and} \quad T = \sqrt{\frac{(b-a)\pi}{g}}.$$

9. (a) In the variable endpoint situation when the right-hand endpoint $x = b$ is restricted to the curve $y = \psi(x)$, verify that the *transversality* condition (4.3–8) can be rewritten as

$$\left[f + (\psi' - y')\frac{\partial f}{\partial y'}\right]_{x=b} = 0.$$

   (b) Derive the form of the comparable condition when the *left*-hand endpoint $x = a$ is restricted to the curve $y = \phi(x)$.

10. (a) Determine the function $y = y(x)$ that minimizes the integral

$$I(y) = \int_0^\pi (2y \sin x + y'^2)dx$$

subject to the single condition $y(0) = 0$. Do first using the natural boundary condition (4.3–8) and compare with a solution procedure that neglects to apply that auxiliary condition.

   (b) How does the minimum value obtained in (a) compare with the minimum value for $I(y)$ that results when the additional restriction $y(\pi) = 0$ is included?

★11. (a) Develop an analogue to Theorem 1 of Section 4.3 for the case of a variable right-hand endpoint when the functional of interest has the form

$$F(y) = \int_a^{x_f} f(x,y,y')dx + h(x_f).$$

In particular, verify that the presence of the term $h(x_f)$ has no effect on the form of the relevant Euler-Lagrange equation and merely leads to the replacement of f by $(f + dh/dx_f)$ within the brackets in (4.3–8).

(b) Show that the analogous result for a variable left-hand endpoint in the case of the functional

$$F(y) = \int_{x_i}^{b} f(x,y,y')dx + h(x_i)$$

is obtained by replacing f with $(f - dh/dx_i)$ in (4.3–8).

12. Find the extremal curve in three-space that joins the origin to the point $(\pi/4, -1, 1)$ and optimizes the integral

$$I(y,z) = \int_0^{\pi/4}\left[y'^2 + z'^2 + 2(yz' - zy') - 3(y^2 + z^2)\right]dx.$$

13. Determine that particular solution of the Euler-Lagrange equation associated with the functional

$$I(y) = \int_0^1 (y''^2 - 48y)dx$$

that also satisfies the boundary conditions $y(0) = y'(0) = y''(0) = 0$ and $y(1) = 1$. Do you anticipate that this solution provides a local maximum or a local minimum for I?

14. Minimize the integral

$$I(u) = \iint_D (x^2u_y^{\,2} + y^2u_x^{\,2})\,dx\,dy$$

where D is the interior of the unit circle $x^2 + y^2 < 1$ and the function $u(x,y)$ satisfies the boundary condition

$$u = x^2 \qquad \text{on} \quad x^2 + y^2 = 1.$$

(The Euler-Lagrange equation can be solved by inspection. Pars (1962) p. 303 demonstrates that the resulting solution provides an absolute minimum for I.)

Section 4.4

15. (a) Verify that under the conditions specified in Section 4.4, the integral formula (4.4–2) gives the length of the curve connecting

the two given points and constrained to lie on the given parametric surface.

(b) Show that in the case of a planar surface, (4.4–2) reduces to a form equivalent to the more familiar expression (4.4–1).

16. (a) Investigate the *geodesics* for the sphere of radius R given parametrically by

$$x = R \cos \phi \sin \theta$$
$$y = R \sin \phi \sin \theta$$
$$z = R \cos \theta$$

in terms of the usual spherical angles ϕ, θ. (It is convenient to let θ be the fundamental independent variable on the spherical surface.)

(b) Demonstrate that the geodesic arcs satisfy

$$\phi = C_1 - \sin^{-1}(C_2 \cot \theta)$$

for suitable constants C_1, C_2, and thus they must perforce also lie on a plane that passes through the center of the sphere, that is, they are *great-circle arcs*.

17. Imagine that light passes from one optically homogeneous medium to another such medium separated from the first by a planar interface, say $y =$ constant. Show that in this situation the fundamental *least-time* relation (4.4–5) implies that the angles of incidence and refraction must be related by Snell's classical law

$$\frac{\sin \theta_1}{v_1} = \frac{\sin \theta_2}{v_2}.$$

18. Consider a particle of mass m moving in the x,y-plane under the sole action of an attractive force directed toward the origin and having a magnitude $|\mathbf{F}| = kr$ where r is the distance of the particle from the origin. Deduce from Hamilton's Principle that the orbit must be an ellipse centered at the origin. (Recall that $\mathbf{F} = -\text{grad } V$ where V is the potential energy of the system.)

19. (a) Reformulate an appropriate boundary-value problem for the differential equation

$$(py'')'' + (qy')' + (\lambda r - s)y = 0 \qquad a < x < b$$

in terms of a variational problem involving the ratio of two functionals similar to (4.4–17). Be specific about the nature of the boundary conditions that $y = y(x)$ must satisfy.

(b) What simple conditions on the coefficient functions p,q,r, and s will ensure that the eigenvalues associated with the original boundary-value problem are nonnegative?

20. Approximate the least eigenvalue of the Sturm-Liouville problem

$$y'' + \lambda y = 0 \qquad 0 < x < 1$$
$$y(0) = 0$$
$$y'(1) = 0$$

by recasting the problem in variational fashion and using approximate solutions of the form

$$\phi_\alpha(x) = x - \frac{x^\alpha}{\alpha} \qquad \alpha > 1.$$

Determine the optimum value of α to two decimal places and find the corresponding eigenvalue approximation. (A calculator will be useful in performing these computations.)

21. Find the "best" cubic polynomial approximation to the lowest-order eigenfunction of the eigenvalue problem (4.4–20). ("Best" here should be interpreted in the sense of yielding the closest approximation to the associated eigenvalue from amongst the class of approximating functions under consideration.)

Section 4.5

22. Consider the problem of optimizing the functional

$$I(y) = \int_a^b f(x,y,y')dx$$

amongst continuously differentiable functions $y = y(x)$ for which $y(a)$, $y(b)$, and the integral

$$J(y) = \int_a^b h(x,y,y')dx$$

all have given prescribed values. Assuming that an extremal solution $\bar{y}(x)$ exists, introduce the *two*-parameter family of "nearby" comparison functions

$$y(x) \equiv \bar{y}(x) + \epsilon_1\eta_1(x) + \epsilon_2\eta_2(x)$$

where η_1, η_2 are arbitrary differentiable functions that vanish at $x = a,b$. Now carry forth the rest of the analysis so as to demonstrate that a *necessary* condition for $\bar{y}(x)$ to be an extremal solution of this basic isoperimetric problem is that it satisfy the Euler-Lagrange equation

$$\frac{\partial \tilde{f}}{\partial y} - \frac{d}{dx}\left(\frac{\partial \tilde{f}}{\partial y'}\right) = 0$$

where $\tilde{f} = \mu_0 f + \mu_1 h$ for certain suitable constants μ_0, μ_1.

23. Assume that the two functionals $I_1(y)$ and $I_2(y)$ are as given by (4.4–16). Analyze the *isoperimetric problem* of finding the minimum value of $I_1(y)$ amongst all continuously differentiable functions $y(x)$ that satisfy the ho-

mogeneous conditions $y(a) = 0 = y(b)$ and also for which $I_2(y) = 1$. In particular, show that the minimizing solution $\tilde{y}(x)$ satisfies the Sturm-Liouville equation (4.4–15) with $\lambda = I_1(\tilde{y})$. (This is yet another approach to the Sturm-Liouville eigenvalue problem initially considered in Chapter 2.)

24. (a) In the *brachistochrone* problem of Section 4.2, we want to include the additional inequality constraint that the descent curve $y(x)$ always satisfies

$$y(x) \geq y_0,$$

where y_0 is a constant which itself is greater than $y_i - \frac{2}{\pi}(b - a)$.

This added condition has the effect of transforming $y(x)$ into a composite curve consisting of a piece of a classic cycloidal path (until the point where it intersects the line $y = y_0$) followed by a portion of this straight line. Formulate this modified brachistochrone problem as a simple *variable endpoint problem* and obtain its solution.

(b) Compare the descent time for this restricted brachistochrone problem with that derived in the original situation.

25. Analyze the control problem of optimizing

$$I(x) = \int_{-\pi/4}^{\pi/4} (x^2 + 11xu + 2u^2)dt$$

subject to the differential equation

$$\dot{x} = u + x$$

and the initial condition $x(-\pi/4) = 8$, where $u = u(t)$ is an appropriate control function. Use the approach outlined in Section 4.5.

5

Analytic function theory and system stability

5.1 Linear systems and transfer functions

In Chapter 1 we saw that the analysis of simple mechanical systems leads to the study of matrix differential equations. The same is true, as one might expect, with simple electrical systems. As elementary examples, consider the RLC networks depicted in Figure 5.1. If, in addition to Ohm's law for voltage drop across a resistor, we also use the corresponding empirical results for capacitors and inductors, namely

$$i_c = C\frac{dv_c}{dt} \qquad \text{and} \qquad v_L = L\frac{di_L}{dt},$$

then the following matrix equations can be easily derived for these configurations:

$$\text{(a)} \quad \begin{bmatrix} L\,di/dt \\ C\,dv_c/dt \end{bmatrix} + \begin{bmatrix} R & 1 \\ -1 & 0 \end{bmatrix} \begin{bmatrix} i \\ v_c \end{bmatrix} = \begin{bmatrix} v \\ 0 \end{bmatrix}$$

$$\text{(b)} \quad \begin{bmatrix} L\,diL/dt \\ C\,dv/dt \end{bmatrix} + \begin{bmatrix} 0 & -1 \\ 1 & 1/R \end{bmatrix} \begin{bmatrix} i_L \\ v \end{bmatrix} = \begin{bmatrix} 0 \\ i \end{bmatrix}.$$

Alternatively, the governing differential equations may be expressed in the uncoupled forms

$$\text{(a)} \quad L\frac{d^2i}{dt^2} + R\frac{di}{dt} + \frac{1}{C}i = \frac{dv}{dt},$$

$$\text{(b)} \quad C\frac{d^2v}{dt^2} + \frac{1}{R}\frac{dv}{dt} + \frac{1}{L}v = \frac{di}{dt}. \qquad \textbf{(5.1–1)}$$

In either approach, the ordinary differential equations that one obtains have constant coefficients and such differential equations are easy to solve. (More complicated passive electrical systems merely give rise to higher-order equations.)

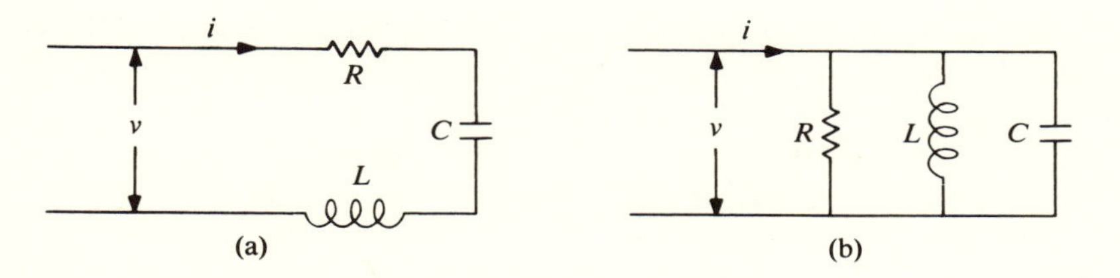

Figure 5.1 Elementary electrical circuits

If we take, say, the first of (5.1–1) and make the standard substitution $i(t) = \alpha e^{st}$, on the left-hand side we find

$$\alpha e^{st}\,[Ls^2 + Rs + 1/C].$$

The quantity within the brackets is often written as $sZ(s)$, where $Z(s)$ is termed the *system impedance*. Those familiar with the Laplace transform

$$\mathcal{L}(f) = F \quad \text{where} \quad F(s) \equiv \int_0^\infty e^{-st} f(t)dt$$

(about which we will have considerably more to say in Chapter 7) will recognize the system impedance as simply the ratio of the Laplace transforms of the input $v(t)$ and the output $i(t)$ when all initial conditions are assumed to be zero. More generally, if $My = f$, where M is the constant-coefficient integro-differential operator associated with a given linear system, then application of the Laplace transform (or its equivalent) leads to

$$Y(s)Z(s) = F(s),$$

where $Z(s)$ is a polynomial whose coefficients depend solely on the parameters of the system.

The above remarks are not restricted to electrical systems. In general, time-invariant linear systems are described by constant-coefficient linear integro-differential operators to which the above analysis applies. In most investigations, however, the construct of interest is the reciprocal $H(s)$ of $Z(s)$. H is called the *transfer function* of the system. In addition to its definition as the ratio of the Laplace transforms of output and input (under rest initial conditions), the transfer function has a noteworthy alternative characterization as the Laplace transform of the system impulse-response. That is, $H(s)$ is the Laplace transform of the (rest) output of the system corresponding to a unit impulse (delta function; see Chapter 9) applied at $t = 0$ (see Zadeh and Desoer (1979), for example).*

The transfer function plays a central role in linear system analysis since the overall behavior of a given system can be determined from knowledge of that system's transfer function. For many systems of interest, even very complicated

---

*For certain stable systems, the transfer function H can also be defined in terms of the steady-state response of the system to sinusoidal inputs.

ones, the transfer function can be determined by relatively simple measurements made on the system.

A satellite attitude-control system

Before we turn to a more detailed discussion of transfer functions and the performance of general systems, let us look briefly at a practical control system suggested by Frederick and Carlson (1971). This system is intended to stabilize an orbiting satellite upon which is rigidly mounted a telescope designed for the long-term exposure photography of several predetermined target stars. The control system senses the response of the satellite to various disturbing torques due in part to extra-satellite sources, such as aerodynamic drag and solar radiation pressure, and in part to intrasatellite sources, such as motor rotation or gas jet operation. Indeed, the system controls the firing of the gas jets to offset these undesirable torques and maintain the satellite's attitude. It also acts to shift the telescope's line-of-sight to allow the photography of the various target stars.

A model of the control system adequate for our purposes is schematically represented by the block diagram of Figure 5.2.

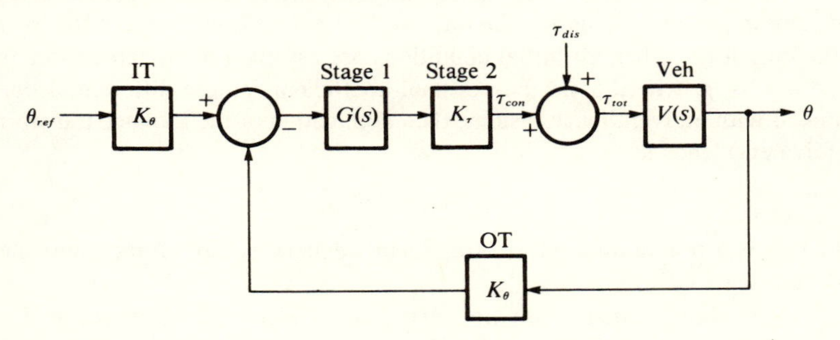

Figure 5.2 Satellite attitude-control system block diagram

The Input Transducer (IT) generates a voltage proportional (with amplification or "gain" K_θ) to an input reference angle θ_{ref}. The Output Transducer (OT), or star tracker, measures the angular deviation θ of the telescope from the specified reference star and provides an output voltage $K_\theta\theta$ proportional to this deviation. When the telescope points precisely at the target star, these voltages are equal. If the voltages are unequal, the situation must be corrected and the difference signal is therefore fed into a controller. Stage 1 of the controller has a transfer function

$$G(s) = K_a(\alpha s + 1 + \beta/s); \qquad \textbf{(5.1–2)}$$

the parameters α and β are termed the "derivative gain" and the "integral gain," respectively. Stage 2, with amplification K_τ, converts the Stage 1 voltage outputs into gas jet control torques. The block labelled Vehicle represents the response of the satellite to the various applied torques. For simplicity, we assume that the vehicle dynamics are described by the transfer function

$$V(s) = \frac{1}{Is^2}, \tag{5.1–3}$$

where I is the moment of inertia of the satellite.

This feedback control system is fairly complicated. It has been designed so that it can damp out undesirable transients as well as respond appropriately even when presented with low-frequency disturbing torques. Crucial to its effective performance, however, are the values of the controller gains α and β. These parameters determine the *stability* of the attitude control system, as we shall see when we return to its analysis in Section 5.5.

5.2 System stability

In many applications, detailed knowledge of the complete behavior of a given system is unnecessary; in such cases it often suffices merely to know the system's response to *bounded* inputs. The following are one form of the usual definitions*:

Definition 1 A time-invariant linear system with a rational transfer function

$$H(s) \equiv \frac{P(s)}{Q(s)}, \tag{5.2–1}$$

where $P(s)$ and $Q(s)$ are polynomials in s with no common factors, is said to be **strictly** (or **asymptotically**) **stable** if all of the zeros of $Q(s)$ lie in the left half of the complex s-plane. For such systems, all free response modes (natural resonances) decay exponentially as $t \to \infty$ and, under the assumption of zero initial conditions, the system response to a bounded input remains bounded.

Definition 2 A system with a rational transfer function as in (5.2–1) is termed (**marginally**) **stable** if $Q(s)$ has no zeros in the right half of the s-plane and any purely imaginary zeros are simple (unrepeated). Although all free responses of such a system are bounded, the system response to a bounded input may become unbounded.

The above definitions apply only to systems with rational transfer functions. In practice, however, this is not a very restrictive assumption. Many systems, even some very complicated ones, have transfer functions satisfying (5.2–1); the complexity manifests itself in the degree of the polynomials $P(s)$ and $Q(s)$. For

*Readers familiar with stability theory for ordinary differential equations should recognize the correspondences between that subject and this material (see Coppel (1965), for example; also see Problem 2.)

example, the satellite attitude control system of the previous section has the reference transfer function (assuming $\tau_{\text{dis}} = 0$)

$$H(s) = \frac{K_\theta K_a K_\tau(\alpha s^2 + s + \beta)}{Is^3 + K_\theta K_a K_\tau(\alpha s^2 + s + \beta)}. \tag{5.2–2}$$

Even if the linear system of interest has a transfer function not of the form (5.2–1), it can generally be represented, with arbitrary accuracy, by an approximating system with such a rational transfer function.* Indeed, for the approximating system the degree of $P(s)$ is generally less than the degree of $Q(s)$. (See Zadeh and Desoer (1979), or comparable references, for details).

In order to apply Definitions 1 and 2 to a given linear system, we need to be able to determine the transfer function of the system. Frequently, with complex systems composed of a number of component parts, a certain amount of so-called "block-diagram algebra" will simplify this task. Utilizing the basic series (or cascade), parallel, and feedback configurations depicted in Figure 5.3, for which the transfer functions are

$$FG, \quad F + G, \quad \text{and} \quad \frac{F}{1 \mp FG},$$

respectively, more complicated systems can be broken down into subsystems. The transfer functions for these subsystems are then obtained before any effort is made

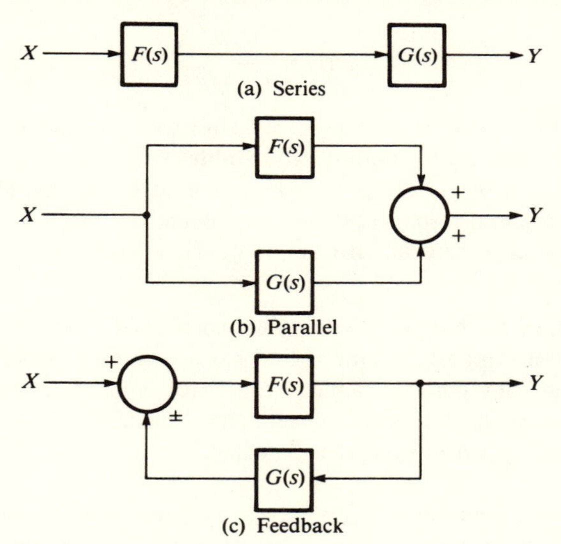

Figure 5.3 Basic configurations commonly encountered as subunits in linear systems

*Unfortunately, we typically cannot measure the quality of approximation in any classical manner; this is especially true if the transfer function of the given system exhibits an infinite number of (pole) singularities. The approximation is generally suitable, however, for stability analyses wherein the general position of the singularities is all that is essential.

to combine the results into the transfer function of the entire system taken as a whole. (In an elementary sense, this is what has been done to arrive at (5.2–2).)

Routh-Hurwitz stability criterion

With the transfer function (assumed rational) of a given system in hand, the stability of the system may be assessed. In view of the earlier definitions, this amounts to establishing the position of the zeros of the polynomial

$$Q(s) \equiv a_0 s^n + a_1 s^{n-1} + \cdots + a_{n-1} s + a_n. \tag{5.2–3}$$

If $n = 1$ or 2, we can calculate the zeros explicitly. For higher-order polynomials it is usually easier to employ an indirect approach that tests for the position of all of the zeros simultaneously. Two variants of such a procedure were developed independently in the late 19th century by Edward Routh (1831–1907) and Adolf Hurwitz (1859–1919). These methods depend upon inequalities involving the so-called *Hurwitz determinants*

$$D_k \equiv \begin{vmatrix} a_1 & a_0 & 0 & 0 & 0 & \cdots & 0 \\ a_3 & a_2 & a_1 & a_0 & 0 & \cdots & 0 \\ \cdot & & & & & & \cdot \\ \cdot & & & & & & \cdot \\ \cdot & & & & & & \cdot \\ a_{2k-1} & a_{2k-2} & \cdot & \cdot & \cdot & \cdots & a_k \end{vmatrix} \qquad k = 1, 2, \ldots, n \tag{5.2–4}$$

(where $a_j = 0$ for $j > n$) associated with the coefficients of $Q(s)$. In its most general (and perhaps most efficient) form the *Routh-Hurwitz criterion* may be stated as

Theorem 1*:

If the polynomial

$$Q(s) = a_0 s^n + a_1 s^{n-1} + \cdots + a_{n-1} s + a_n$$

has real coefficients, with $a_0 > 0$, then *any one* of the following conditions is necessary and sufficient for every zero of $Q(s)$ to have negative real part:

(i) $a_n > 0,\quad a_{n-2} > 0,\quad a_{n-4} > 0, \ldots;\quad D_1 > 0,\quad D_3 > 0, \ldots$

(ii) $a_n > 0,\quad a_{n-2} > 0,\quad a_{n-4} > 0, \ldots;\quad D_2 > 0,\quad D_4 > 0, \ldots$

(iii) $a_n > 0,\quad a_{n-1} > 0,\quad a_{n-3} > 0, \ldots;\quad D_1 > 0,\quad D_3 > 0, \ldots$

(iv) $a_n > 0,\quad a_{n-1} > 0,\quad a_{n-3} > 0, \ldots;\quad D_2 > 0,\quad D_4 > 0, \ldots.$

Proof: See Gantmacher (1959) or Coppel (1965). Kaplan (1962) provides a proof of the less general classical result. □

*The version of the result we give here is due to Liénard and Chipart (1914). It avoids redundancies usually present in the classical statement of the Routh-Hurwitz criterion (see the discussion of Zadeh and Desoer (1979), pp. 419 ff.). For extensions to complex polynomials, see the appendix of Coppel (1965) or Marden (1966).

In the terminology of Definition 1, Theorem 1 furnishes a test for the *strict* stability of a linear system with a rational (real) transfer function of the form (5.2–1). If any of the coefficients a_k or determinants D_k are negative we are assured that the system is unstable as $t \to \infty$. However, if some of the coefficients and/or determinants are merely zero while the remainder are positive, the given system *may* be marginally stable. Two particular cases of marginal stability are of special interest:

(I) $a_n = 0$ and $D_n = 0$; all other a_k and D_k positive

(II) all a_k nonnegative; $D_n = D_{n-1} = 0$, but all other D_k positive.

Case (I) corresponds to the situation wherein the origin $s = 0$ is a simple zero of the nth degree polynomial $Q(s)$; all other zeros lie in the left half of the complex s-plane. In case (II), all zeros of $Q(s)$ have negative real parts with the exception of a single complex-conjugate pair situated on the imaginary s-axis. In both these cases, Definition 2 ensures that the underlying linear system is indeed marginally stable.

If $Q(s)$ has additional zeros on the imaginary s-axis, other Hurwitz determinants may vanish, and it is typically more difficult to decide the stability/instability question by a cursory inspection of the D_k. The above special examples, however, do suggest the existence of a single simple criterion that should be useful in the design of stable linear systems, particularly those for which the overall system transfer functions are readily expressible in analytic form. Such turns out to be the case, and we express the result as follows:

Theorem 2:

Let a linear system have a real transfer function of the form (5.2–1) in which $Q(s)$ is an nth degree polynomial whose coefficients depend continuously on a real parameter μ. If the system is known to be strictly stable for $\mu_0 < \mu < \mu_1$ and at most marginally stable when $\mu = \mu_0, \mu_1$, then $D_n(\mu_0) = 0 = D_n(\mu_1)$ where D_n is the nth Hurwitz determinant associated with $Q(s)$.

Proof: See Problem 7. □

We will have occasion to apply this theorem in our further analysis of the satellite control system of the previous section.

An alternative stability criterion

For a practical problem in which the coefficients of the denominator $Q(s)$ of the system transfer function have specified numerical values, the calculation of the various Hurwitz determinants can be systematized in algorithmic fashion by setting up a so-called *Routh array* (see Frederick and Carlson (1971), Coppel (1965), Kaplan (1962) or any of a number of other references). This can be particularly advantageous when the degree of $Q(s)$ is large.

A rather different method for obtaining information about the location of the zeros of $Q(s)$ is embodied in

Theorem 3:

If the polynomial $Q(s)$ does not vanish on a simple closed contour C in the complex s-plane, then the number of zeros of $Q(s)$ within C is given by

$$N = \frac{1}{2\pi i} \oint_C \frac{Q'(s)}{Q(s)}\, ds. \qquad \textbf{(5.2–5)*}$$

By choosing the curve C carefully we can, in theory, determine if all of the zeros of $Q(s)$ are in the left half of the complex s-plane, and hence ascertain whether the system of interest is stable.

In practical terms, an approach based upon (5.2–5) turns out to be a very useful alternative procedure for stability calculations and constitutes a basic technique in systems design and analysis efforts. For linear control systems with feedback the method gives rise to the important Nyquist Stability Criterion, which we shall discuss in Section 5.5. It is appropriate, therefore, to lay some groundwork that will allow the appreciation of the relation (5.2–5) as an applied tool, of value in a variety of contexts. At the same time we can provide a convenient overview of related material that will be useful in other applications to be considered later in this text. The next sections of this chapter and Appendix 4 are devoted to these tasks.

5.3 Analytic functions

The feature that sets $Q(s)$ apart, and indeed engenders the applicability of a relation such as (5.2–5), is the fact that Q is a well-behaved function of a complex variable. Most readers are undoubtedly acquainted with the foundations of the theory of such functions, and perhaps may have studied the subject using a standard reference such as Churchill (1974), Pennisi (1976), or Wylie (1975), for example. For completeness, however, we have included the basic definitions and properties associated with complex function theory in Appendix 4 at the end of the book. We also now remind the reader of the nature of those functions belonging to that important subclass: analytic functions.

Differentiation

If $w = f(z)$ is a complex function of the complex variable $z = x + iy$ defined in some neighborhood of a point z_0, the derivative of f at z_0, denoted variously by

*This result is implicit in the *Argument Principle*, which we discuss in detail in Section 5.5. Also see Churchill (1974), p. 296 or Wylie (1975), p. 829, for example.

$dw/dz|_{z=z_0}$, $w'(z_0)$, or $f'(z_0)$, is given by

$$f'(z_0) \equiv \lim_{\Delta z \to 0} \frac{f(z_0 + \Delta z) - f(z_0)}{\Delta z}. \tag{5.3–1}$$

In one sense, there is nothing new in this definition owing to the formal identity of (5.3–1) with that associated with real-variable differential calculus. Consequently, all the familiar differentiation formulas for sums, products, quotients, powers, and composites of functions of a single real-variable are equally valid for functions of a complex variable.

On the other hand, since $\Delta z = \Delta x + i\,\Delta y$, some important differences do crop up, not unlike those that occur in the calculus of several variables; there is no *a priori given specific* manner in which Δz in (5.3–1) should approach zero. Indeed, for the derivative of $f(z)$ to exist at a point it is necessary that the limit of the above difference quotient be *independent* of the precise way in which $\Delta z \to 0$. The function $f(z) = \bar{z}$, for example, is nowhere differentiable since

$$\begin{aligned}\frac{f(z + \Delta z) - f(z)}{\Delta z} &= \frac{(x + \Delta x) - i(y + \Delta y) - (x - iy)}{\Delta x + i\,\Delta y} \\ &= \frac{\Delta x - i\,\Delta y}{\Delta x + i\,\Delta y} \\ &= \begin{cases} 1 & \text{if } \Delta y = 0 \\ -1 & \text{if } \Delta x = 0 \end{cases}\end{aligned}$$

so that the limit of the difference quotient as $\Delta z \to 0$ does not exist.

If we set $f(z) = w = u(x,y) + iv(x,y)$ as in (A4–2) and investigate in general the implications of the definition (5.3–1), we find

$$\begin{aligned} f'(z) &= \lim_{\Delta z \to 0} \frac{u(x + \Delta x, y + \Delta y) + iv(x + \Delta x, y + \Delta y) - u(x,y) - iv(x,y)}{\Delta x + i\,\Delta y} \\ &= \frac{\partial u}{\partial x} + i\frac{\partial v}{\partial x} \qquad \text{if } \Delta y = 0 \qquad (5.3\text{–}2) \\ &= \frac{\partial v}{\partial y} - i\frac{\partial u}{\partial y} \qquad \text{if } \Delta x = 0. \end{aligned}$$

(We only need to look at approach from these two directions.) In addition to providing formulas for calculating the derivative $f'(z)$ in terms of partial derivatives of the component functions $u(x,y)$, $v(x,y)$, the expressions (5.3–2) lead to the celebrated *necessary* condition for differentiability of complex functions, the *Cauchy-Riemann equations*:

$$\frac{\partial u}{\partial x} = \frac{\partial v}{\partial y}, \qquad \frac{\partial u}{\partial y} = -\frac{\partial v}{\partial x}. \tag{5.3–3}$$

(See Problem 9 for the polar form of these equations.) As most readers know, with the addition of modest assumptions regarding continuity, these equations, named after their discoverer Augustin Cauchy (1789–1857) and their principal early exploiter Georg Riemann (1826–1866), turn out to be *sufficient* also.

Theorem 1:

Let z_0 be an interior point of a region in which $w = u + iv = f(z)$ is a single-valued and continuous function of the complex variable $z = x + iy$. Then $f'(z)$ exists and is continuous at $z = z_0$ if $u(x,y)$, $v(x,y)$ have continuous first partial derivatives satisfying the Cauchy-Riemann $(C-R)$ equations (5.3–3) in some neighborhood of the point z_0.

Analyticity

This classical result prompts the following definition.

Definition 1 The complex function $f(z)$ is termed **analytic** at a point z_0 if $f'(z)$ exists and is continuous* throughout some neighborhood of z_0.

It is important to note that, whereas differentiability is a *point* concept, analyticity is a *neighborhood* concept. The specific example $f(z) = |z|^2$ illustrates this nicely; it has a derivative at $z = 0$, but nowhere else, and is thus not analytic. (Also see Problem 10.)

Definition 2 A point at which a complex function $f(z)$ is analytic is called a **regular point** of f. Non-regular points are termed **singular points** or *singularities* of f. A function analytic at every point of a region R is said to be **analytic in R.** If the region is the whole of the complex z-plane, the function is called **entire**.

The definitions of the complex counterparts of the elementary functions of real-variable calculus make them analytic functions, with the expected derivatives, throughout the complex plane. As examples we mention

(i) polynomials in z,

(ii) rational functions (except at the finite number of points where the denominator vanishes),

(iii) rational powers of the form $z^{m/n}$ with $n > 1$ (except along, say, some ray $\arg z = \alpha$, $|z| \geq 0$),

(iv) the exponential function $e^z \equiv e^x (\cos y + i \sin y)$, **(5.3–4)**

(v) simple combinations of exponential functions such as the trigonometric and hyperbolic functions (except at zeros of the denominators).

As suggested by the proviso appearing in the case of rational powers, however, some functions require special treatment in order to ensure single-valuedness.

*Most authors do not require continuity of $f'(z)$ in their definition of analyticity; the continuity is then proved subsequently. The end result, of course, is the same, and no loss of generality ensues from this more restrictive definition.

An archetypal example is the multivalued inverse of the exponential function, $w = \ln z$. Since $e^w = z$, using the polar representation (A4–1) we find

$$e^{u + iv} = re^{i\theta}$$

and hence $e^u = r, v = \theta$. It follows that $w = \ln r + i\theta$, or expressed perhaps more informatively,

$$\ln z = \ln|z| + i \arg z. \qquad \textbf{(5.3–5)}$$

It is clear that the origin $z = 0$ must be avoided in order that $\ln z$ be well-defined. If $\arg z$ is likewise restricted, say, so that $\delta < \arg z < \delta + 2\pi$, then (5.3–5) defines a single-valued analytic function, with derivative $1/z$, as may be verified. The choice $\delta = -\pi$ leads to what is usually termed the *principal branch* of the multivalued logarithm function. Each point of the negative real axis is a singular point of this branch and this line of singularities itself is called a *branch cut* for the function. This branch cut joins the origin $z = 0$ with the point at infinity, thereby uniting the so-called *branch points*.

These results for the logarithm function have implications for the general power function z^α where the exponent α is an arbitrary complex number, since

$$z^\alpha \equiv e^{\alpha \ln z}. \qquad \textbf{(5.3–6)*}$$

Naturally we must restrict z to ensure a well-defined, single-valued analytic function. The conditions $|z| > 0$ and $-\pi < \arg z < \pi$ again give rise to the principal branch.

5.4 Complex integration

In this section we continue our review of helpful background material and remind the reader of those results fundamental to the analysis of integrals of analytic functions. We begin by noting that line integrals in the complex plane can be defined in the following manner. Assume that we have a continuous function f of the complex variable z and a contour starting at $z = \alpha$ and ending at $z = \beta$, along which we wish to integrate f. We partition C into n component contours C_k $(k = 1,2, \ldots, n)$ by the points $\alpha = z_0, z_1, z_2, \ldots, z_n = \beta$ and set $\Delta z_k \equiv z_k - z_{k-1}$ with $\Delta(n) \equiv \max_k |\Delta z_k|$. In the limit of small Δ (large n), the sum $\sum_k f(z_k)\Delta z_k$ yields the desired line integral. Indeed,

$$\lim_{\Delta\to 0} \sum_{k=1}^{n} f(z_k)\Delta z_k \equiv \int_C f(z)dz. \qquad \textbf{(5.4–1)}^\dagger$$

*This relation is equivalent to the generalized de Moivre result discussed in connection with Property 7 of Appendix 4.

†If $\alpha = \beta$ and C is a closed contour, the notation $\oint_C f(z)dz$ will typically be employed. Unless otherwise noted, the usual interpretation will be made that the contour is traversed in the *counterclockwise* direction.

If we desire to exhibit the endpoints of this particular line integral, then we will use the notation $\int_{\alpha\,C}^{\beta} f(z)dz$. In general, however, we must specify the contour of integration or our line integrals will be ambiguous, since most line integrals of interest are path-dependent.

If we introduce real and imaginary parts of z and f, then we can express the line integral (5.4–1) as the (complex) sum of two real integrals, namely

$$\int_C f(z)dz = \int_C (u\,dx - v\,dy) + i\int_C (v\,dx + u\,dy). \qquad \textbf{(5.4–2)}$$

Alternatively, since $|dz| = ds$, the arclength infinitesimal, the definition (5.4–1) leads to

Property 1

$$\left|\int_C f(z)dz\right| \leq \int_C |f(z)|\,|dz| \leq ML$$

where $M = \max|f(z)|$ on C and $L =$ length of C.

Other useful properties of line integrals that the reader may easily establish are:

Property 2

$$\int_{\alpha \atop C}^{\beta} f(z)dz = -\int_{\beta \atop C}^{\alpha} f(z)dz$$

Property 3

$$\int_{\alpha \atop C}^{\beta} [\lambda f(z) + \mu g(z)]dz = \lambda\int_{\alpha \atop C}^{\beta} f(z)dz + \mu\int_{\alpha \atop C}^{\beta} g(z)dz$$

Property 4

$$\int_{\alpha \atop C}^{\gamma} f(z)dz + \int_{\gamma \atop C}^{\beta} f(z)dz = \int_{\alpha \atop C}^{\beta} f(z)dz.$$

Here f is assumed to be continuous along the contour C, γ is some intermediate point on the contour, and λ, μ are arbitrary complex numbers.

An especially important line integral is that obtained from integrating the function $f(z) = 1/(z - z_0)$ in counterclockwise fashion along the closed circular contour $|z - z_0| = \rho$ (z_0 and $\rho > 0$ arbitrary). If we parameterize the path of integration C by setting $z = z_0 + \rho\, e^{i\varphi}$, then

$$\oint_C \frac{dz}{z - z_0} = \int_{\varphi_0}^{\varphi_0 + 2\pi} \frac{\rho i\, e^{i\varphi} d\varphi}{\rho e^{i\varphi}} = 2\pi i. \qquad \textbf{(5.4–3)}$$

This result, as most readers know, plays a fundamental role in complex integration theory.*

The Cauchy-Goursat theorem

One of the essential results in vector calculus is the *Divergence Theorem*, usually best remembered in its three-dimensional form (take $\eta = 1$ in (4.3–17)). A useful two-dimensional version of that theorem, however, is:

Divergence Theorem:

Let R be the simply-connected closed region consisting of points within and on a closed contour C. Assume $\mathbf{F}$ and div $\mathbf{F}$ are continuous throughout R. Then

$$\oint_C \mathbf{F} \cdot \mathbf{e}_n \, ds = \iint_R \operatorname{div} \mathbf{F} \, dx \, dy \tag{5.4–4}$$

where $\mathbf{e}_n$ is the *outward* normal (unit vector) to the contour C.

If we let the vector function $\mathbf{F}$ in the Divergence Theorem be given by $\mathbf{F} = Q(x,y)\mathbf{e}_x - P(x,y)\mathbf{e}_y$ where $P, Q, \dfrac{\partial P}{\partial y}$, and $\dfrac{\partial Q}{\partial x}$ are continuous throughout R, (5.4–4) can be written alternatively as

$$\oint_C (P \, dx + Q \, dy) = \iint_R \left(\frac{\partial Q}{\partial x} - \frac{\partial P}{\partial y} \right) dx \, dy. \tag{5.4–5}$$

In this form, we recognize the result as *Green's Theorem* for line integrals in the plane. As a ready consequence, we note

Property 5

If $\dfrac{\partial Q}{\partial x} = \dfrac{\partial P}{\partial y}$ throughout R, then

$$\oint_C (P \, dx + Q \, dy) = 0$$

and the line integral $\int_\alpha^\beta (P \, dx + Q \, dy)$ is path independent for all α, β in R.

---

*If we divide the path C into two component contours and reverse the direction of integration of one of them, this result also serves to vividly illustrate the path dependence of line integrals.

The relation (5.4–5) can be used as the basis for a simple proof of the classical

Cauchy-Goursat Theorem:

Let $f(z)$ be an analytic function of the complex variable z throughout a region R of the complex z-plane. Then

$$\oint_C f(z)dz = 0$$

for every closed contour C that lies in R and has its interior completely within R.

Proof: By virtue of (5.4–2), (5.4–5)

$$\begin{aligned}\oint_C f(z)dz &= \oint_C (u\,dx - v\,dy) + i\oint_C (v\,dx + dy)\\ &= -\iint_D \left(\frac{\partial v}{\partial x} + \frac{\partial u}{\partial y}\right)dx\,dy + i\iint_D \left(\frac{\partial u}{\partial x} - \frac{\partial v}{\partial y}\right)dx\,dy\\ &= 0.\end{aligned}$$

Here we have denoted by D the region interior to C and used the Cauchy-Riemann equations (5.3–3). □

The ease with which our proof is accomplished is somewhat deceptive. We have avoided the difficulties that occur with more general versions of this important result by assuming considerable smoothness for the curves C (see Appendix 4).

The Cauchy-Goursat* theorem gives rise to two important corollaries that will be of considerable value in the rest of the chapter:

Corollary 1:

$$\oint_C f(z)dz = \oint_{C'} f(z)dz$$

if the closed contour C can be continuously deformed into the closed contour C' without passing through a point where $f(z)$ is not analytic.

Corollary 2:
The integral $\int_\alpha^\beta f(z)dz$ is path independent in any simply-connected region throughout which $f(z)$ is analytic.

*The name of Édouard Goursat (1858–1936) is included since he was the first to establish that merely the existence, rather than the continuity, of the derivative of f was needed for the result to be true.

The path independence of line integrals of analytic functions detailed in this second result allows the proof of the fact that such definite integrals are actually analytic functions of their upper limits. Indeed, with comparable effort the following can be established:

Theorem 1:

Let $f(z)$ be a continuous function throughout a region R with the property that

$$\oint_C f(z)dz = 0$$

for every closed contour C interior to R. Then, for arbitrary z_0 in R,

$$F(z) = \int_{z_0}^{z} f(s)ds$$

defines an analytic function whose derivative is $f(z)$.

Proof: See Churchill (1974), for example. □

We will use this result shortly to establish the converse to the Cauchy-Goursat theorem due to Morera. Meanwhile, we note that combining this result with that earlier theorem leads directly to:

Corollary:

If $f(z)$ is analytic throughout a region R with the property that

$$\oint_C f(z)dz = 0$$

for every closed contour C interior to R, then

$$\int_{z_1}^{z_2} f'(z)dz = f(z_2) - f(z_1).$$

Thus the Fundamental Theorem of the Integral Calculus is equally valid for well-behaved complex functions.

The Cauchy-integral formula

Now we turn to another basic result, namely the *Cauchy Integral Formula*, which we prove for the sake of illustrating a valuable technique.

Cauchy Integral Formula:

Let $f(z)$ be an analytic function of the complex variable z throughout a region R of the complex z-plane. Then for any (simple) closed contour C that lies in R and has its interior completely within R,

$$\frac{1}{2\pi i}\oint_C \frac{f(z)}{z - z_0}\,dz = \begin{cases} f(z_0) \\ 0 \end{cases} \qquad \textbf{(5.4–6)*}$$

according as z_0 is interior or exterior to C.

Proof: If z_0 is exterior to C, the result follows immediately upon applying the Cauchy-Goursat theorem to the analytic integrand. If z_0 is within C, we shift the contour of the desired integral, using Corollary 1 of the Cauchy-Goursat theorem, to the circle $|z - z_0| = \rho$ which, for small enough ρ, is interior to C. We denote this new contour by C'. Owing to our earlier calculation (5.4–3), we find then that

$$\begin{aligned}\frac{1}{2\pi i}\oint_C \frac{f(z)}{z - z_0}\,dz &= \frac{1}{2\pi i}\oint_{C'} \frac{f(z)}{z - z_0}\,dz \\ &= \frac{f(z_0)}{2\pi i}\oint_{C'} \frac{dz}{z - z_0} + \frac{1}{2\pi i}\oint_{C'} \frac{f(z) - f(z_0)}{z - z_0}\,dz \\ &= f(z_0) + I.\end{aligned}$$

The modulus of I can be bounded using Property 1 on p. 139. Indeed

$$|I| \leq M(\rho)$$

where $M(\rho) = \max |f(z) - f(z_0)|$ on C'. Since $M(\rho) \to 0$ as $\rho \to 0$, I must vanish and the proof is complete. □

The consequences of the Cauchy Integral Formula are legion. With modest effort we can show, for example, that

Theorem 2:

Under the same hypotheses as the Cauchy Integral Formula, if z_0 is interior to C, then

$$f^{(n)}(z_0) = \frac{n!}{2\pi i}\oint_C \frac{f(z)\,dz}{(z - z_0)^{n+1}} \qquad \textbf{(5.4–7)}$$

for $n = 0,1,2, \ldots$.

*Recall our convention that the contour is traversed in the counterclockwise, or what is usually termed *positive*, direction.

From this we immediately conclude that

Corollary 1:
If $f(z)$ is analytic at a point, then $f^{(n)}(z)$ is also analytic at that point for arbitrary positive integral n.

The case $n = 1$ of this last result is sufficient for the proof of

Morera's Theorem:

Let $f(z)$ be a continuous function throughout a region R with the property that

$$\oint_C f(z)dz = 0$$

for every closed contour C interior to R. Then $f(z)$ is analytic throughout R.

Proof: By virtue of Theorem 1, for arbitrary z_0 in R,

$$F(z) = \int_{z_0}^{z} f(s)ds$$

defines an analytic function whose derivative is $f(z)$. But now we know $F'(z) = f(z)$ is analytic also. □

Another important result engendered by Theorem 2 is

Corollary 2:
If $f(z)$ is analytic within and on the circle $|z - z_0| = \rho$, then the *Cauchy inequalities*

$$|f^{(n)}(z_0)| \leq \frac{n!M}{\rho^n} \qquad \textbf{(5.4–8)}$$

are valid for $n = 0,1,2, \ldots$, where M is the maximum value of $|f(z)|$ on the circular contour.

When $n = 0$, we see that $|f(z_0)|$ is bounded by the maximum assumed by $|f(z)|$ on the circle $|z - z_0| = \rho$. If $n = 1$ (5.4–8) shows that an upper bound on the modulus of the derivative $|f'(z_0)|$ is directly proportional to M and inversely proportional to ρ. The first observation is fundamental in the demonstration of the

Maximum Modulus Principle:

If $f(z)$ is analytic and nonconstant in a (open) region R, then $|f(z)|$ cannot have a maximum in R.

The second special case noted leads to the proof of

Liouville's Theorem:

If $f(z)$ is analytic throughout the complex plane (that is, $f(z)$ is an entire function) and bounded, then $f(z)$ is constant.

We remark that, in view of Liouville's Theorem, the interesting analytic functions are those that either have singularities or grow unboundedly large as $|z| \to \infty$, or both.

5.5 The argument principle and its consequences

We are now ready to investigate some function theoretic results that bring us closer to our earlier stability problem of Section 5.2. We begin by considering the logarithmic derivative of an analytic function. To be precise, let $f(z)$ be analytic within and on a closed contour C, and assume that $f(z)$ vanishes at the points $z = z_k$ $(k = 1,2, \ldots, n)$ inside C, but not on C itself. On the one hand, if we imagine C arbitrarily partitioned into component contours by the points $\alpha = \sigma_0, \sigma_1, \ldots, \sigma_{m-1}, \sigma_m = \beta$, then by virtue of the corollary to Theorem 1 of the previous section,

$$\begin{aligned}
\frac{1}{2\pi i}\oint_C \frac{d}{dz}[\ln f(z)]\,dz &= \frac{1}{2\pi i}\sum_{i=1}^{m}\int_{\sigma_{i-1}}^{\sigma_i}\frac{d}{dz}[\ln f(z)]dz \\
&= \frac{1}{2\pi i}\sum_{i=1}^{m}[\ln f(\sigma_i) - \ln f(\sigma_{i-1})] \\
&= \frac{1}{2\pi i}[\ln f(\beta) - \ln f(\alpha)] \\
&= \frac{1}{2\pi i}[\ln|f(\beta)| - \ln|f(\alpha)|] + \frac{1}{2\pi}[\arg f(\beta) - \arg f(\alpha)] \\
&= \frac{1}{2\pi}\Delta_C \arg f(z). \qquad \textbf{(5.5–1)}
\end{aligned}$$

This last expression takes cognizance of the fact that, since C is closed, $\beta = \alpha$ and, although $|f(z)|$ undergoes no net change as z makes a complete circit around C, arg $f(z)$ does. This latter change, of course, is independent of the choice of reference (starting/ending) point and so can be given a generic designation. On the other hand, we also have

$$\begin{aligned}
\frac{1}{2\pi i}\oint_C \frac{d}{dz}[\ln f(z)]\,dz &= \frac{1}{2\pi i}\oint_C \frac{f'(z)}{f(z)}\,dz \\
&= \sum_{k=1}^{n}\frac{1}{2\pi i}\oint_{C_k}\frac{f'(z)}{f(z)}\,dz \qquad \textbf{(5.5–2)}
\end{aligned}$$

where the C_k ($k = 1,2, \ldots, n$) are small (nonoverlapping) circular contours centered at each of the zeros z_k of $f(z)$, respectively. (Here we have used a natural extension to Corollary 1 of the Cauchy-Goursat Theorem).

The contour integrals on the right-hand side of (5.5–2) can be easily evaluated. In the neighborhood of each point $z = z_k$ where f vanishes, the function must have the form

$$f(z) = (z - z_k)^{p_k} g_k(z)$$

where $g_k(z)$ is analytic and $g_k(z_k) \neq 0$. (The constant p_k is called the order or multiplicity of the zero z_k. It must, of course, be integral, lest some derivative of $f(z)$ not exist at z_k in violation of the analyticity of f.) It then follows that

$$\frac{f'(z)}{f(z)} = \frac{p_k}{(z - z_k)} + \frac{g_k'(z)}{g_k(z)},$$

and therefore, using (5.4–3),

$$\frac{1}{2\pi i} \oint_{C_k} \frac{f'(z)}{f(z)}\, dz = p_k. \qquad \textbf{(5.5–3)}$$

(For small enough contours C_k, there are no contributions from the terms involving $g_k(z)$ since $g_k(z)$ is nonvanishing within C_k by continuity. Recall Property 12 of Appendix 4.) If we combine the results embodied in (5.5–1), (5.5–2), and (5.5–3), we obtain the

Argument Principle:*

Let $f(z)$ be a nontrivial complex function, analytic within and on a closed contour C and nonvanishing on C. Then

$$\frac{1}{2\pi} \Delta_C \arg f(z) = \sum_{k=1}^{n} p_k \qquad \textbf{(5.5–4)}$$

= total number of zeros of $f(z)$, counting multiplicity, within C.

Expressed geometrically, the argument principle says that the total number of zeros of an analytic function within a closed contour C is equal to the (net) number of times the point $w = f(z)$ "winds around" the origin $w = 0$ in the positive (counter-clockwise) direction while z is making one circuit around C in the positive direction.

Applications to stability

It is easy to see that our earlier assertion (5.2–5) is merely a special case of (5.5–4) above. In order to carry out the desired stability calculations, then, we make the

*This is a restricted form of the result that usually goes by this name. In the more general case of a function with pole singularities, the right-hand side of (5.5–4) is the total number of zeros less the total number of poles of f (zeros of $1/f$) within C.

associations $z = s$ and $f(z) = Q(s)$ and choose for our contour C the union of the semicircle in the *right* half-plane of arbitrarily large radius R with the connecting portion of the imaginary axis (see Figure 5.4).

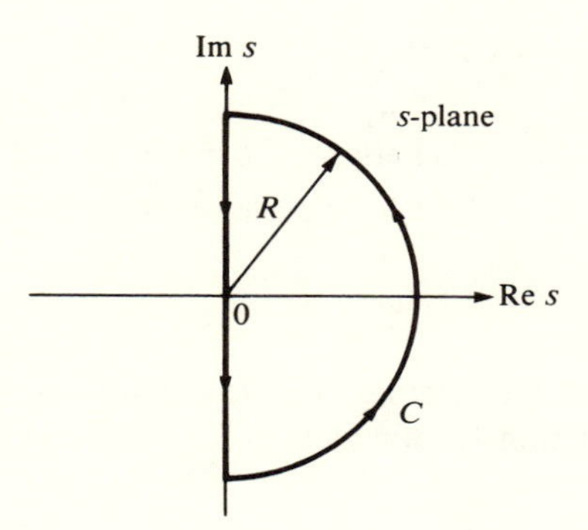

Figure 5.4 Appropriate contour in the s plane for stability calculations

If $Q(s)$ has no purely imaginary zeros, we can apply the Argument Principle with this fundamental contour C to ascertain if all of the zeros of $Q(s)$ lie in the left half of the complex s-plane. This will determine if the rational transfer function having $Q(s)$ as its denominator belongs to a strictly stable system. The following serve to illustrate the nature of these calculations:

Example I

$$Q(s) = s^2 - 4$$

$$Q(i\omega) = -(\omega^2 + 4); \qquad Q(Re^{i\theta}) \sim R^2 e^{2i\theta}$$

No purely imaginary zeros

On infinite semicircle: Δ arg $= 2(\pi/2) - 2(-\pi/2) = 2\pi$
On imaginary axis: Δ arg $= 0$

Number of zeros in *RHP* = 1; system unstable. □

Example II

$$Q(s) = s^4 + 2s^3 + s^2 + 3s + 2$$

$$Q(i\omega) = \omega^4 - \omega^2 + 2 + i\omega(3 - 2\omega^2); \qquad Q(Re^{i\theta}) \sim R^4 e^{4i\theta}$$

No purely imaginary zeros

On infinite semicircle: Δ arg $= 4(\pi/2) - 4(-\pi/2) = 4\pi$
On imaginary axis: Δ arg $= 0$

Number of zeros in *RHP* = 2; system unstable. □

Example III

$$Q(s) = s^5 + 5s^4 + 11s^3 + 14s^2 + 10s + 4$$

$$Q(i\omega) = 5\omega^4 - 14\omega^2 + 4 + i\omega(\omega^4 - 11\omega^2 + 10); \qquad Q(Re^{i\theta}) \sim R^5 e^{5i\theta}$$

No purely imaginary zeros

On infinite semicircle: $\Delta \arg = 5(\pi/2) - 5(-\pi/2) = 5\pi$

On imaginary axis: $\Delta \arg = -5\pi/2 - (5\pi/2) = -5\pi$

Number of zeros in $RHP = 0$; system strictly stable. □

Turning now to the satellite attitude control system discussed in Section 5.1, with the reference transfer function (5.2–2) we see that the denominator of interest* can be written as

$$Q(s) = s^3 + a(\alpha s^2 + s + \beta) \tag{5.5–5}$$

where $a \equiv K_\theta K_a K_\tau / I$. An analysis based upon the argument principle yields

$$Q(i\omega) = a(\beta - \alpha\omega^2) + i\omega(a - \omega^2); \qquad Q(Re^{i\theta}) \sim R^3 e^{3i\theta}$$

No purely imaginary zeros unless $a = 0$, $\beta = 0$, or $\beta = \alpha a$

On infinite semicircle: $\Delta \arg = 3(\pi/2) - 3(-\pi/2) = 3\pi$

On imaginary axis: $\Delta \arg = \begin{cases} -3\pi & \text{if } \alpha a > \beta > 0 \text{ with } \alpha > 0 \\ -\pi, \pi, \text{ or } 3\pi & \text{otherwise } (a\beta \neq 0, \alpha a^2) \end{cases}$

The control system is strictly stable, therefore, only when both the derivative and integral gains α and β are positive and satisfy $\alpha a > \beta$.

In their consideration of this control system, Frederick and Carlson (1971) suggest some practical values for the various system parameters:

$$K_\theta = 0.20 \text{ volts/rad}$$
$$K_a = 1.25$$
$$K_\tau = 0.10 \text{ ft lbs/volt}$$
$$I = 10.0 \text{ slug ft}^2.$$

These specific values lead numerically to $a = 0.0025$. If $\alpha = 20$, for example, then the system is strictly stable for $0 < \beta < 0.05$. However, as Frederick and Carlson point out, due to the presence of the differentiator, Stage 1 of the controller under discussion is not physically realizable. From a mathematical point of view, this difficulty manifests itself in the order of the polynomial in the numerator of (5.1–2), when this expression is rewritten as a rational function, being greater than the order of the polynomial in the denominator.

*The disturbance transfer function relating τ_{dis} and θ has the same denominator (see Problem 21).

As an alternative component transfer function that avoids this difficulty we consider

$$G(s) = k\frac{\alpha s^2 + s + \beta}{\alpha s^2 + \gamma s + \beta} \tag{5.5–6}$$

with $k > 0$ and $\gamma > 1$. Substituting (5.5–6) for (5.1–2) leads to a new overall system transfer function, in place of (5.2–2), whose stability properties must be investigated. It is not difficult to show that in this case the resulting control system is strictly stable if $K_\theta K_\tau k\alpha > \gamma\beta I > 0$, that is, using the specific numerical values given earlier, whenever $k\alpha > 500\ \beta\gamma$ (see Problem 22).

Nyquist's diagrams and criteria

Oftentimes the stability analyses carried out above can be simplified by making a plot of the contour Γ that $Q(s)$ describes in the w-plane as s traverses the fundamental contour C in the s-plane. The angular changes in arg Q are then easily determined in general.

Harry Nyquist, an American Telephone and Telegraph Company engineer, was one of the first to suggest the utility of these diagrams and related constructs for the design as well as the analysis of feedback systems. Expressed in modern terms, Nyquist (1932) recognized that for rational functions of the form (5.2–1), that is,

$$H(s) \equiv \frac{P(s)}{Q(s)},$$

where $P(s)$ and $Q(s)$ are polynomials in s with no common factors, owing to the Argument Principle, the relationship

$$\frac{1}{2\pi}\Delta_C \arg H(s) = N_P - N_Q \tag{5.5–7}$$

= total number of zeros of $P(s)$ within C *less* the total number of zeros of $Q(s)$ within C

must hold. Since the zeros of Q are the pole singularities of H*, the right-hand side of (5.5–7) is merely the *difference* between the number of zeros and the number of poles of $H(s)$ within C (recall the footnote on p. 146). The left-hand side of this expression, of course, can be viewed as the net number of times Γ, the image contour in the w-plane of C under the mapping $w = H(s)$, winds around the origin while s traverses the original contour in the positive direction.

For stability calculations C becomes the fundamental contour depicted in Figure 5.4 and the resulting image contour Γ is termed the *Nyquist diagram*. We want to close this chapter by recasting the above observation regarding the interpre-

*That is, the (isolated) zeros of $1/H$

tation of (5.5–7) in terms of the behavior of a Nyquist diagram into a useful test for stability.

Consider the simple single-loop feedback system with rational transfer functions $F(s)$ and $G(s)$ illustrated in Figure 5.3(c) on p. 132. The overall feedback system transfer function is $F/(1 + FG)$, so that the system is strictly stable only if no zeros of $1 + FG$ lie in the right half-plane or upon the imaginary s-axis. If we make the association $w = FG$ and apply (5.5–7) to the function $H(s) = 1 + w$, we see that the strict stability of the system necessitates a balance between the "winding number" for the Nyquist diagram of $H(s)$ and the number of poles of this complex function that lie in the right half-plane. Since the poles of w and of $1 + w$ are equivalent, we can combine these statements so as to yield the following:

Nyquist Stability Criterion:*

Let $F(s)$ and $G(s)$ be rational functions of s and assume that in the product FG the only cancellations that occur involve poles and zeros in the left half of the complex s-plane. If $w = FG$ has N poles in the right half-plane, then the single-loop feedback system of Figure 5.3(c) is strictly stable *if and only if* the Nyquist diagram in the w-plane encircles the point $w = -1$ in the *negative* or *clockwise* sense exactly N times.

Note that no assumption needs to be made here relative to the stability of either of the component subsystems with transfer functions F and G.

Example As an illustration of the utility of the Nyquist criterion, let us investigate the stability of a single-loop linear feedback system with a product transfer function FG given by

$$F(s)\,G(s) = \frac{ks}{s^2 - s + 1}$$

where k is a real constant. It is a simple matter to ascertain that the Nyquist diagram associated with $w = FG$ is the circle $|w + \tfrac{1}{2}k| = \tfrac{1}{2}k$, covered twice in the negative direction (see the sketch in Figure 5.5). Since the product transfer function has 2 poles in the right half of the complex s-plane, the Nyquist criterion then shows that $k > 1$ is necessary (and sufficient) for the strict stability of the overall system. □

In this example, the determination of the Nyquist diagram was facilitated by two features of the product transfer function:

(i) $F(s)G(s) \to 0$ as $|s| \to \infty$ so that the image of the semicircular portion of the fundamental contour C was the origin $w = 0$ and the entire Nyquist diagram was bounded;

*For an extremely general version of this classic result under minimal assumptions, see Desoer (1965).

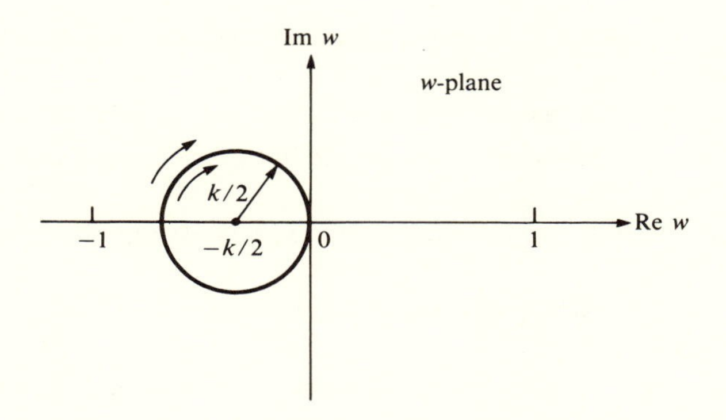

Figure 5.5 Nyquist diagram for the product transfer function $F(s)G(s) = ks\,(s^2 - s + 1)$

(ii) $F(\bar{s})G(\bar{s}) = \overline{F(s)G(s)}$ and thus the Nyquist diagram was symmetric with respect to the real w axis.

The first feature is shared by virtually all realizable systems, the second by all real systems. Owing principally to these features, therefore, the Nyquist diagram can be sketched almost automatically for the most general of systems (see Problems 23 and 24 for two additional illustrative examples). As a consequence, the Nyquist criterion has proved over the years to be an extremely effective tool in the analysis and design of an impressive variety of time-invariant linear controllers.

References

Churchill, R. V., J. W. Brown, and R. F. Verhey (1974): *Complex Variables and Applications*, 3rd ed., McGraw-Hill, New York.

Coppel, W. A. (1965): *Stability and Asymptotic Behavior of Differential Equations*, D. C. Heath, Boston; Chapter IV, Appendix.

Desoer, C. A. (1965): "A General Formulation of the Nyquist Criterion," *IEEE Trans. Circuit Theory* CT–12: 230–234.

Frederick, D. K. and A. B. Carlson (1971): *Linear Systems in Communication and Control*, Wiley, New York; Chapters 9, 10, 11.

Gantmacher, F. R. (1959): *Theory of Matrices, Vol. II*, transl. from the Russian, Chelsea, New York; Chapter XV.

Kaplan, W. (1962): *Operational Methods for Linear Systems*, Addison-Wesley, Reading, Mass.; Chapters 1, 2, 7.

Liénard, A. and M. H. Chipart (1914): "Sur le signe de la partie réelle des racines d'une équation algébrique," *J. Math. Pures Appl.* (6) 10: 291–346.

Lorentz, G. G. (1966): *Approximation of Functions*, Holt, Rinehart and Winston, New York.

Marden, M. (1966): "Geometry of Polynomials," *Math. Surveys #3*, Revision of 1949 ed., American Mathematical Society, Providence, R.I.; Chapter 9.

Newcomb, R. W. (1968): *Concepts of Linear Systems and Controls*, Brooks/Cole, Belmont, Calif.

Nyquist, H. (1932): "Regeneration Theory," *Bell Syst. Tech, J.* 11: 126–147.

Pennisi, L., L. I. Gordon and S. Lasher (1976): *Elements of Complex Variables*, Holt, Rinehart and Winston, New York.

Silverman, R. A. (1974): *Complex Analysis with Applications*, Prentice-Hall, Englewood Cliffs, N.J.

Wylie, C. R. (1975): *Advanced Engineering Mathematics*, 4th ed., McGraw-Hill, New York; Chapters 15, 16, 17.

Zadeh, L. A. and C. A. Desoer (1979): *Linear System Theory*, 2nd ed., Krieger, Huntington, N.Y.; McGraw-Hill, New York (1963).

Problems

Sections 5.1, 5.2

1. Consider the *RLC* network depicted below.

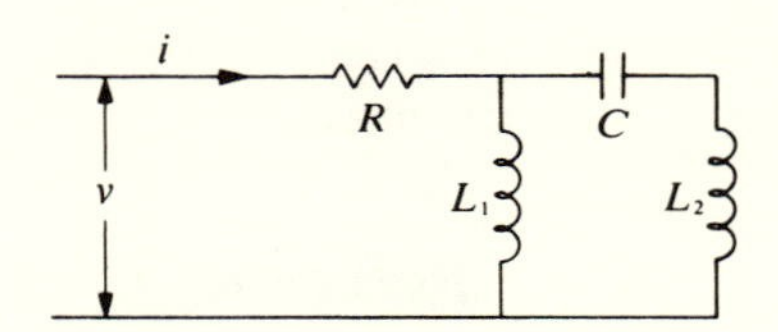

   (a) Derive both a first-order system of differential equations (in matrix form) as well as a single higher-order differential equation that governs the behavior of this network.

   (b) Determine the system transfer function, taking $v(t)$ as input and $i(t)$ as output, and discuss its relation to the characteristic polynomial associated with the appropriate matrix developed in part (a).

2. Illustrate the correlation between the stability theory of linear systems and related concepts in ordinary differential equations by analyzing the response of the constant-coefficient nonhomogeneous equation

$$\frac{d^2y}{dt^2} + a\frac{dy}{dt} + by = ce^{pt}$$

   for various values of the input parameters p and c.

3. Let a time-invariant linear system be described by the equation

$$\dot{\mathbf{y}} + A\mathbf{y} = \mathbf{f}$$

where A is the matrix

$$A = \begin{bmatrix} 4 & 1 & 2 \\ 2 & 3 & 2 \\ -2 & -2 & -1 \end{bmatrix}.$$

Find the transfer functions associated with the output variables $y_i(t)$, $i = 1,2,3$, and discuss the system's stability.

4. The following is a block diagram for a simple linear control system.

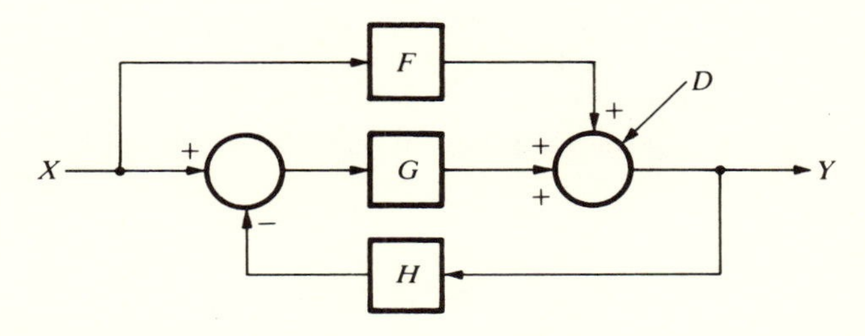

If the transfer functions of the respective blocks are F, G, and H, find the system transfer function when

(a) X is arbitrary, but $D = 0$ (reference transfer function);
(b) D is arbitrary, but $X = 0$ (disturbance transfer function).

How should these results be combined to provide the general system output when *both* X and D are arbitrary?

5. (a) Demonstrate the equivalence of the various conditions in Theorem 1 of Section 5.2 for polynomials of degree less than or equal to 4.
(b) Show that the polynomial

$$Q(s) = s^4 + 3s^3 + 5s^2 + 9s + 6$$

has no right half-plane zeros.

6. The stability of the feedback system (see the accompanying figure) is dependent upon the nature of the feedback transfer function $H(s)$.

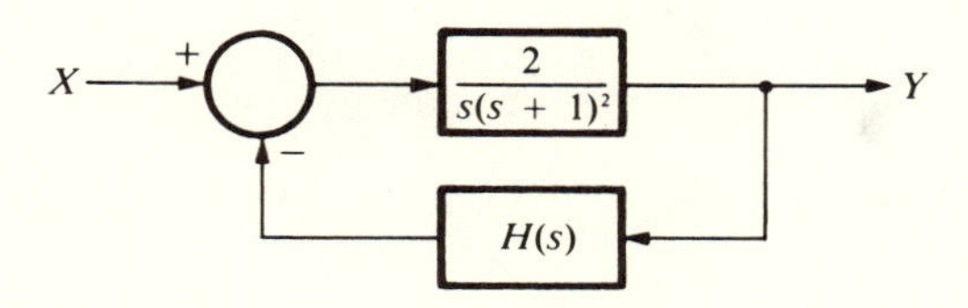

(a) Use the Routh-Hurwitz criterion to verify that the system is unstable if $H(s) \equiv 1$.
(b) Find a simple rational transfer function $H(s)$ of the form (5.2–1) (with the degree of $P(s)$ less than the degree of $Q(s)$) such that $\lim_{s \to 0} H(s) = 1$ but the overall system is strictly stable in general.

★7. Show that if an nth degree polynomial with real coefficients has one or more pairs of purely imaginary zeros and/or a zero at the origin, then the nth Hurwitz determinant D_n must vanish. Use this result to establish the validity of Theorem 2 of Section 5.2.

8. The overall system transfer function of a linear controller is

$$H(s) = \frac{k(s - 1)}{k(s - 1) + (s + 1)(s + 10)}$$

where k is an adjustable system parameter.

(a) Determine the range of real k for which the controller is strictly stable.

(b) Carry out the same analysis for the system with overall transfer function

$$H(s) = \frac{k(s + 100)^2}{k(s + 100)^2 + s^2(s + 10)} .$$

[Hint: Recall Theorem 2 of Section 5.2.]

Sections 5.3, 5.4

9. Let the function $f(z) = u(x,y) + iv(x,y)$ be analytic. Rewrite z in polar form as $re^{i\theta}$ and hence express $f(z)$ as $U(r,\theta) + iV(r,\theta)$. Show that the Cauchy-Riemann equations (5.3–3) become

$$r\frac{\partial U}{\partial r} = \frac{\partial V}{\partial \theta}, \quad r\frac{\partial V}{\partial r} = -\frac{\partial U}{\partial \theta},$$

and

$$f'(z) = e^{-i\theta}\left[\frac{\partial U}{\partial r} + i\frac{\partial V}{\partial r}\right]$$

$$= \frac{e^{-i\theta}}{r}\left[\frac{\partial V}{\partial \theta} - i\frac{\partial U}{\partial \theta}\right].$$

10. Given two differentiable functions $u(x,y)$, $v(x,y)$ of the real variables x,y, form the complex function $w = u(x,y) + iv(x,y)$. Assume that w is analytic. Prove that, if the substitutions $x = ½(\bar{z} + z)$, $y = \frac{i}{2}(\bar{z} - z)$ are made, the dependence of w upon $\bar{z}$ disappears and w becomes a function of z alone.

11. (a) Demonstrate that the exponential function e^z is zero-free for all finite z.

(b) Find all the zeros of $f(z) = \cosh z - i$.

12. Let $f(z) = 1/(1 + z^2)$ and denote by C and C' the two circular arcs centered on $z = i$ beginning at the origin $z = 0$, ending at $z = 2i$ and lying in the half-planes Re $z \geq 0$, Re $z \leq 0$, respectively.

(a) Use Property 1 of Section 5.4 to show that

$$\left| \int_{0 \atop C}^{2i} \frac{dz}{1 + z^2} \right| \leq \pi.$$

(b) Evaluate $\displaystyle\int_0^{2i} \frac{dz}{1 + z^2}$ along both of the contours C and C'.

13. Let C be the closed circular contour $|z - z_0| = \rho$ (z_0 and $\rho > 0$ arbitrary). Verify that

$$\oint_C (z - z_0)^n dz = 0$$

for all integers $n \neq -1$.

14. Establish the two Corollaries of the Cauchy-Goursat Theorem of Section 5.4.

15. (a) Find all possible values of the integral

$$\int_{0 \atop C}^{1} \frac{dz}{2 - z}$$

where C is a contour joining the origin $z = 0$ to $z = 1$. Are there any restrictions that must be placed on the choice of C?
(b) Repeat part (a) for the function $1/(2 - z)^2$.

16. Let C_1, C_2, and C_3 be the circles of radius 2 and centered at $z = -2, 0$, and 2, respectively. Evaluate

$$\oint \frac{dz}{1 - z^2}$$

along each of these closed contours.

★17. State and prove a Minimum Modulus Principle for nonconstant analytic functions.

18. Assume $f(z)$ is an entire function which for all sufficiently large z satisfies the inequality

$$|f(z)| \leq M|z|^\alpha$$

where M and α are positive constants. Show that $f(z)$ is a polynomial of degree $n \leq \alpha$.

★19. Use Liouville's Theorem to extend the result of Problem 11, Chapter 3 to the case of coefficients that are analytic throughout the entire complex plane.

Section 5.5

20. Employ the Argument Principle to determine the number of right half-plane zeros of
    (a) $Q(s) = s^4 + s^3 + s^2 + 10s + 10$
    (b) $Q(s) = s^4 + 3s^3 + 5s^2 + 9s + 6.$

21. The reference transfer function of the satellite attitude control system of Section 5.1 (see also Figure 5.2) is given by (5.2–2). Determine the disturbance transfer function ($\theta_{\text{ref}} = 0$) for this system and verify that both this function and that of (5.2–2) have the same denominators when expressed as rational functions as in (5.2–1).

22. (a) Derive the reference transfer function for the satellite attitude control system of Section 5.1 in the case when Stage 1 of the controller is replaced by an alternative component with transfer function given by (5.5–6).
    (b) Show that this new control system is strictly stable if $K_\theta K_\tau k\alpha > \beta I\gamma > 0$.

23. Consider a single-loop linear feedback system as depicted in Figure 5.3(c) on p. 132. Assume the product transfer function FG has the form

$$F(s)G(s) = \frac{k}{(s+1)(s+2)}$$

where k is a real constant.
    (a) Sketch the Nyquist diagram for this product function and employ the Nyquist criterion to show that the overall system is strictly stable if $k > -2$.
    (b) Derive the result in alternative fashion by using the Argument Principle directly to ascertain the position of the zeros of $1 + FG$.

24. (a) In the manner of part(a) of Problem 23, determine the values of the real parameter k for which the linear feedback system with product transfer function

$$F(s)G(s) = \frac{k}{(s+1)(s+2)(s+3)}$$

is strictly stable.
    (b) Use Hurwitz determinants to obtain the same result. (Recall Theorem 2 of Section 5.2 on p. 134.)

6

Conformal mapping

6.1 The kinematics of Euler/potential flows

In Chapter 2 we discussed one aspect of the fascinating field of fluid mechanics. In this section we begin consideration of another important area of that applications-rich subject. Our current interest is in fluid flows whose vector velocities can be described in terms of velocity potentials satisfying Laplace's equation. Such *potential flows* can be conveniently studied in two dimensions using analytic function theory.

Our starting point is the assumption that the motion of the fluid is characterized by a vector velocity $\mathbf{v}(\mathbf{x};t)$ satisfying Euler's equation

$$\frac{\partial \mathbf{v}}{\partial t} + (\mathbf{v} \cdot \nabla)\mathbf{v} = -\frac{1}{\rho} \nabla p. \qquad \textbf{(6.1–1)}$$

(We encountered the one-dimensional version of this equation earlier as (2.1–3).) In this expression $p(\mathbf{x};t)$ is the pressure at the point $\mathbf{x}$ within the fluid and ρ is the fluid density. Accompanying (6.1–1), which expresses conservation of momentum in the absence of external forces, is a "continuity" equation. This equation reflects the conservation of mass and, consequent to the Divergence Theorem, can be given the form

$$\frac{\partial \rho}{\partial t} + \operatorname{div}(\rho \mathbf{v}) = 0. \qquad \textbf{(6.1–2)}$$

For incompressible fluids this expression reduces to

$$\operatorname{div} \mathbf{v} = 0. \qquad \textbf{(6.1–3)}$$

In flow problems it is convenient to associate with a closed contour C the line integral

$$\Gamma(C) = \oint_C \mathbf{v} \cdot \mathbf{ds},$$

termed the circulation around C. In view of *Stokes' Theorem**, the circulation around C can be expressed alternatively as the surface integral of the normal component of the curl of the vector velocity. If the circulation is zero for every closed contour C bounding a surface lying entirely within the flow field, then the fluid motion is said to be *curl-free* or *irrotational*. For such flows it is easy to show that there exists a scalar function $\phi(\mathbf{x};t)$, called the *velocity potential*, with the property that

$$\mathbf{v} = \text{grad}\ \phi. \tag{6.1–4}$$

Equation (6.1–3) implies that ϕ satisfies Laplace's equation and hence, if suitably smooth, is harmonic.

For potential flows, the earlier equation of motion (6.1–1) may be rewritten, using a well-known vector identity†, as

$$\frac{\partial \phi}{\partial t} + \tfrac{1}{2}|\nabla\phi|^2 + \frac{p}{\rho} = \text{constant}. \tag{6.1–5}$$

Here the function of time t obtained upon integration over the space variables has been incorporated into the pressure p, since it changes nothing in the flow. Equation (6.1–5) is *Bernoulli's equation*; it determines the pressure p in terms of the velocity potential ϕ.

Two-dimensional potential flows

Since frictional forces have been neglected, the idealized model of fluid flow inherent in (6.1–3), (6.1–4), (6.1–5) is not without shortcomings. Nevertheless, consideration of specific potential flows can often be of substantial qualitative (and sometimes even quantitative) value in given practical situations. Low speed fluid flow around smooth airfoils is one case in point, particularly if the motion is steady. In the rest of this chapter we therefore drop any explicit dependence upon the time t.

In two dimensions it is customary to denote the x and y components of the vector velocity $\mathbf{v}$, by u and v respectively. Equation (6.1–3) then assumes the form

$$u_x + v_y = 0.$$

An implication of this relation is that there exists a function ψ whose partial derivatives satisfy

$$\frac{\partial \psi}{\partial y} = u \qquad \text{and} \qquad \frac{\partial \psi}{\partial x} = -v.$$

---

*Stokes' Theorem (see Wylie (1975), p. 678, for example) asserts that

$$\oint_C \mathbf{F}\cdot d\mathbf{s} = \iint_R \mathbf{e}_n \cdot (\text{curl}\ \mathbf{F})\, dS\,,$$

where $\mathbf{e}_n$ denotes the appropriate unit normal to any surface R bounded by C (perhaps nonplanar in three dimensions) and dS is the area element on R. Compare with the Divergence Theorem (5.4–4).
†$(\mathbf{v}\cdot\nabla)\mathbf{v} = \tfrac{1}{2}\nabla(\mathbf{v}\cdot\mathbf{v}) - \mathbf{v}\times \text{curl}\ \mathbf{v}$.

The function ψ is called the *stream function*, and curves along which ψ = constant are called *streamlines*. The streamlines are tangent to the velocity field since along each of them

$$\mathbf{v} \cdot \left(\frac{\partial \psi}{\partial x}, \frac{\partial \psi}{\partial y} \right) = 0.$$

Therefore, the fluid actually flows along these curves where ψ is constant.

In the case of potential flows, owing to (6.1–4), the velocity potential ϕ and the stream function ψ are related by the Cauchy-Riemann equations (5.3–3), that is, $\partial\phi/\partial x = \partial\psi/\partial y$ and $\partial\phi/\partial y = -\partial\psi/\partial x$. The functions ϕ and ψ are thus the real and imaginary parts, respectively, of an analytic function

$$w(z) = \phi(x,y) + i\,\psi(x,y) \qquad \textbf{(6.1–6)}$$

of the complex variable $z = x + iy$. The function $w(z)$ is often termed the *complex (velocity) potential*. Its derivative satisfies

$$\begin{aligned} \frac{dw}{dz} &= \frac{\partial \phi}{\partial x} - i\frac{\partial \phi}{\partial y} \\ &= u - iv; \end{aligned}$$

that is, its complex conjugate is the vector velocity written in complex notation. Reversing these steps, if the complex velocity function

$$u + iv \equiv qe^{i\theta} \text{ with } q = (u^2 + v^2)^{1/2}$$

is known in terms of ϕ and ψ, then formally

$$z = \int^{w} \left(\frac{1}{\zeta} \right) dw \qquad \textbf{(6.1–7)}$$

where

$$\zeta \equiv u - iv = qe^{-i\theta}. \qquad \textbf{(6.1–8)*}$$

These last two expressions relate the coordinates of physical space to the components of the vector velocity and will be particularly useful in the analyses we will eventually carry out in this chapter.

An important feature of the flows we are considering is that the region where the fluid motion takes place, the *flow field*, is bounded by level curves of ψ. These streamlines ensue from two sources. Some are associated with "fixed" boundaries occasioned by obstacles such as plates, cylinders, wings (or, more precisely their two-dimensional equivalents). Others are the streamlines arising from the "free" boundaries that occur at fluid interfaces, as in cavity flows and jets. Along those latter streamlines it is customary to assume that the speed $q = |\nabla\phi|$ remains constant.[†]

---

*Historically ζ has often been used to designate the *reciprocal* of the complex velocity function rather than its conjugate.

[†]The reasoning is as follows: in steady flow the pressure must be continuous across free boundaries. For wakes the fluid on the "outside" of the dividing streamline is at rest. At a liquid-gas interface, the "external" density is negligible in comparison with the "internal" density. In either case, Bernoulli's equation (6.1–5) suggests that the assumption q = constant is appropriate.

6.2 Simple examples of ideal plane flows

Example The most elementary example of a two-dimensional potential flow is provided by the complex potential $w = U_0 z$, where U_0 is a positive constant. For this case

$$\phi = U_0 x, \quad \psi = U_0 y \qquad \text{and} \qquad u = U_0, \quad v = 0.$$

As illustrated in Figure 6.1, all the streamlines are lines y = constant. The flow is uniform, parallel to the x-axis, and has constant speed U_0. □

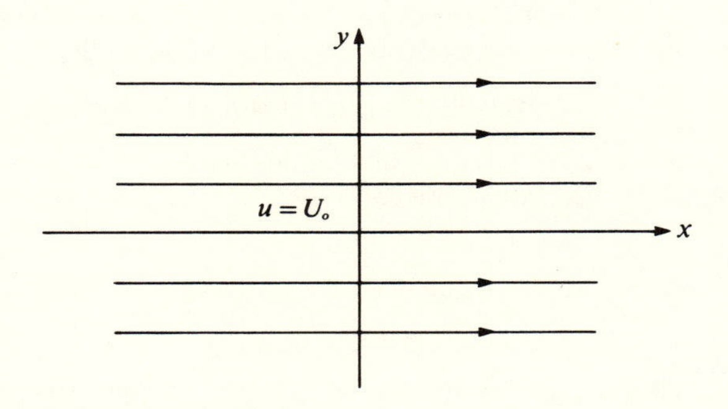

Figure 6.1 Uniform flow with speed U_0

Example Somewhat more generally, if $w = U_0 z^\beta$ with $U_0 > 0$ and $\beta > 1/2$, we find

$$\phi = U_0 |z|^\beta \cos(\beta \arg z), \qquad \psi = U_0 |z|^\beta \sin(\beta \arg z)$$

and

$$u = \beta U_0 |z|^{\beta-1} \cos[(\beta - 1)\arg z], \qquad v = -\beta U_0 |z|^{\beta-1} \sin[(\beta - 1)\arg z].$$

Here the flow is "between" two walls making an angle $\alpha = \pi/\beta$ with each other (see Figure 6.2 where flow fields for two representative cases are depicted. A

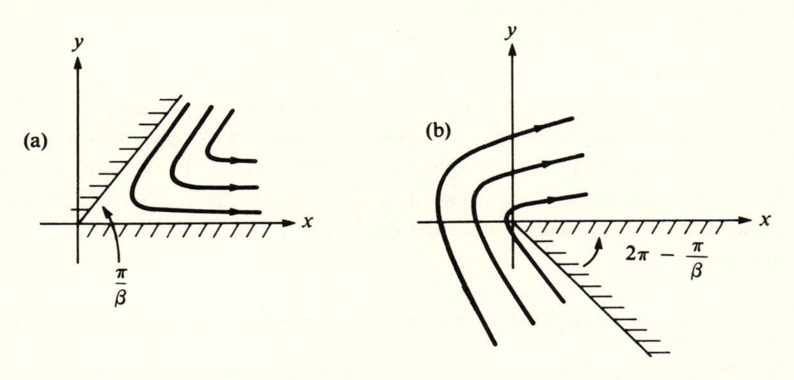

Figure 6.2 Potential flow "between" rigid walls (a) $\beta \geq 1$ (b) $1 > \beta > ½$

different, but obviously related, flow in the exterior of these infinite "wedges" is considered in Problem 1). It should be noted that when $\beta > 1$, the fluid velocity vanishes at $z = 0$; the origin is thus a *stagnation point* in this case. On the other hand, the velocity is unbounded at the origin if $\beta < 1$, that is, if the wedge corner is convex. □

Two other intuitively appealing potentials are provided by the logarithm function and the dipole function $(z/a + a/z)$. A particularly interesting flow occurs when we combine them (see Shinbrot (1973), Meyer (1971), Vallentine (1967), or the classic Milne-Thomson (1968)).

Example Consider the complex potential

$$w(z) = U_0 a \left(\frac{z}{a} + \frac{a}{z}\right) + \frac{Ai}{2\pi} \ln \frac{z}{a} \tag{6.2–1}$$

with a and U_0 positive, A real. An easy calculation determines the stream function

$$\psi = U_0 y \left(1 - \left|\frac{a}{z}\right|^2\right) + \frac{A}{2\pi} \ln \left|\frac{z}{a}\right|;$$

note that the circle $z = a$ is the streamline $\psi = 0$. We also observe that

$$\zeta = U_0 \left(1 - \frac{a^2}{z^2}\right) + \frac{Ai}{2\pi z},$$

which implies $\lim_{z\to\infty} \zeta = U_0$. Therefore, if $A = 0$, the fluid motion given by (6.2–1) is symmetric flow past a circle (circular cylinder) of radius a (see Figure 6.3(a)).

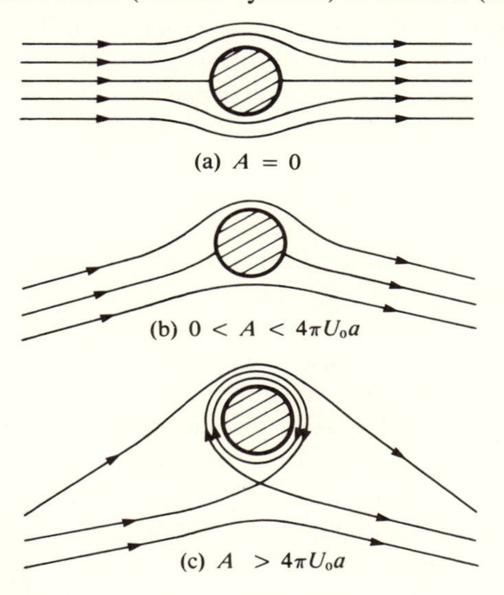

Figure 6.3 Potential flow, with and without circulation, past a circular cylinder

For nonvanishing A, there is circulation of the fluid around the cylinder. Indeed, for any closed contour C surrounding the circle $|z| = a$ we find

$$\begin{aligned}\Gamma(C) &= \oint_C \nabla\phi \cdot \mathbf{ds} = \oint_C (u\,dx + v\,dy) = \text{Re}\oint_C \zeta\,dz \\ &= \text{Re}\oint_{|z|=a} \zeta\,dz = \text{Re}\oint_{|z|=a}\left[U_0\left(1 - \frac{a^2}{z^2}\right) + \frac{Ai}{2\pi z}\right]dz \\ &= -A.\end{aligned}$$

(Here Re stands for the real part of the expression that follows; we have also used Corollary 1 of the Cauchy-Goursat Theorem of Section 5.4.) The stagnation points, at $z = \pm a$ when $A = 0$, are now to be found at

$$z = -\frac{Ai}{4\pi U_0} \pm \frac{1}{2}\left[4a^2 - \left(\frac{A}{2\pi U_0}\right)^2\right]^{1/2}.$$

If $|A| < 4\pi U_0 a$, they are symmetrically situated on the circle $|z| = a$. For larger values of $|A|$, the two points of vanishing velocity lie on the imaginary axis; only one of them, however, is in the flow field. The flow pattern associated with this complex velocity potential is sketched in Figure 6.3 on p. 161 for three illustrative values of the parameter A. □

Cavitation

The above are examples of flows of incompressible ideal fluids along and around various simple obstacles. In order to describe the flow of an ideal fluid past a more general profile P, we need a complex velocity potential $w(z) = \phi + i\psi$ that satisfies at least the following three conditions:

(i) $w(z)$ is analytic in the exterior of the profile P;
(ii) $\zeta = dw/dz = qe^{-i\theta}$ is singlevalued* in the exterior of P and approaches a given (constant) value as $|z| \to \infty$;
(iii) Im w = constant on P, that is, P is a streamline.

It is possible to show that for "reasonable" profiles of finite extent, complex velocity potentials exist that satisfy these assumptions (see Shinbrot (1973) for details). In fact, as the potential function with the free parameter A given by (6.2–1) suggests, there can be a host of permissible potentials associated with a particular profile. The problem is not one of existence, therefore, it is one of uniqueness.

If a given profile P is smooth, the additional assumption that $w'(z)$ is continuous *on* P, that is, u and v are continuous, assures uniqueness. If P lacks smoothness at even a single point, however, a condition of a different sort is needed. In 1910 Nikolai Joukowski (1847–1921) suggested a possible way out of this dilemma for wing profiles, that is, bodies shaped as in Figure 6.4, which are smooth

*$w(z)$ itself may not be single-valued and, indeed, cannot be when circulation is present.

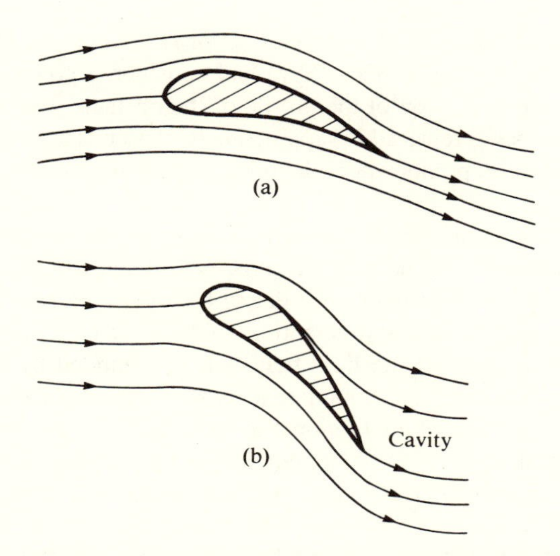

Figure 6.4 (a) Potential flow past an airfoil with K-J condition satisfied
(b) Similar airfoil at high angle-of-attack with streamline detachment

except for a single sharp downstream corner or cusp. His auxiliary hypothesis is often called the

Kutta-Joukowski Condition*:

At the trailing edge of a wing profile, the velocity is bounded.

What the K-J condition does, in essence, is sort out from amongst all theoretically allowable flows past the wing profile, that one for which the associated velocity *everywhere* along P is finite. In practice the condition can be applied by adjusting the circulation around P (see Problem 3), and for thin airfoils at small angles of attack this "correct" value of circulation leads to lift forces that agree remarkably well with measured values.

The inherent attractiveness of the K-J condition notwithstanding, the hypothesis has no theoretical justification. Moreover, the imposition of the K-J condition in combination with hypothesis (iii) on p. 162 ceases to appear so reasonable for high angles of attack. More plausible is the assumption that the flow detaches from the fixed boundary and creates a cavity (see Figure 6.4(b)). (Recall that we earlier suggested this possibility in our discussion of free streamlines.) Historically, the concept of free-boundaries actually predates Joukowski's work with airfoils. Already in the late 1860's, in attempts to overcome some of the

*Working independently, both the Russian Joukowski and the German M. Wilhelm Kutta (1867–1944) (perhaps known best for his contributions to the theory of numerical solution of differential equations), were concerned with the relationships between lifting force and circulation. This condition, as well as a companion result that in two-dimensional flows around finite bodies the lift is independent of the specific shape of the body, is usually credited to both researchers.

inconsistencies of classical fluid dynamics, both Hermann von Helmholtz (1821–1894) and Gustav Kirchhoff (1824–1887) had considered potential flows past obstacles in which only a portion of the profile of the obstacle was taken as a streamline, the remainder being replaced by appropriately chosen free boundaries separating from the profile at those points where the body form makes an abrupt (downstream) turn.

In our further considerations we will allow for the presence of free boundaries in this natural manner and analyze some significant potential flows involving both cavitation and jets. However, in order to be able to carry out our analyses in closed form, we will restrict our attention to flow past *rectilinear* profiles (see Figure 6.5). This means that the flow field will be bounded by streamlines along which either q = constant (free boundaries) or θ = constant (fixed boundaries). It follows that the region of fluid motion for such flows is readily depicted in the ζ-plane (recall equation (6.1–8)). With equal ease we can describe the flow field in the w-plane. If a functional relationship between these two regions can be found, then, using (6.1–7), we can relate these results to the desired profile(s) in physical space (the z-plane) and hence determine the underlying complex velocity potential $w(z)$ for the problem.

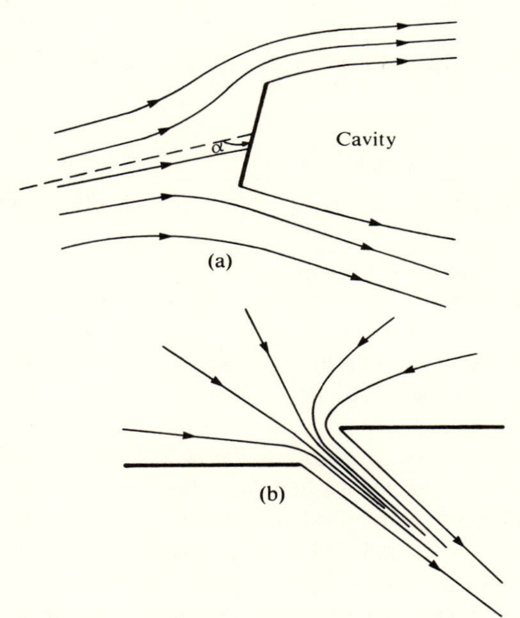

Figure 6.5 Cavity (a) and jet (b) flows associated with motion of an ideal fluid past flat plates

Since the ζ-plane is commonly termed the *hodograph plane*, the procedure outlined above is called the *hodograph method*. Transformations involving analytic functions and so-called conformal mappings, particularly of the Schwarz-Christoffel variety, play an essential role in these applied analyses. Hence, before applying the hodograph method to specific configurations of practical interest, we devote the next several sections to a careful consideration of general analytic transformations and these special conformal mappings.

6.3 The nature of analytic transformations

Basic results

Analytic functions $w = f(z)$ give rise to transformations (mappings) with special properties. To deduce one renowned result we recall from (5.3–1) that

$$f'(z_0) \equiv \lim_{\Delta z \to 0} \frac{\Delta w}{\Delta z} \tag{6.3–1}$$

where $\Delta w \equiv f(z_0 + \Delta z) - f(z_0)$. It follows that

$$\lim_{\Delta z \to 0} \left| \frac{\Delta w}{\Delta z} \right| = |f'(z_0)| \tag{6.3–2}$$

and, if $f'(z_0) \neq 0$,

$$\lim_{\Delta z \to 0} \arg \left(\frac{\Delta w}{\Delta z} \right) = \arg f'(z_0) \tag{6.3–3}$$

(the fundamental range is implied for all angles). Letting z approach z_0 along a smooth curve C through z_0 will lead to a corresponding approach of w to $f(z_0)$ along the image curve $f(C)$. Moreover, the inclinations of the tangent lines to C and $f(C)$ will be related by (6.3–3). More specifically, if $z \to z_0$ along C

$$\lim_{\Delta z \to 0} \arg \Delta w = \lim_{\Delta z \to 0} \arg \Delta z + \arg f'(z_0).$$

In summary, we have

> Property 1
> Let a smooth curve C pass through a point z_0 in the complex z-plane. If a complex function $f(z)$ is analytic at z_0 and $f'(z_0) \neq 0$, then the directed tangent to the curve C at z_0 is rotated through the angle $\arg f'(z_0)$ by the transformation $w = f(z)$.

Since the angle of rotation is the same for all smooth curves through z_0, being determined solely by the specific function $f(z)$ and the point z_0, these transformations preserve the angles *between* smooth curves intersecting at a point. Transformations that are angle-preserving in both magnitude and sense are said to be *conformal*.* Therefore, in view of Property 1, we have

*If merely the magnitude of the angle is preserved, the transformation is called *isogonal*. The mapping $w = \bar{z}$ is a simple example of an isogonal transformation that is not conformal.

Theorem 1:

At each point where the complex function $f(z)$ is analytic and $f'(z) \neq 0$, the mapping $w = f(z)$ is conformal.

An important application of Theorem 1 is provided by considering families of orthogonal curves.*

Example The preservation of orthogonality can be nicely illustrated with the function $w = f(z) = z^2$. In this case the straight lines $x = a$ and $y = b$ are transformed into the orthogonal families of parabolas

$$v^2 = 4a^2(a^2 - u) \qquad \text{and} \qquad v^2 = 4b^2(b^2 + u),$$

respectively. (Note that the y-axis maps into the negative half of the u-axis, the x-axis into the positive half and the origin is a point of nonorthogonality.) The inverse mapping $z = \sqrt{w}$ is likewise conformal except, of course, at the origin. Here the lines $u = c$ and $v = d$ are mapped into the rectangular hyperbolas

$$x^2 - y^2 = c, \qquad 2xy = d.$$

□

One's appreciation of the nature of a regular analytic transformation is enhanced when he or she is able to view the mapping in both directions, that is, with w as a function of z and conversely, with z as a function of w. The following theorem gives the fundamental theoretical result in this regard:

Theorem 2:

If the complex function $f(z)$ is analytic, the transformation $w = f(z)$ will have a single-valued inverse (denoted by $z = f^{-1}(w)$) in the neighborhood of any regular point where $f'(z) \neq 0$. Moreover,

$$\frac{df^{-1}(w)}{dw} = \frac{1}{f'(z)},$$

and thus the *local* inverse function is analytic.

Proof: The essential step in verifying this result involves recognizing that the pair of relations $u = u(x,y)$ and $v = v(x,y)$ can be solved uniquely for $x = x(u,v)$ and $y = y(u,v)$ in the neighborhood of any point where the Jacobian J of the transformation is nonvanishing. But, using the Cauchy-Riemann equations (5.3–3),

$$J\begin{pmatrix} u & v \\ x & y \end{pmatrix} = \begin{vmatrix} \dfrac{\partial u}{\partial x} & \dfrac{\partial u}{\partial y} \\[2ex] \dfrac{\partial v}{\partial x} & \dfrac{\partial v}{\partial y} \end{vmatrix} = \left(\frac{\partial u}{\partial x}\right)^2 + \left(\frac{\partial v}{\partial x}\right)^2$$

*Recall that in most of the commonly employed coordinate systems, the coordinate curves (surfaces) of one family are orthogonal to the coordinate curves (surfaces) of other families.

$$= \left| \frac{\partial u}{\partial x} + i \frac{\partial v}{\partial x} \right|^2 = |f'(z)|^2. \qquad \textbf{(6.3–4)}$$

It should be clear from the discussion thus far that in our consideration of the analytic transformation $w = f(z)$, the points where $f'(z)$ vanishes are exceptional. Such points are generally termed *critical points*. In the neighborhood of such a point, the behavior of the mapping function can be analyzed as follows. Let n be the order of the zero of $f'(z)$ at a given critical point z_0 so that

$$f'(z) \equiv (z - z_0)^n g(z)$$

where $g(z)$ is analytic in the neighborhood of z_0 and $g(z_0) \neq 0$. Then by virtue of the Corollary of Theorem 1 of Section 5.4

$$f(z) - f(z_0) = \int_{z_0}^{z} (s - z_0)^n g(s)ds$$

$$= \frac{(z - z_0)^{n+1}}{n + 1} g(z) - \int_{z_0}^{z} \frac{(s - z_0)^{n+1}}{n + 1} g'(s)ds.$$

This implies that

$$\lim_{z \to z_0} \frac{f(z) - f(z_0)}{(z - z_0)^{n+1}} = \frac{g(z_0)}{n + 1} - \lim_{z \to z_0} \int_{z_0}^{z} \left(\frac{s - z_0}{z - z_0} \right)^{n+1} \frac{g'(s)ds}{(n + 1)}$$

$$= \frac{g(z_0)}{n + 1}, \qquad \textbf{(6.3–5)}$$

since the limit of the integral on the right-hand side of this last relation vanishes by Property 1 of Section 5.4. Therefore, as a complement to (6.3–1), (6.3–3), and our earlier Property 1 of this section, we find that

$$\lim_{\Delta z \to 0} \arg \Delta w = (n + 1) \lim_{\Delta z \to 0} \arg \Delta z + \arg g(z_0)$$

and the following:

> Property 2
> If a complex function $f(z)$ is analytic at a point z_0 where $f'(z)$ has an nth order zero, then under the transformation $w = f(z)$, the angle at z_0 between any two smooth curves passing through this point is multiplied by the factor $(n + 1)$.

The case $n = 0$ of this property is merely Theorem 1 on p. 166.

Transformation of harmonic functions

A twice continuously differentiable function φ that satisfies Laplace's equation

$$\Delta \varphi \equiv \text{div}(\text{grad } \varphi) = 0$$

is termed *harmonic*. Fundamental to our earlier analysis is the fact that, as a consequence of the Cauchy-Riemann equations, the real and imaginary parts of analytic functions are each harmonic. In this way, two-dimensional harmonic function theory and complex analysis are inseparably interwoven. However, exploiting this relationship proves to be less fruitful for our purposes than does considering the way in which a harmonic function behaves when its independent variables are subjected to an analytic transformation, that is, considering the result of composing harmonic functions and analytic functions. The key fact is the following:

Theorem 3:

If $w = f(z)$ is an analytic function, in the neighborhood of a point z_0 where $f'(z_0) \neq 0$ the analytic local inverse function $z = f^{-1}(w)$ transforms an arbitrary harmonic function of x and y into a harmonic function of u and v.

Proof: If $\Phi(u,v) \equiv \varphi[x(u,v),y(u,v)]$, then a simple calculation shows

$$\Delta_{x,y}\varphi = |f'(z)|^2 \, \Delta_{u,v}\Phi,$$

and the desired conclusion clearly ensues. □

Also of importance to our considerations is

Theorem 4:

Under the conditions of Theorem 3,

(i) curves along which φ = constant are transformed into curves along which Φ = constant;

(ii) curves along which $\dfrac{\partial\varphi}{\partial n} = 0$ are transformed into curves along which $\dfrac{\partial\Phi}{\partial n} = 0$;*

(iii) data along other curves is transformed appropriately; in particular,

$$\frac{\partial\varphi}{\partial n_z} = |f'(z)| \frac{\partial\Phi}{\partial n_w}.$$

Proof: Conclusion (i) is obvious; (ii) is a consequence of the fact that curves along which $\partial\rho/\partial n = 0$ are orthogonal to curves along which φ = constant. For (iii) see Problem 5. □

The two results we have just stated engender the solutions to some very complicated boundary-value problems for Laplace's equation in two-dimensions that arise not only in fluid dynamics, but also in electrostatics, heat conduction, and other fields where equilibrium phenomena occur. In many cases, the boundary-

---

* $\dfrac{\partial\varphi}{\partial n}$ designates the derivative in the direction normal to the curve.

value problem is not amenable to forthright attack. However, if a conformal transformation (or, equivalently, a series of such transformations) can be found that maps the domain in which the solution is sought onto a domain for which the corresponding boundary-value problem can be solved, then the solution in the new domain can be pulled back to the original domain by means of the inverse transformation. We now illustrate this powerful technique.

Example We wish to find the solution to the so-called "mixed" boundary-value problem

$$\Delta\varphi = 0 \qquad x > 0, \quad y > 0$$

where

$$\frac{\partial\varphi(x,0)}{\partial y} = 0 \quad \text{for } 0 < x < 1,$$

$$\varphi(x,0) = 0 \quad \text{for } x > 1,$$

and

$$\frac{\partial\varphi(0,y)}{\partial x} = 0 \quad \text{for } 0 < y < 1,$$

$$\varphi(0,y) = 1 \quad \text{for } y > 1.$$

The transformation $w = \sin^{-1} z^2$ maps the domain in question (the first quadrant of the z-plane) conformally onto the semi-infinite strip $|u| < \pi/2$, $v > 0$.* The transformed boundary-value problem is

$$\Delta\Phi = 0 \qquad -\pi/2 < u < \pi/2, \quad v > 0$$

with

$$\Phi(-\pi/2,v) = 1, \qquad \Phi(\pi/2,v) = 0, \qquad v > 0$$

$$\frac{\partial\Phi(u,0)}{\partial v} = 0, \qquad -\pi/2 < u < \pi/2.$$

Since this latter problem has the obvious solution

$$\Phi(u,v) = \frac{1}{2} - \frac{u}{\pi},$$

it follows that

$$\varphi(x,y) = \frac{1}{2} - \frac{1}{\pi}\operatorname{Re}(\sin^{-1} z^2) \tag{6.3–6}$$

$$= \frac{1}{2} - \frac{1}{\pi}\sin^{-1}\frac{1}{2}\left[\sqrt{(x^2 - y^2 + 1)^2 + 4x^2y^2} - \sqrt{(x^2 - y^2 - 1)^2 + 4x^2y^2}\right] \tag{6.3–7}$$

solves the original problem. □

*The conformal mapping appropriate for this problem can be "discovered" using the procedure of Schwarz and Christoffel (see Section 6.4).

Similar techniques will serve us well in the future. It will be helpful, therefore, to have a suitable repertory of basic conformal mappings at our disposal.

The linear fractional transformation

We conclude this section with a brief discussion of the simplest, yet very useful, class of mapping functions: *the linear fractional transformations* (or Möbius transformations, named after the 19th century German geometer). These mappings have the form

$$w = \frac{az + b}{cz + d} \tag{6.3–8}$$

where $a,b,c,$and d are arbitrary complex constants satisfying $ad - bc \neq 0$. (This proviso prevents the transformation from reducing to the trivial function $w =$ constant.) Since

$$\frac{dw}{dz} = \frac{ad - bc}{(cz + d)^2}$$

never vanishes, the linear fractional transformation is conformal except for $z = -d/c$; it is also one-to-one. It is convenient to associate the "value" $w = \infty$ with $z = -d/c$ and the value $w = a/c$ with $z = \infty$. When $c = 0$ we agree that $w = \infty$ corresponds to $z = \infty$.*

The bilinear transformation (6.3–8) includes three special cases:

(i) the *translation* $w = z + \lambda$,

(ii) the *rotation/dilation* $w = \mu z$, and

(iii) the *inversion* $w = 1/z$.

The general transformation is actually a composition of transformations of these transparent types, a fact that one can often exploit when investigating the characteristics of the function (6.3–8). Indeed, if $c \neq 0$,

$$w = \frac{bc - ad}{c^2}\left(\frac{1}{z + d/c}\right) + \frac{a}{c},$$

while if $c = 0$, d cannot vanish, and

$$w = \frac{a}{d}z + \frac{b}{d}.$$

Since circles are transformed into circles by each of these special mappings (straight lines counting as circles of arbitrarily large radius; see Problem 9), it follows that

---

*In complex analysis it is customary to accord the symbol ∞ many of the characteristics of other complex numbers. Geometrically we envisage the "point" at ∞ as lying at the ends of every line in the plane, while analytically we combine it in the natural way with complex numbers. In particular, $z/\infty = 0$ and $z/0 = \infty$ if $z \neq 0$.

Theorem 5:

If $ad - bc \neq 0$, the general linear fractional transformation

$$w = \frac{az + b}{cz + d}$$

maps (generalized) circles onto (generalized) circles.

A distinguishing, and easily verified, characteristic of linear fractional mappings is given by (see Problem 10):

Property 3
Under a linear fractional transformation, the cross ratio $[(z_1 - z_2)(z_3 - z_4)] / [(z_1 - z_4)(z_3 - z_2)]$ of any four points is invariant.

Since the general transformation (6.3–8) actually depends on only three constants (the ratios of any 3 of a, b, c, and d to the fourth), specifying the images of any three distinct points determines a unique linear fractional mapping. Property 3 with $z_4 = z$, $w_4 = w$ can then be used advantageously to ascertain the transformation.

Example If we seek the linear fractional mapping satisfying

$$w(i) = 0, \qquad w(-i) = \infty, \qquad \text{and} \qquad w(0) = -i,$$

then, using the invariance of the cross ratio *in formal fashion*, we find

$$\frac{(0 - \infty)(-i - w)}{(0 - w)(-i - \infty)} = \frac{(i + i)(0 - z)}{(i - z)(0 + i)} .$$

Upon simplifying, we obtain

$$\frac{i + w}{w} = \frac{2z}{z - i},$$

or

$$w = i\frac{z - i}{z + i} . \qquad \textbf{(6.3–9)}$$

For later reference, we note that this particular analytic function conformally maps the upper half-plane $\text{Im } z > 0$ onto the interior of the unit circle $|w| < 1$, and takes the point $z = i$ into the origin $w = 0$ with $w'(i) = ½$. □

6.4 The Schwarz-Christoffel transformation

The transformation given by (6.3–9) is just one example of a wide class of conformal mappings that take simply-connected regions onto the interior of the unit circle.

Indeed, the existence of such transformations is guaranteed by the

Riemann Mapping Theorem:

Let R be a simply-connected region in the z-plane. Either R is the entire complex plane or there exists an analytic function $w = f(z)$ that maps R one-to-one onto the interior of the unit circle $|w| < 1$. The mapping is uniquely determined by specifying that $f(z_0) = 0$ and $f'(z_0) > 0$ for some given point z_0 in R.

(See Henrici (1974), p. 380; Silverman (1974), p. 199.) Although nontrivial to establish, it is at least intuitively clear that for a sufficiently smooth boundary C to the region R, the mapping function assured by this important theorem takes C onto $|w| = 1$. Therefore, when either analyzing given transformations or generating new ones for specified domains, one must take care to ensure that not only the interiors but also the boundaries coincide appropriately.

Example Both of the functions

$$\text{(I)}\ w = e^z$$

and

$$\text{(II)}\ w = \frac{1 + \sinh(z/2)}{1 - \sinh(z/2)}$$

are regular analytic in the semi-infinite strip $|y| < \pi$, $x < 0$ and map this simply-connected region one-to-one *into* the unit circle $|w| < 1$ (see Figure 6.6). Transformation I, however, does not provide an *onto* mapping. Moreover, only a portion of the boundary ($x = 0$, $|y| \leq \pi$) is mapped onto the circumference of the circle $|w| = 1$; the remainder ($|y| = \pi$, $x \leq 0$) is mapped into that part of the negative real-axis ($v = 0$, $-1 \leq u < 0$) *interior* to the unit circle. Transformation (II) does not exhibit such features; the mapping is onto and the boundary correspondence is complete. □

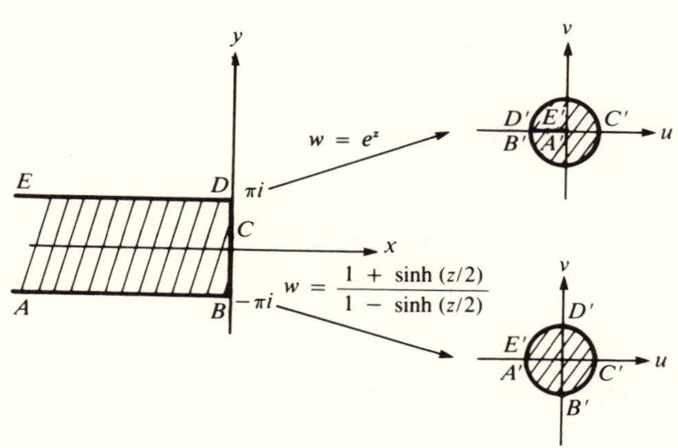

Figure 6.6 Mappings of a semi-infinite strip into the unit circle

In the remainder of the chapter our primary interest is in the mapping of simply-connected regions which, like the one just considered, are bounded by a finite number of line segments. Fortunately there is a well-developed theory, due originally to H. A. Schwarz (1843–1921) and E. B. Christoffel (1829–1900), for determining mapping functions associated with such regions. For simplicity, we will analyze the transformation in the reverse direction, and seek in rather intuitive fashion to determine mappings of the upper half-plane (which we know can be mapped conformally onto the unit circle by means of a linear fractional transformation; see Problem 12) onto these "polygonal" regions. Initially, we assume that the polygon is closed and, of course, is not self-intersecting; an open polygon can be viewed as a limiting configuration with one or more vertices at ∞. The geometry pertinent to our considerations is displayed in Figure 6.7.

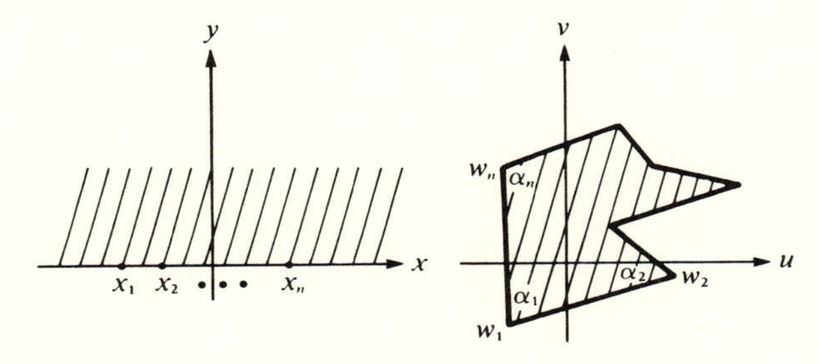

Figure 6.7 Geometry for Schwartz-Christoffel transformations

Let us begin with the simple mapping function

$$w = w_1 + k(z - x_1)^{\alpha_1/\pi}$$

where x_1 is real, k is constant, and $0 < \alpha_1 \leq 2\pi$. This transformation is conformal in the upper half-plane and, provided we restrict $\arg(z - x_1)$ appropriately, one-to-one there also. More important, however, since

$$\frac{dw}{dz} = \frac{\alpha_1}{\pi} k(z - x_1)^{(\alpha_1 - \pi)/\pi}, \tag{6.4–1}$$

the transformation maps every open segment of the x-axis containing x_1 into a *corner* at $w = w_1$ with an *interior* angle α_1.*

The relation (6.4–1) suggests how to generate a single transformation with analogous mapping properties at a finite number of specified points $x_1, x_2, \cdots, x_n$ on the real z-axis. Indeed, the product

---

*Note that from (6.4–1)

$$\lim_{\Delta z \to 0} \arg \Delta w = \lim_{\Delta z \to 0} \arg \Delta z + \frac{\alpha_1 - \pi}{\pi} \lim_{\Delta z \to 0} \arg(z - x_1) + \arg k,$$

so that as z goes through x_1 from left to right, $\arg w$ changes by the amount

$$\frac{\alpha_1 - \pi}{\pi}(-\pi) = \pi - \alpha_1.$$

$$\frac{dw}{dz} = K(z - x_1)^{(\alpha_1 - \pi)/\pi} (z - x_2)^{(\alpha_2 - \pi)/\pi} \cdots (z - x_n)^{(\alpha_n - \pi)/\pi}$$

$$\equiv K \prod_{i=1}^{n} (z - x_i)^{(\alpha_i - \pi)/\pi} \tag{6.4–2}$$

appears to have the desired properties. However, to complete the mapping of the upper half-plane onto the interior of a given polygonal region with interior angles $\alpha_1, \alpha_2, \ldots, \alpha_n$, we have to make sure that not only the angles but also the lengths of the sides of the polygon have the correct values. And this is not an easy task.

Integrating (6.4–2) yields the Schwarz-Christoffel transformation

$$w = K \int_{z_0}^{z} \prod_{i=1}^{n} (z - x_i)^{(\alpha_i - \pi)/\pi} dz + C, \tag{6.4–3}$$

a function which, once z_0 has been selected, has three adjustable parameters: C, $\arg K$, and $|K|$. Thus, if the integral in (6.4–3) maps the x-axis onto a polygon at least *similar* to the desired one, then the constants C and K can be chosen so that the resulting translation, rotation, and dilation make the two polygons coincide. For triangles, once α_1, α_2, and α_3 have been specified, the integral in (6.4–3) automatically does its job, independent of the choice of x_1, x_2, and x_3. For a polygon of n sides, on the other hand, $(n-3)$ additional conditions, beyond the specification of the α_i, are needed to determine the mapping to within similarity. (For example, all rectangles have the same angles.) The upshot of this is that only *three* of the pre-image points on the x-axis associated with corners of the polygon can be assigned arbitrarily;* the conditions of similarity then determine the remaining $(n - 3)$ x_is. Sometimes considerable ingenuity is required to effect the calculation of these x_is and hence to determine completely the transformation (6.4–3).

In practice, the integrals that arise in the Schwarz-Christoffel transformation are often rather complex. Some simplification can be effected by choosing one of the three assignable pre-image points x_i to be the point at ∞. When this is done, the corresponding factor in the representation (6.4–3) is eliminated (see Churchill (1974) or Silverman (1974) for details).

Analogous to the case when one of the x_is is infinite is the situation wherein the image(s) of one (or more) of the x_is, that is, one (or more) of the vertices of the vertices of the polygon, lies at infinity. Here a simple limiting argument suffices to show that the formula (6.4–3) for the Schwarz-Christoffel transformation remains valid. The angle between two rays "with vertex at infinity" merely needs to be interpreted (defined) as the negative of the angle between the two rays at their *finite* point of intersection. (Again Silverman (1974), for example, has the details; so also does Henrici (1974).)

The above discussion has been necessarily brief. Therefore we now treat a number of illustrative examples that should help to clarify these notions.

---

*An identical, but of course deeper, result is valid for simply-connected regions bounded by more general curves; see Silverman (1974), pp. 209 f., for example.

Triangles: regular and degenerate

Example 1 When the polygonal image region is a regular triangle with vertices at the points w_1, w_2, w_3, the transformation (6.4–3) assumes the form

$$w = K\int_{z_0}^{z} (z - x_1)^{(\alpha_1-\pi)/\pi}(z - x_2)^{(\alpha_2-\pi)/\pi}(z - x_3)^{-(\alpha_1+\alpha_2)/\pi}\,dz + C.$$

If x_1, x_2, and x_3 are all chosen to be finite, the integral in this expression does not generally represent an elementary function. If $x_3 = \infty$, the mapping function becomes

$$w = K\int_{z_0}^{z} (z - x_1)^{(\alpha_1-\pi)/\pi}(z - x_2)^{(\alpha_2-\pi)/\pi}\,dz + C. \tag{6.4–4}$$

The integral in this latter equation is an elliptic integral when the triangle is either equilateral or a right triangle with one of its other angles equal to $\pi/4$ or $\pi/3$ (see Problem 16).

Example 2 As suggested in Figure 6.8, the Schwarz-Christoffel transformation of the upper half-plane onto the semi-infinite strip $|u| < \pi/2$, $v < 0$ can be obtained in formal fashion as the limit of mappings onto a sequence of triangles of increasing size. If $x_1 = -1$, $x_2 = 0$, and $x_3 = 1$ with $w_1 = -\pi/2$, $w_2 = \pi/2 - iA$, and $w_3 = \pi/2$, then in the limit $A \to +\infty$, we find (choosing $z_0 = 1$)

$$\begin{aligned} w &= K\int_1^z \frac{dz}{z\sqrt{z^2-1}} + \pi/2 \\ &= K[\sec^{-1}z - \sec^{-1}1] + \pi/2. \end{aligned}$$

Now $\sec^{-1}1 = 0$ and $\sec^{-1}(-1) = \pi$. It follows that $K = -1$, whence

$$w = \pi/2 - \sec^{-1}z$$

or, inverting,

$$z = \frac{1}{\sin w}.$$

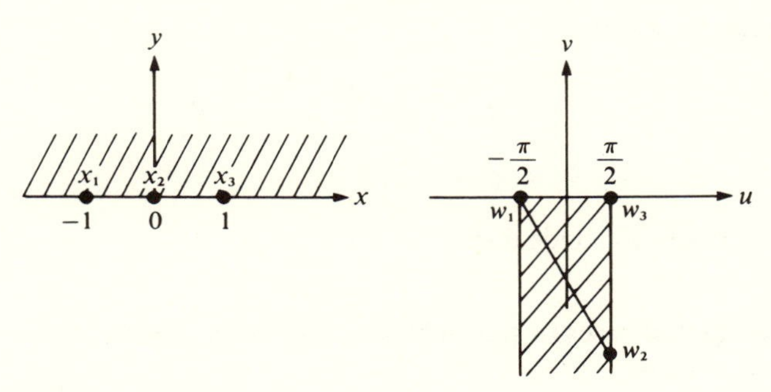

Figure 6.8 Schwarz-Christoffel mapping onto a semi-infinite strip

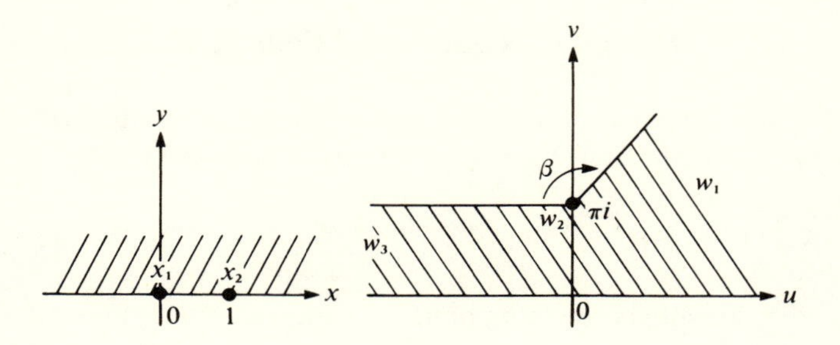

Figure 6.9 Schwarz-Christoffel mapping onto a funnel

Example 3 As a final "triangle" consider the funnel-shaped region of Figure 6.9. If the x_is, "corners," and angles have the values given in the table

i	x_i	w_i	α_i
1	0	∞	$\beta - \pi$
2	1	πi	$2\pi - \beta$
3	∞	∞	0

then the requisite mapping from the upper half-plane onto the interior of this degenerate triangle is given by

$$\begin{aligned} w &= K \int_1^z \left(\frac{z-1}{z}\right)^{(\pi-\beta)/\pi} \frac{dz}{z} + \pi i \\ &= K \int_{1/z}^1 (1 - t)^{(\pi-\beta)/\pi} \frac{dt}{t} + \pi i. \end{aligned} \tag{6.4–5}$$

To determine the constant K, we note that when $z = x > 1$, the real part of w should be negative while the imaginary part should equal π. On the other hand, if $z = x < 0$, Im w should be zero. The first observation leads to $K < 0$. Utilizing the second observation, we find

$$\begin{aligned} 0 = \operatorname{Im} w(x) &= \pi + K \operatorname{Im} \int_{-1/|x|}^1 (1 - t)^{(\pi-\beta)/\pi} \frac{dt}{t} \\ &= \pi + K\pi, \end{aligned}$$

from which we easily conclude that $K = -1$.

Three special cases of this Schwarz-Christoffel transformation are of particular interest, namely

(i) $\beta = 0$; $\qquad w = \pi i + 1 - \dfrac{1}{z} - \ln z$

(ii) $\beta = \pi/2;\qquad w = \pi i + 2\sqrt{\dfrac{z-1}{z}} - \ln\left[2z - 1 + 2\sqrt{z^2 - z}\right]$

(iii) $\beta = \pi;\qquad w = \pi i - \ln z.$

The forms of the mapping functions given here should be compared with those developed in Problem 17 where a different selection of x_is is employed.

Quadrilaterals: regular and degenerate

Example 1 One of the classic Schwarz-Christoffel mappings is that associated with the rectangle. If the x_is and corresponding vertices have the values $x_1 = -1$, $x_2 = 0$, $x_3 = \lambda$, $x_4 = \infty$ and $w_1 = ai$, $w_2 = 0$, $w_3 = b$, $w_4 = b + ai$, respectively, with λ, a, b all positive, then the mapping function (6.4–3) has the form

$$w = A\int_0^z \frac{dz}{\sqrt{z(z+1)(\lambda - z)}}.$$

To match the points x_i with their corresponding corners we choose, in the notation of (3.3–6),

$$\begin{aligned} a &= \frac{A}{\sqrt{\lambda}}\int_0^1 \left(1 + \frac{t}{\lambda}\right)^{-1/2} t^{-1/2}(1-t)^{-1/2}\,dt \\ &= \frac{A\pi}{\sqrt{\lambda}}\,{}_2F_1\left(\frac{1}{2}, \frac{1}{2}; 1; \frac{1}{\lambda}\right) \end{aligned}$$

and

$$\begin{aligned} b &= A\int_0^1 (1 + \lambda t)^{-1/2} t^{-1/2}(1-t)^{-1/2}\,dt \\ &= A\pi\,{}_2F_1\left(\frac{1}{2}, \frac{1}{2}; 1; -\lambda\right). \end{aligned}$$

These hypergeometric functions are related to the Legendre functions of order $-1/2$ and hence to elliptic integrals (see Abramowitz and Stegun (1965), for example).

When the rectangle is a square, the specific values of the various functions can be readily determined. We find that $\lambda = 1$ and hence

$$\begin{aligned} a = b &= A\pi\,{}_2F_1\left(\frac{1}{2}, \frac{1}{2}; 1; -1\right) \\ &= A\,\frac{\Gamma^2\left(\frac{1}{4}\right)}{2\sqrt{2\pi}} \\ &\doteq 2.622A \end{aligned}$$

since $\Gamma(1/4) \doteq 3.626$.

Example 2 For a second example of a mapping onto a "quadrilateral," consider the infinite strip depicted in Figure 6.10. The choices $x_1 = \infty$, $x_2 = -1$, $x_3 = 0$ and $x_4 = \lambda > 0$ (arbitrary) with $w_1 = \infty$, $w_2 = \pi i$, $w_3 = \infty$, and $w_4 = 0$ lead to

$$\begin{aligned} w &= K \int_\lambda^z \frac{dz}{z} \\ &= K \ln(z/\lambda). \end{aligned}$$

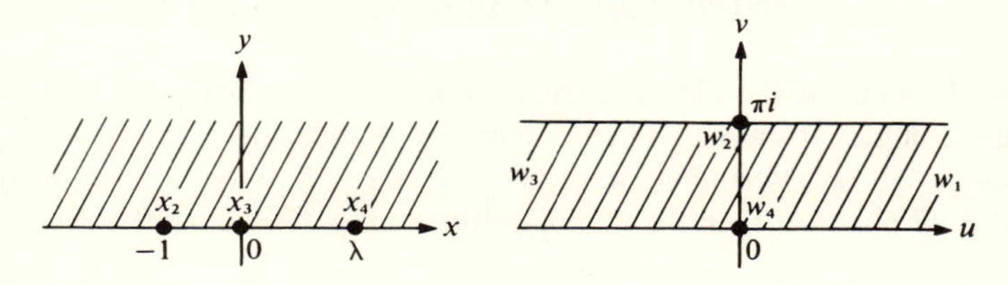

Figure 6.10 Schwarz-Christoffel mapping onto an infinite strip

Setting $z = x < 0$ determines K as 1, and therefore, because $w(-1) = \pi i$, λ must also be unity. The mapping thus obtained, namely

$$w = \ln z,$$

should be compared with (iii) of the "triangular" example 3 on p. 176, since the polygonal image domains are identical.

Example 3 We conclude our selection of illustrative transformations by considering the Schwarz-Christoffel mapping, which takes the upper half-plane onto the bifurcated strip domain shown in Figure 6.11. Assuming the values given in the table

i	x_i	w_i	α_i
1	λ	α	0
2	0	$k\pi i$	2π
3	1	∞	0
4	∞	∞	0

with $\lambda < 0$ (arbitrary) and $0 < k < 1$, we derive

$$\begin{aligned} w &= K \int_0^z \frac{z\,dz}{(z - \lambda)(z - 1)} + k\pi i \\ &= \frac{K}{1-\lambda}\left[\ln(z - 1) - \lambda \ln\left(1 - \frac{z}{\lambda}\right)\right] + \pi i\left(k - \frac{K}{1-\lambda}\right). \end{aligned}$$

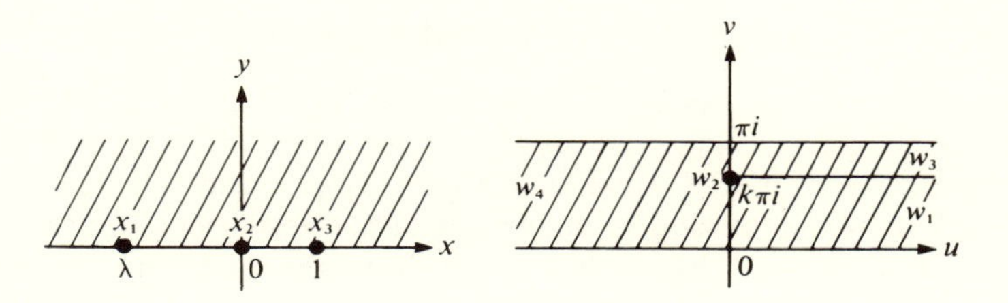

Figure 6.11 Schwarz-Christoffel mapping onto an infinite strip with bifurcation

In the now familiar manner, we set $z = x > 1$ to find, owing to boundary correspondence, that

$$k - \frac{K}{1-\lambda} = 1.$$

Similarly, $z = x < \lambda$ leads to

$$k = \frac{K\lambda}{1 - \lambda}.$$

From these relations we determine $\lambda = k/(k-1)$ and $K = -1$. Thus the desired transformation has the final form

$$w = \pi i - (1 - k)\ln(z - 1) - k\ln\left(1 + \frac{1-k}{k}z\right).$$

When $k = 0$, we obtain

$$w = \pi i - \ln(z-1),$$

which we recognize as yet another mapping of the upper half-plane onto the infinite strip of width π.

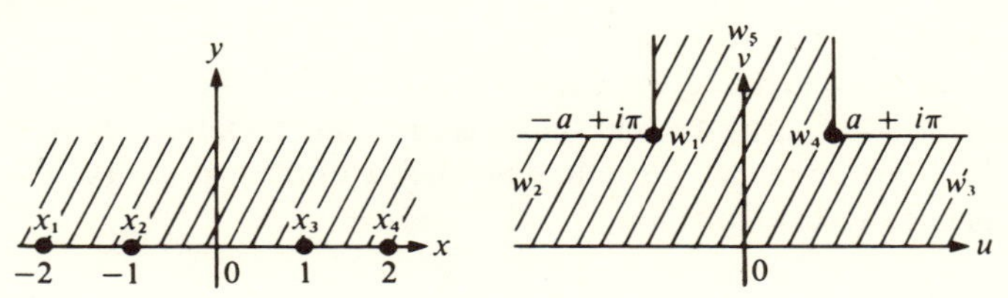

Figure 6.12 Schwarz-Christoffel mapping onto a degenerate pentagon

In leaving this topic we note that R. A. Silverman, in his 1974 text, treats a slightly more general example than the one just discussed, while Churchill (1974) considers other related mappings and, in an appendix, provides analytic transformations for a number of degenerate polygonal domains of 3 or 4 sides. A thorough discussion of both theoretical and applied conformal mapping, including a careful analysis of a modification of the Schwarz-Christoffel transformation to handle rounded corners, appears in Henrici (1974). Finally, for the ardent student, the dictionary of Kober (1952), as its title implies, contains a wealth of information on specific conformal mappings of various regions.

6.5 Cavity and jet flows

Now we return to fluid-dynamical considerations, as we are prepared to analyze two somewhat more complicated cases of potential flow. Both examples exhibit free streamlines and, indeed, were among the examples that Helmholtz and Kirchhoff used in 1868–9 to suggest the efficacy of the free-streamline concept.* We will take a more modern approach to these illustrative problems, however, and employ the hodograph method outlined in Section 6.2, in which we seek to determine the complex velocity potential appropriate for the given profile. Before continuing, therefore, readers may find it helpful to review the notation introduced in the initial sections of this chapter.

Flow past a flat plate

Consider the case of uniform flow of an ideal fluid impinging "at an angle α" upon a fixed flat plate of length L.† Under the assumption of streamline detachment, a cavity is formed behind the plate, and the flow field in physical space (the z-plane) resembles that sketched in Figure 6.5(a). If we assume unit speed at ∞, it follows that $q = 1$ on the free streamlines extending from the ends of the flat plate to ∞; whereas along the plate itself the argument of the complex velocity remains constant.

If we take $\alpha = \pi/2$ and assume that there is only one stagnation point (in the center of the plate, owing to symmetry considerations), the region of fluid motion in the z- and ζ-planes is as depicted in Figure 6.13. Letting $\psi = 0$ correspond to the combined streamline *DABCD* formed from the two free boundaries and the fixed flat plate and, without loss of generality, choosing the origin in the (ϕ,ψ)-plane as the image of the stagnation point at B, the flow field in potential space becomes the slit domain of Figure 6.13(c), that is, the complement of the positive real-axis.

In order to determine the underlying complex velocity potential $w(z)$ for this problem, first we need to find the analytic transformation relating the flow regions in the ζ- and w-planes. Recalling the nature of the exponential map, we see that the associated inverse function

$$\tau = i \ln \zeta \tag{6.5–1}$$

will take the flow field in the ζ-plane conformally onto the interior of the semi-infinite strip in the τ-plane (see Figure 6.8) with $\tau_A = -\pi/2$, $\tau_B = \infty$, and $\tau_C = \pi/2$. In the "triangular" Example 2 of Section 6.4 we used the Schwarz-Christoffel procedure to show that this strip is itself conformally equivalent to the upper half of, say, the t-plane under the transformation

$$t = \frac{1}{\sin \tau}. \tag{6.5–2}$$

*Robertson (1965) provides an excellent historical account of the development of free-streamline flow analyses.

†The angle α is measured between the direction of the uniform flow at negative infinity and the plate.

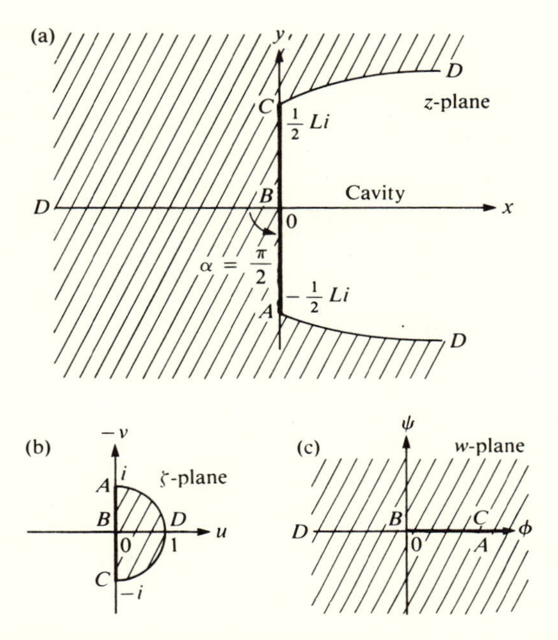

Figure 6.13 Flow field in (a) physical, (b) velocity, and (c) potential space for uniform flow past a vertical flat plate

(Here $t_A = -1$, $t_B = 0$, and $t_C = 1$.) Furthermore, the upper half of the t-plane can be easily mapped onto the desired slit domain in the w-plane using the function

$$w = K t^2 \tag{6.5–3}$$

where K is a positive constant, yet to be evaluated.

It is a simple matter to compose the mappings given by (6.5–1)–(6.5–3). We find

$$w = \frac{K}{\sin^2 (i \ln \zeta)},$$

or, inverting,

$$\zeta = \sqrt{1 - \frac{K}{w}} - i \sqrt{\frac{K}{w}}. \tag{6.5–4}$$

This expression is the sought-for analytic relation between ζ and w.

Continuing with the hodograph method, we substitute (6.5–4) into (6.1–7) to obtain $z(w)$ as

$$\begin{aligned} z &= \int_0^w \left(\frac{1}{\zeta}\right) dw \\ &= \int_0^w \left(\sqrt{1 - \frac{K}{w}} + i \sqrt{\frac{K}{w}}\right) dw \\ &= K\left[\sqrt{W}\sqrt{W-1} - \ln(\sqrt{W} + \sqrt{W-1}) + 2i\sqrt{W} + i\pi/2\right] \end{aligned} \tag{6.5–5}$$

where $W \equiv w/K$. Associating $z = \frac{1}{2}Li$ with $W = 1$ gives

$$\frac{1}{2}Li = K(2i + i\pi/2),$$

and thus

$$K = \frac{L}{4 + \pi}.$$

The evaluation of K completes the determination of z as a function of w. Although we cannot calculate it explicitly, the inverse of the expression (6.5–5) then yields the desired velocity potential $w(z)$.

We note that the free streamlines surrounding the cavity behind the flat plate can be expressed parametrically as

$$\left.\begin{aligned} x &= \frac{L}{4 + \pi}\left[t\sqrt{t^2 - 1} - \ln(t + \sqrt{t^2 - 1})\right] \\ y &= \pm\frac{L}{4 + \pi}[\tfrac{1}{2}\pi + 2t]. \end{aligned}\right\} \quad t \geq 1$$

In this model, then, these streamlines are essentially parabolic for large x and the resulting cavity is of arbitrarily large cross section.*

In the more general case where the flow at negative infinity is directed at an angle $\alpha \neq \pi/2$ to the flat plate, the principal alteration that needs to be made in the above analysis is in the positions of the various images of the point D. To be precise

$$\zeta_D = e^{i(\pi/2 - \alpha)}, \qquad \tau_D = \alpha - \pi/2, \qquad t_D = -(1/\cos\alpha).$$

Thus the analogue of (6.5–3) must be

$$w = K\left[\frac{t(1 + \cos\alpha)}{1 + t\cos\alpha}\right]^2$$

and hence

$$\zeta = \sqrt{1 - F^2(w)} + iF(w)$$

where

$$F(w) \equiv \cos\alpha - (1 + \cos\alpha)\sqrt{\frac{K}{w}}.$$

The remainder of the derivation then follows in a fashion comparable to that given in the case of "normal" incidence, although the requisite integrals generally cannot be evaluated in closed form.

*In actual flows, there is substantial wake underpressure (pressure in the cavity less than that in the oncoming undisturbed flow), so the cavity behind the plate is of finite size. A number of improvements to the Helmholtz-Kirchhoff model have been proposed to make it accord with these facts (see Birkhoff and Zarantonello (1957) or Robertson (1965) for details).

Jet flow from a slot

As our second example, we will analyze the case of ideal fluid flow past two semi-infinite flat plates positioned to form a slot-like orifice, as in Figure 6.14. Again assuming streamline detachment, the fluid passing through this slot forms a symmetrical jet not unlike that shown in a popular television commercial for antifreeze. With no external forces present, the jet is bounded by free streamlines upon which the speed q is constant, taken to be unity. The flow diagram in the ζ-plane is therefore identical to that given earlier in the flat-plate-plus-cavity problem. (The origin $\zeta = 0$ corresponds to the stagnation point, which is now at negative infinity.) The region of fluid motion in the w-plane, however, is markedly different; in the current problem it is an infinite horizontal strip of width $K\pi$, yet to be determined. We position the strip so that $-\frac{1}{2}K\pi \leq \psi \leq \frac{1}{2}K\pi$ and associate the bottom edge of the orifice ($z = -\frac{1}{2}Li$) with the point $w = -\frac{1}{2}K\pi i$.

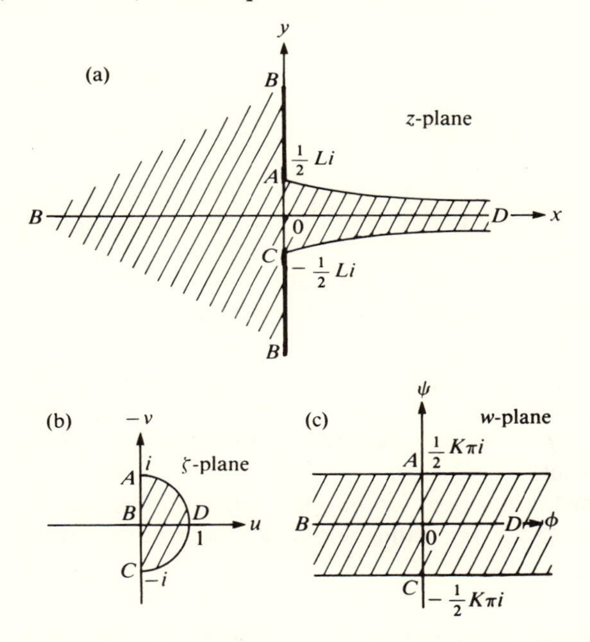

Figure 6.14 Flow field is (a) physical, (b) velocity, and (c) potential space for a symmetrical fit

In analogy with the steps taken in the previous example, we employ the function

$$\tau = i \ln \zeta$$

to map the ζ-plane flow field conformally onto the interior of the semi-infinite strip of Figure 6.8. This degenerate triangle is then transformed as before onto the upper half of the t-plane by means of the mapping function

$$t = \frac{1}{\sin \tau}.$$

However, in this case, the analytic transformation

$$w = K(\ln t - \tfrac{1}{2}\pi i),$$

is the appropriate final mapping function of the trio.

Carrying out the hodograph procedure, we compose the above mappings to determine

$$\zeta = \sqrt{1 + e^{-2w/K}} - e^{-w/K}.$$

Employing (6.1–7) we then calculate for this configuration

$$\begin{aligned} z &= \int_{-\frac{1}{2}K\pi i}^{w} \left(\frac{1}{\zeta}\right) dw - \frac{1}{2}Li \\ &= \int_{-\frac{1}{2}K\pi i}^{w} \left(\sqrt{1 + e^{-2w/K}} + e^{-w/K}\right) dw - \frac{1}{2}Li \\ &= K\left[\ln\left(\frac{1 + \sqrt{1 + W}}{\sqrt{W}}\right) - \sqrt{W} - \sqrt{1 + W}\right] \\ &\quad + i\left[K\left(\frac{1}{2}\pi + 1\right) - \frac{1}{2}L\right] \end{aligned} \qquad \textbf{(6.5–6)}$$

where $W \equiv e^{-2w/K}$. Since $w = \tfrac{1}{2}K\pi i$ (that is $W = e^{-\pi i}$) corresponds to the upper edge of the orifice ($z = \tfrac{1}{2}Li$), K must be such that

$$\frac{1}{2}Li = K\left[\frac{1}{2}\pi i + i\right] + i\left[K\left(\frac{1}{2}\pi + 1\right) - \frac{1}{2}L\right].$$

This yields $K = L/(2 + \pi)$, which, when substituted in (6.5–6), gives the (inverse of the) desired velocity potential.

Parametric equations for the upper and lower free streamlines z_u and z_l can be derived from (6.5–6) by letting $w = K(s \pm \tfrac{1}{2}\pi i)$, respectively, with $s > 0$. It follows that

$$\left.\begin{matrix} z_u \\ z_l \end{matrix}\right\} = \frac{L}{2 + \pi}\left[\ln(1 + \sqrt{1 - e^{-2s}}) + s \pm \tfrac{1}{2}\pi i \pm ie^{-s} - \sqrt{1 - e^{-2s}}\right].$$

If we define the width of the jet as $d(s) \equiv \operatorname{Im}(z_u - z_l)$, then

$$d(s) = \frac{L(\pi + 2e^{-s})}{2 + \pi}.$$

The ratio of the minimum jet width (which occurs here in the limit of large s) to the distance between the plate edges, that is, the width of the orifice, is customarily designated the *contraction coefficient* C_c. Accordingly,

$$C_c = \frac{d(\infty)}{L} = \frac{\pi}{2 + \pi} \doteq 0.611$$

for this symmetrical jet, a value that agrees well with the comparable quantity in the case of non-ideal (real) fluids.* (See Robertson (1965) for an informative discussion of the relation between the contraction coefficients for two- and three-dimensional jet flows.)

Just as in the case of the flat plate, there is also a somewhat more general case here that can be formulated with comparable ease. If the upper plate is shifted to the right or left, then the free jet no longer emerges horizontally from the orifice; its direction, however, can be specified in terms of the angle α between the (parallel) jet flow at positive infinity and the vertical (the plate direction). All the associated flow diagrams remain essentially unchanged, except for the position of the various images of the point D. Since the analogy with the discussion of our earlier example is complete, we leave the details to the interested reader.

Other cases of interest

An investigation similar to that carried out above can be applied to other two-dimensional jet/cavity flows involving nozzles, plates, wedges, and the like. Indeed, Helmholtz himself used the free-streamline model to analyze the potential flow associated with a jet *entering* the space between two symmetric plane-parallel walls: the so-called *Borda mouthpiece*.† (The general case of symmetric "orifice" walls making an angle other than $\pi/2$ or π with the direction of the jet was subsequently studied extensively by von Mises.) More recent contributions of other researchers such as Bolbleff with symmetric wedges, Morton with asymmetric wedges, and von Mises (again) with unsymmetric jets and other plane effluxes, for example, could be mentioned, but even cursory analysis of their work would take us rather far afield. The references of Robertson and Birkhoff and Zarantonello, cited earlier, contain substantial accounts of these and related efforts.

For our purpose it suffices to observe that in all of the flows just mentioned, the region of fluid motion in physical space is bounded by a combination of rectilinear arcs and free streamlines. This means that the flow field in velocity space (the ζ-plane) is bounded by circular arcs concentric with the origin and radial straight lines. In such cases the logarithmic transformation will map the relevant domain onto the interior of a polygon, the Schwarz-Christoffel method will apply, and the hodograph procedure can, in theory, be employed successfully.

References

Abramowitz, M. and I. A. Stegun (1965): *Handbook of Mathematical Functions*, Dover, New York; NBS Applied Math Series 55, U.S. Dept. of Commerce, Washington, D.C. (1964); Chapters 8, 17.

*In actual flows, the free jet first contracts in cross section, then expands. The minimum cross-section is termed the *vena contracta*, and C_c^2 measures the ratio of the area of the vena contracta to that of the orifice.

†Named in honor of J. C. Borda (1733–99) who in 1766 used momentum considerations to show that the contraction coefficient for this configuration equals ½.

Birkhoff, G. and E. H. Zarantonello (1957): *Jets, Wakes, and Cavities*, Academic Press, New York.

Churchill, R. V., J. W. Brown, and R. F. Verhey (1974): *Complex Variables and Applications*, 3rd ed., McGraw-Hill, New York; Chapters 4, 8, 9, 10.

Dettman, J. W. (1969): *Mathematical Methods in Physics and Engineering*, 2nd ed., McGraw-Hill, New York; Chapter 7.

Henrici, P. (1974): *Applied and Computational Complex Analysis, Vol. 1*, Wiley-Interscience, New York; Chapter 5.

Kober, H. (1952): *Dictionary of Conformal Representations*, Dover, New York.

Meyer, R. E. (1971): *Introduction to Mathematical Fluid Dynamics*, Wiley-Interscience, New York.

Milne-Thomson, L. M. (1968): *Theoretical Hydrodynamics*, 4th ed., Macmillan, New York.

Robertson, J. M. (1965): *Hydrodynamics in Theory and Application*, Prentice-Hall, Englewood Cliffs, N.J.

Shinbrot, M. (1973): *Lectures on Fluid Mechanics*, Gordon and Breach, New York.

Silverman, R. A. (1974): *Complex Analysis with Applications*, Prentice-Hall, Englewood Cliffs, N.J.; Chapters 8, 13, 14, 15.

Valentine, H. R. (1967): *Applied Hydrodynamics*, 2nd ed., Plenum Press, New York.

Wylie, C. R. (1975): *Advanced Engineering Mathematics*, 4th ed., McGraw-Hill, New York; Chapter 18.

Problems

Sections 6.1, 6.2

1. Consider the two-dimensional ideal fluid flow given by the complex velocity potential

$$w(z) = -U_0\,(z\,e^{-i\pi})^{\beta} \qquad |\arg z - \pi| < \pi/\beta$$

where U_0 and β are fixed constants satisfying $U_0 > 0$, $1 \le \beta \le 2$.

(a) Verify that this function describes flow of a fluid impinging upon a symmetric infinite wedge with interior angle $\alpha = 2\pi(\beta - 1)/\beta$.

(b) Sketch the flow field and some representative streamlines.

2. A *source* is defined to be a point at which fluid is created at a constant rate. Let

$$w(z) = C \ln (z^2 - h^2),$$

where C and h are positive constants, be the complex potential associated with a two-dimensional potential flow.

(a) Show that this function gives rise to ideal fluid flow with a source in the presence of a rigid wall.

(b) Sketch the flow field and some representative streamlines.

★3. The classic Joukowski airfoil is the image of the unit circle $|t| = 1$ under the mapping

$$z = f(t) = t - a + \frac{(1 - a)^2}{t - a} \qquad 0 < a < 1.$$

(a) Verify that the function, with free real parameter A,

$$w(z) = U_0 \left[k(z) + \frac{1}{k(z)} \right] + \frac{Ai}{2\pi} \ln k(z),$$

where $k(f(t)) = te^{i\beta}$ with β real, that is,

$$e^{-i\beta}k(z) = a + \frac{1}{2}z + \frac{1}{2}[z^2 - 4(1 - a)^2]^{1/2},$$

satisfies conditions (i), (ii) and (iii) of Section 6.2 and thus is a complex velocity potential for ideal fluid flow past the Joukowski airfoil.

(b) Determine the appropriate value of the constant A so that the resulting flow satisfies the Kutta-Joukowski condition.

Section 6.3

4. Consider the analytic transformation $w = 3z - z^3$.

(a) Determine the critical points $z_0 = x_0 + iy_0$ of this function and exhibit the effect of nonconformality by sketching the image of the lines x = constant and y = constant in the neighborhood of these exceptional points. [Hint: Let $x = x_0 + \epsilon$, $y = y_0 + \delta$ where $|\epsilon|$, $|\delta|$ are small.]

(b) Find the locus of points at which infinitesimal segments are rotated 90° in the positive sense.

(c) Using (6.3–2), determine the curve(s) along which mapping by this function leaves scale sizes unchanged, that is, the "magnification" of infinitesimal lengths is equal to 1.

5. Let $w = f(z)$ be an analytic function, conformal at a point z_0. Denote by C a smooth curve passing through z_0 and by $f(C)$ the image curve of C under the mapping f. If $\varphi(x,y)$ is a differentiable function defined in the neighborhood of z_0 and $\phi(u,v) \equiv \varphi[x(u,v),y(u,v)]$, then show that

$$\left.\frac{\partial \varphi}{\partial s_z}\right|_{z_0} = |f'(z_0)| \left.\frac{\partial \phi}{\partial s_w}\right|_{w_0} \quad \text{and} \quad \left.\frac{\partial \varphi}{\partial n_z}\right|_{z_0} = |f'(z_0)| \left.\frac{\partial \phi}{\partial n_w}\right|_{w_0},$$

where the tangential and normal derivatives are calculated with respect to the curves C and $f(C)$ as appropriate.

6. For the example of Section 6.3, demonstrate that (6.3–6) and (6.3–7) are

equivalent and hence verify that the latter is a solution to the given boundary-value problem.

7. (a) Satisfy yourself that $\rho(x,y) = \dfrac{1}{\pi}\tan^{-1}\left\{\dfrac{2y}{x^2+y^2-1}\right\}$ is a solution to the boundary-value problem

$$\Delta\rho = 0 \qquad -\infty < x < \infty, \quad y > 0$$

$$\rho(x,0) = \begin{cases} 1 & |x| < 1 \\ 0 & |x| > 1. \end{cases}$$

(b) Combine the result of part (a) with the mapping function of the example in Section 6.3 to effect the solution of

$$\begin{aligned} \Delta\rho &= 0 && |x| < \pi/2, \quad y > 0 \\ \rho(\pm\,\pi/2,y) &= 1 && y > 0 \\ \rho(x,0) &= 0 && -\pi/2 < x < \pi/2. \end{aligned}$$

8. A *fixed point* of a transformation $w = f(z)$ is a point z_0 such that $z_0 = f(z_0)$. Determine the fixed points of the general linear fractional transformation given by (6.3–8). Can there ever be no fixed points? Precisely one? two? three?

★9. Consider the three canonical linear transformation mappings (i) the translation $w = z + \lambda$, (ii) the rotation/dilation $w = \mu z$, and (iii) the inversion $w = 1/z$, where λ and μ are arbitrary complex constants. Show that each of these special mappings transforms arbitrary circles into circles (with straight lines counting as circles of unbounded radius).

10. Establish Property 3 of Section 6.3.

11. (a) Find the linear fractional transformation that maps the points $z = -i$, 0, and i onto the points $w = -1$, i, and 1, respectively.

(b) Determine the images of the lines $x =$ constant and $y =$ constant under this transformation.

★12. (a) Verify that the most general linear fractional transformation that maps the upper half of the z-plane onto the interior of the unit circle in the w-plane has the form

$$w = e^{i\alpha}\,\frac{z-\lambda}{z-\bar{\lambda}}$$

where α is real and Im $\lambda > 0$.

(b) Use the result of part (a) to determine the most general linear fractional transformation that maps the upper half of the z-plane onto the lower half of the w-plane.

13. Find the most general bilinear transformation that maps the interior of the unit circle $|z| < 1$ onto the interior of the unit circle $|w| < 1$ and determine the fixed points of the mapping.

14. Solve the boundary-value problem

$$\Delta\varphi(r,\theta) = 0 \qquad |z| = r < 1,$$

$$\varphi(1,\theta) = \begin{cases} 0 & 0 < \theta < \pi/2 \\ 1 & \pi/2 < \theta < \pi, \end{cases}$$

$$\frac{\partial\varphi(1,\theta)}{\partial r} = 0 \qquad -\pi < \theta < 0.$$

[Hint: Recall the example of Section 6.3.]

Section 6.4

15. Let $z = -\coth\dfrac{w}{2} = -\dfrac{\cosh(w/2)}{\sinh(w/2)}$. Analyze the mapping function $w = f(z)$ inverse to this transformation and, in particular, describe the image, in the w-plane, of the upper half of the complex z-plane.

16. Analyze the Schwarz-Christoffel transformation (6.4–4) in the case that $\alpha_1 = \pi/4$ and $\alpha_2 = \pi/2$. Show that the complex constants can be selected so that the upper half-plane with $x_1 = 0$, $x_2 = 1$ *and* $x_3 = \infty$ maps onto the interior of the right triangle with $w_1 = 0$, $w_2 = a$, and $w_3 = (1 + i)a$, where $a > 0$.

17. (a) Determine the Schwarz-Christoffel transformation that maps the upper half-plane with $x_1 = \infty$, $x_2 = -1$, and $x_3 = 0$ onto the interior of the funnel region, with the corners indexed as shown in Figure 6.9 on p. 176.
(b) Explicitly evaluate the integral involved in the special cases $\beta = 0$, $\beta = \pi/2$, and $\beta = \pi$.

18. For appropriate choices of the constants K and C, the transformation

$$w = K\int_0^z \frac{\sqrt{4 - t^2}}{1 - t^2}\,dt + C$$

maps the upper half of the z-plane onto the interior of the domain in the w-plane depicted in Figure 6.12 on p. 179, with the images of the points $x_1 = -2$, $x_2 = -1$, $x_3 = 1$, $x_4 = 2$, and $x_5 = \infty$ as shown.
(a) Carefully analyze this transformation and determine the correct values of K,C.
(b) Find the numerical value of $2a$, the cross-sectional width of the upper channel.

19. Consider the analytic function

$$w = f(z) = \frac{1}{2}\left(z + \frac{1}{z}\right).$$

(a) Onto what domain does this function map the interior of the unit circle $|z| < 1$?
(b) Where is the exterior region $|z| > 1$ mapped by $f(z)$?

20. Describe in detail the image of the unit circle $|z| < 1$ under the analytic transformation

$$w = \int_0^z \frac{dz}{\sqrt{1 - z^4}}.$$

Section 6.5

21. Show that, in the limit of large L, the expression (6.5–5) leads, in essence, to the velocity potential associated with uniform flow impinging normally upon an infinite flat plate. (Compare Problem 1.)

★22. The velocity potential for ideal fluid flow past a symmetric infinite wedge with interior angle α could be obtained by a limiting process from the potential function appropriate to a finite wedge with cavity. Alternatively, it can be derived directly using the hodograph method. Sketch the flow field for this configuration in physical, velocity, and potential space, and then employ this *latter* approach to determine $w(z)$. (Compare Problem 1.) [Hint: The flow field in velocity space (the ζ-plane) is interior to an infinite wedge of angle α.]

23. (a) The case of uniform flow impinging "at an angle α" upon a flat plate of finite extent was briefly discussed in Section 6.5. Investigate the nature of that portion of the streamline $\psi = 0$ with $\phi < 0$ for this situation and demonstrate that it has the appropriate direction for both $\phi > 0^-$ and $\phi \to -\infty$.

 (b) Perform the comparable analysis for a mid-jet streamline in the case of two-dimensional jet flow that makes an angle α with the fixed-boundary plates.

24. Show that, with a suitable translation, the expression (6.5–6) leads, in the limit of large L, to the velocity potential appropriate for uniform flow.

7

Integral transforms

7.1 Mixed boundary-value problems

Models of real phenomena frequently lead to simple partial differential equations. When the physical situation does not change with time, the governing equation often turns out to be either *Poisson's equation*

$$\Delta\phi = f(x)$$

or the homogeneous special case, *Laplace's equation*

$$\Delta\phi = 0. \qquad \textbf{(7.1–1)}$$

We encountered this latter equation in our study in Chapter 6 of potential flows associated with ideal fluids (also see the discussion leading up to (4.4–14)). As most readers know, Laplace's equation also occurs in the analysis of equilibrium phenomena, which arise in a wide variety of other physical contexts including electrostatics, elastostatics, and diffusion.

In most mathematical problems of physical interest, the desired (potential) function must satisfy not only the equation (7.1–1) in some specified region but must also exhibit certain prescribed behavior along the curve or surface that bounds the region. In other words, the desired function is a (the) solution of a boundary-value problem in multidimensional space. (Recall that we discussed one-dimensional BVP's in Chapter 2.) Occasionally, the boundary conditions are "uniform," that is to say, the value of the function alone or of its normal derivative alone is prescribed along the *entire* boundary. More often, the appropriate problem is of *mixed type* wherein some linear combination $\alpha(s)\phi + \beta(s)\dfrac{\partial\phi}{\partial n}$ is specified a priori on the boundary (the earlier example of Section 6.3 was of this type).

In this chapter we consider in some detail a rather general class of boundary value problems of mixed type and investigate an "operational" procedure for the solution of such problems. Our point-of-departure is the Mellin transform, which arises naturally in the course of our analysis. Although this integral transform is probably not as familiar to most readers as the Fourier or Laplace transforms, it actually contains each of them as special cases and is closely related to other common transforms (see Oberhettinger (1974); also Titchmarsh (1959), Sneddon (1972)). In view of this, and because much of the analysis of Mellin transforms can be carried out in a comfortable analytic function-theoretic framework, we opt for this nonstandard, but quite reasonable, approach. (For those readers who are interested, additional background information on the important Fourier and Laplace transforms is given in Appendix 6.)

A class of mixed problems

Let us assume that the mathematical problem we are interested in solving for $\phi(x,y)$ is equivalent (possibly after a conformal transformation)* to

$$\begin{aligned} &\Delta\phi = 0 && -\infty < x < \infty,\ y > 0 \\ &\phi(x,0) = f(x) && x > 0 \\ &\frac{\partial\phi}{\partial y}(x,0) = 0 && x < 0. \end{aligned} \qquad \textbf{(7.1–2)}$$

This is akin to what Sneddon (1966) terms the "first basic (plane) problem," and includes both the simple situation mentioned above, which we treated in Section 6.3, as well as an array of more significant problems, one of which we will consider later.

As an aid to our intuition, we will think of the mixed boundary-value problem (7.1–2) as a mathematical description of heat conduction, wherein $\phi(x,y)$ is the steady-state (i.e., time independent) temperature of a semi-infinite plate.† On one half of the edge $y = 0$ the temperature is specified; the other half of the edge is insulated, that is, there is no heat flow across this portion of the boundary.

In analyzing the above canonical problem it is important to note that the region in which we seek the solution is unbounded. Mathematically, therefore, as far as uniqueness is concerned, this mixed boundary-value problem is *not* well-posed. Indeed, the nontrivial function

$$\phi(x,y) = [\sqrt{x^2 + y^2} - x]^{1/2},$$

for example, satisfies the homogeneous version ($f=0$) of (7.1–2). Fortunately, such "extraneous" solutions can be removed from consideration by the physically natural assumption of boundedness. In future examples, therefore, we assume that we are interested only in that harmonic function which not only takes on the boundary conditions of (7.1–2), but also remains bounded as $\sqrt{x^2 + y^2} \to \infty$.

Although the function $f(x)$ which appears in the statement of (7.1–2) should be, say, continuous and bounded, it is otherwise arbitrary. As a consequence, this mixed problem cannot be treated as was the simple example of Section 6.3.

*Theorems 3 and 4 of Section 6.3 were concerned with analytic transformations of two-dimensional boundary-value problems for Laplace's equation. Henrici (1974) discusses this process of *conformal transplantation* at some length.

†The classic study of Carslaw and Jaeger (1959) details a number of steady temperature applications that fall into the class of mixed problems under consideration in this section. See also Özisik (1968) or analogous books in other subject areas.

Instead, we apply the time-honored technique of *separation of variables* in order to take advantage of the linearity of Laplace's equation. The separated "solutions" will then be superposed to produce an appropriate solution.

If we express the Laplacian in cartesian coordinates and write ϕ as a product $X(x)Y(y)$, we are led to

$$\frac{X''}{X} = -\frac{Y''}{Y} = -\beta^2,$$

where β is the separation parameter. From this it follows that

$$X(x) = A(\beta)\cos \beta x + B(\beta)\sin \beta x$$

and

$$Y(y) = C(\beta)\, e^{\beta y} + D(\beta)\, e^{-\beta y},$$

where A, B, C, D are yet to be determined functions of β. The boundedness of the solution for large $|x|$ mandates that β be real. Assuming, without loss of generality, that β is nonnegative, boundedness for large y then implies $C(\beta) \equiv 0$. Incorporating D into A and B and superposing "solutions" by integration, as we have done by summation earlier, we thus obtain

$$\phi(x,y) = \int_0^\infty [A(\beta)\cos \beta x + B(\beta)\sin \beta x]\, \mathrm{e}^{-\beta y}\, d\beta \qquad \textbf{(7.1–3)}$$

as a candidate for the desired solution of (7.1–2).

If (7.1–3) is to satisfy the given boundary conditions, A and B must be such that

$$\int_0^\infty [A(\beta)\cos \beta x + B(\beta)\sin \beta x]\, \mathrm{d}\beta = f(x) \qquad \textbf{(7.1–4)}$$

for positive x, while

$$\int_0^\infty [\beta A(\beta)\cos \beta x + \beta B(\beta)\sin \beta x]\, d\beta = 0$$

when x is negative. Changing x into $-x$ in this latter relation, however, gives us

$$\int_0^\infty [\beta A(\beta)\cos \beta x - \beta B(\beta)\sin \beta x]\, d\beta = 0 \qquad x > 0. \qquad \textbf{(7.1–5)}$$

Now it is not difficult to convince ourselves that the two expressions (7.1–4) and (7.1–5), which both must hold for arbitrary positive x, are unfortunately incompatible for general $f(x)$.

The failure of the above procedure to provide a viable representation for the solution of the given problem (7.1–2) is not due to the method we employed. Rather the fault lies at the very first step when we chose to separate variables in cartesian coordinates. We see in retrospect that, owing to the nature of the boundary conditions, what is needed is a coordinate system in which the negative x-axis is more clearly differentiated from the positive x-axis. We will therefore try a similar approach with a set of independent variables having the desired feature, such as polar coordinates.

Accordingly, we write ϕ as $R(r)\Theta(\theta)$ and apply separation of variables in polar coordinates to Laplace's equation, to find that

$$r^2 \frac{R''}{R} + r\frac{R'}{R} = -\frac{\Theta''}{\Theta} = \lambda^2,$$

where λ is now the separation parameter. The resulting ordinary differential equations are again easy to solve, and yield

$$\Theta(\theta) = A(\lambda)\cos\lambda\theta + B(\lambda)\sin\lambda\theta$$

and

$$R(r) = C(\lambda)\, r^{\lambda} + D(\lambda)\, r^{-\lambda}$$

as the separated solutions. Taking $Re\lambda \geq 0$, the boundedness condition once again implies $C(\lambda) \equiv 0$, and hence, incorporating D into A and B as before, we are led to

$$\phi(r,\theta) = \int_{\Gamma} [A(\lambda)\cos\lambda\theta + B(\lambda)\sin\lambda\theta]\, r^{-\lambda}\, d\lambda \qquad \textbf{(7.1–6)}$$

as a putative representation for the solution to (7.1–2). In this expression, Γ is a contour along which λ has nonnegative real part.

In applying the boundary conditions of (7.1–2) to (7.1–6) we first observe that the second condition is equivalent to

$$\frac{\partial\phi}{\partial\theta}(r,\pi) = 0.$$

Obviously, this condition will be trivially satisfied if the parametric functions A and B are related by

$$B(\lambda) = A(\lambda)\tan\lambda\pi. \qquad \textbf{(7.1–7)}$$

The other boundary condition of (7.1–2) implies that

$$\int_{\Gamma} A(\lambda)\, r^{-\lambda}\, d\lambda = f(r) \qquad \textbf{(7.1–8)}$$

If we can find $A(\lambda)$ (and an appropriate contour Γ), which makes (7.1–8) an identity, then we can completely determine the desired solution using (7.1–7).

Once the contour Γ is specified, (7.1–8) is seen to be an integral equation of the first kind (see Chapter 10), and the problem of inverting this expression to find $A(\lambda)$ for given $f(r)$ appears formidable. To our good fortune, however, when Γ is a straight line parallel to the imaginary λ axis, the equation has been extensively studied. In this case, the expression (7.1–8) is related to the so-called Mellin transform, and the associated methodology does indeed permit the determination of $A(\lambda)$ and hence the resolution of the posed problem. It is therefore appropriate that we become more familiar with these integral transforms.

7.2 The Mellin transform

The inversion formula

Although an inversion formula for (7.1–8) in the case of the infinite vertical contour discussed above was derived by Riemann in the course of his fundamental investigations in number theory*, the first rigorous analysis of the matter was given by R. H. Mellin (1854–1933), and so the associated integral transform bears his name. The following theorem suggests the pair-wise relations valid for this transform; it also illustrates the importance of analytic function theory in these considerations.

Theorem 1:

Let $\mathscr{F}(s)$ be a regular analytic function of the complex variable $s = \sigma + i\omega$ in some strip $a < \sigma < b$. Assume

(i) $\int_{\infty}^{\infty} |\mathscr{F}(\sigma + i\omega)| d\omega < \infty$ for fixed $\sigma, < \sigma < b$;

(ii) $\mathscr{F}(s) \to 0$ uniformly as $|\omega| \to \infty$ in the strip $a + \delta \leq \sigma \leq b - \delta$, where $\delta > 0$ is arbitrary.

For $x > 0$, define

$$f(x) \equiv \frac{1}{2\pi i} \int_{\sigma - i\infty}^{\sigma + i\infty} x^{-s} \mathscr{F}(s)\, ds\,. \qquad \textbf{(7.2–1)}$$

Then

$$\mathscr{F}(s) = \int_0^{\infty} x^{s-1} f(x) dx \qquad a < \sigma < b. \qquad \textbf{(7.2–2)}$$

Proof: Our demonstration follows that of Courant and Hilbert (1953); see also Sneddon (1966). We begin by observing that, in view of hypothesis (ii) and Cauchy's theorem, $f(x)$ is independent of σ. For given $s = \sigma + i\omega$ in the strip, therefore, we can choose σ_1, σ_2 such that $a < \sigma_1 < \sigma < \sigma_2 < b$ and

$$\begin{aligned}
\int_0^{\infty} x^{s-1} f(x) dx &= \int_0^1 x^{s-1} f(x) dx + \int_1^{\infty} x^{s-1} f(x) dx \\
&= \frac{1}{2\pi i} \int_0^1 x^{s-1} \left[\int_{\sigma_1 - i\infty}^{\sigma_1 + i\infty} x^{-s_1} \mathscr{F}(s_1) ds_1 \right] dx \\
&\quad + \frac{1}{2\pi i} \int_1^{\infty} x^{s-1} \left[\int_{\sigma_2 - i\infty}^{\sigma_2 + i\infty} x^{-s_2} \mathscr{F}(s_2) ds_2 \right] dx \\
&= I_1 + I_2.
\end{aligned}$$

*And also earlier (1815) by Simeon Poisson as part of some practical studies of heat conduction in solids.

Notice that the integrands of the two subsidiary integrals I_1 and I_2 are (absolutely) integrable when the integration is performed in the reverse order. Indeed,

$$\frac{1}{i}\int_{\sigma_1-i\infty}^{\sigma_1+i\infty}\left[\int_0^1\left|x^{s-s_1-1}\,\mathcal{F}(s_1)\right|dx\right]ds_1 = \int_{-\infty}^{\infty}\left|\mathcal{F}(s_1)\right|d\omega\int_0^1 x^{\sigma-\sigma_1-1}dx$$

$$= \frac{1}{\sigma-\sigma_1}\int_{-\infty}^{\infty}\left|\mathcal{F}(s_1)\right|d\omega$$

and

$$\frac{1}{i}\int_{\sigma_2-i\infty}^{\sigma_2+i\infty}\left[\int_1^{\infty}\left|x^{s-s_2-1}\,\mathcal{F}(s_2)\right|dx\right]ds_2 = \int_{-\infty}^{\infty}\left|\mathcal{F}(s_2)\right|d\omega\int_1^{\infty} x^{\sigma-\sigma_2-1}\,dx$$

$$= \frac{1}{\sigma_2-\sigma}\int_{-\infty}^{\infty}\left|\mathcal{F}(s_2)\right|d\omega$$

(recall the choice of σ_1, σ_2). Since both these expressions are, by assumption, finite, the celebrated theorem of Fubini* applies, and we may interchange the orders of integration in I_1 and I_2. Thus, for I_1 we find

$$I_1 = \frac{1}{2\pi i}\int_{\sigma_1-i\infty}^{\sigma_1+i\infty}\mathcal{F}(s_1)\left[\int_0^1 x^{s-s_1-1}\,dx\right]ds_1$$

$$= \frac{1}{2\pi i}\int_{\sigma_1-i\infty}^{\sigma_1+i\infty}\frac{\mathcal{F}(s_1)}{s-s_1}\,ds_1,$$

while for I_2 we obtain the analogous expression

$$I_2 = \frac{1}{2\pi i}\int_{\sigma_2-i\infty}^{\sigma_2+i\infty}\mathcal{F}(s_2)\left[\int_1^{\infty} x^{s-s_2-1}\,dx\right]ds_2$$

$$= \frac{1}{2\pi i}\int_{\sigma_2-i\infty}^{\sigma_2+i\infty}\frac{\mathcal{F}(s_2)}{s_2-s}\,ds_2.$$

It follows from these two relations that

$$\int_0^{\infty} x^{s-1}f(x)dx = \frac{1}{2\pi i}\int_C\frac{\mathcal{F}(z)}{z-s}\,dz$$

where C is the union of the two infinite rectilinear contours detailed above, the right-most traversed from bottom to top, the left-most from top to bottom. However, by virtue of hypothesis (ii), the two disjoint portions of C can be joined at infinity to form a closed contour, and we can apply the Cauchy integral formula. Since s is interior to C, the result we seek then follows readily. □

*Actually it is the extension of Fubini's theorem due to Tonelli and Hobson that we invoke here; for a concise statement of this generalization see Apostol (1974), for example.

Actually, the hypotheses of the theorem we have just established are unnecessarily strong. The conclusion is still valid if, in addition to (i), $\mathscr{F}(s)$ is merely continuously differentiable (Titchmarsh (1959), for example, has the details).

The Mellin transform pair

The relations (7.2–1) and (7.2–2), taken together, constitute the Mellin transform pair. Given $f(x)$ with $x > 0$, we say that

$$M[f(x)] \equiv \mathscr{F}(s) = \int_0^\infty x^{s-1} f(x)dx \qquad \textbf{(7.2–2')}$$

is the *Mellin transform* of f with complex parameter $s = \sigma + i\omega$. Conversely, the formula

$$f(x) = \frac{1}{2\pi i}\int_{\sigma-i\infty}^{\sigma+i\infty} x^{-s}\,\mathscr{F}(s)ds \qquad \textbf{(7.2–1')}$$

defines the *inverse Mellin transform* of a complex function $\mathscr{F}(s)$. The theorem proved above shows that under appropriate conditions, the Mellin transform "undoes" what the inverse Mellin transform "does." (This is just what is needed to invert (7.1–8).) However, in order to complete the correspondence, we need a converse to Theorem 1 which shows that the inverse Mellin transform likewise "undoes" what the Mellin transform "does." At the heart of the result in the reverse direction lies the famous

Riemann-Lebesgue Theorem:

If $\displaystyle\int_{-\infty}^{\infty} \left|f(x)\right| dx < \infty,$

then

$$\lim_{\lambda\to\infty}\int_{-\infty}^{\infty} e^{i\lambda x} f(x)dx = 0.*$$

Proof: The proof is classical. For arbitrary $\epsilon > 0$ we determine $X(\epsilon) > 0$ such that

$$\int_{-\infty}^{-X} \left|f(x)\right| dx < \epsilon/4 \qquad \text{and} \qquad \int_{X}^{\infty} \left|f(x)\right| dx < \epsilon/4.$$

*Alternatively,

$$\lim_{\lambda\to\infty}\int_{-\infty}^{\infty} f(x)\cos\lambda\, x\, dx = 0 \qquad \text{and} \qquad \lim_{\lambda\to\infty}\int_{-\infty}^{\infty} f(x)\sin\lambda\, x\, dx = 0.$$

Moreover, all these results are equally valid if the domain of integration is merely the finite or semi-infinite interval (a,b).

Then

$$\left|\int_{-\infty}^{\infty} e^{i\lambda x} f(x)dx\right| \leq \epsilon/2 + \left|\int_{-X}^{X} e^{i\lambda x} f(x)dx\right|.$$

Now we select a continuously differentiable function $g(x)$ with the property that

$$\int_{-X}^{X} \left| f(x) - g(x) \right| dx < \epsilon/4.$$

[This can be done since the continuously differentiable functions are dense in the space of functions integrable over $(-X,X)$.] Observe that for $g(x)$ we have

$$\left|\int_{-X}^{X} e^{i\lambda x} g(x)dx\right| = \left|\left(\frac{e^{i\lambda x} g(x)}{i\lambda}\right)_{-X}^{X} - \frac{1}{i\lambda}\int_{-X}^{X} e^{i\lambda x} g'(x)dx\right|,$$

so this expression can be made arbitrarily small, say less than $\epsilon/4$, by choosing λ sufficiently large. It follows then that as $\lambda \to \infty$,

$$\begin{aligned}\left|\int_{-\infty}^{\infty} e^{i\lambda x} f(x)dx\right| &\leq \epsilon/2 + \left|\int_{-X}^{X} e^{i\lambda x}(f(x) - g(x))dx + \int_{-X}^{X} e^{i\lambda x} g(x)dx\right| \\ &\leq \epsilon/2 + \int_{-X}^{X} |f(x) - g(x)|dx + \left|\int_{-X}^{X} e^{i\lambda x} g(x)dx\right| \\ &\leq \epsilon/2 + \epsilon/4 + \epsilon 4 \\ &= \epsilon,\end{aligned}$$

which, since ϵ is arbitrary, completes the demonstration. □

It is now a relatively easy matter to establish the converse result for Mellin transforms.

Theorem 2:

Let $f(x)$ be a continuously differentiable function of a positive real variable which satisfies

$$\int_{0}^{\infty} x^{\sigma-1}|f(x)|dx < \infty \qquad \textbf{(7.2–3)}$$

for some real σ. For complex $s = \sigma + i\omega$, define

$$\mathscr{F}(s) \equiv \int_{0}^{\infty} x^{s-1} f(x)dx.$$

Then

$$f(x) = \frac{1}{2\pi i}\int_{\sigma-i\infty}^{\sigma+i\infty} x^{-s}\,\mathscr{F}(s)ds.$$

Proof: In view of the definition of $\mathscr{F}(s)$, we have for positive x

$$I \equiv \frac{1}{2\pi i}\int_{\sigma-i\infty}^{\sigma+i\infty} x^{-s}\,\mathscr{F}(s)ds = \frac{1}{2\pi i}\lim_{\lambda\to\infty}\int_{\sigma-i\lambda}^{\sigma+i\lambda} x^{-s}\left[\int_0^{\infty} u^{s-1} f(u)du\right]ds$$

$$= \frac{1}{2\pi i}\lim_{\lambda\to\infty}\int_0^{\infty} f(u)\left[\int_{\sigma-i\lambda}^{\sigma+i\lambda}\left(\frac{u}{x}\right)^s ds\right]\frac{du}{u}$$

$$= \frac{1}{\pi}\lim_{\lambda\to\infty}\int_0^{\infty} f(u)\left(\frac{u}{x}\right)^{\sigma}\frac{\sin[\lambda\ln(u/x)]}{\ln(u/x)}\frac{du}{u}$$

$$= \frac{1}{\pi}\lim_{\lambda\to\infty}\int_0^{\infty} f(xv)v_{\sigma}\frac{\sin(\lambda\ln v)}{\ln v}\frac{dv}{v}.$$

The integrability condition on f, and thence Fubini's theorem, justifies interchanging the order of integration. Owing to the Riemann-Lebesgue theorem, however, this last relation (which is equivalent to the so-called *Fourier single-integral formula*) can be further simplified. Indeed, in view of (7.2–3), it follows that the sole contribution to the above integral, in the limit of large λ, comes from the vicinity of $v = 1$. Therefore, a valid alternative expression for I is

$$I = \frac{1}{\pi}\lim_{\lambda\to\infty}\int_{1-\delta}^{1+\delta} f(xv)v^{\sigma}\frac{\sin(\lambda\ln v)}{\ln v}\frac{dv}{v},$$

where δ is some arbitrary positive parameter less than unity.

Now we are ready to take advantage of the continuous differentiability of f. Adding and subtracting $f(x)$ within the integrand, we rewrite I as $I_1 + I_2$ with

$$I_1 \equiv \frac{1}{\pi}\lim_{\lambda\to\infty} f(x)\int_{1-\delta}^{1+\delta}\frac{\sin(\lambda\ln v)}{\ln v}\frac{dv}{v}$$

and

$$I_2 \equiv \frac{1}{\pi}\lim_{\lambda\to\infty}\int_{1-\delta}^{1+\delta}[f(xv)v^{\sigma} - f(x)]\frac{\sin(\lambda\ln v)}{\ln v}\frac{dv}{v}$$

$$= \frac{1}{\pi}\lim_{\lambda\to\infty}\int_{\ln(1-\delta)}^{\ln(1+\delta)}\left[\frac{f(xe^u)e^{\sigma u} - f(x)}{u}\right]\sin\lambda u\,du.$$

The term within brackets in this last expression, however, satisfies the conditions of the Riemann-Lebesgue theorem, and thus I_2 vanishes in the limit of large λ. To conclude our proof, we make an obvious change of variables and write

$$I = I_1 = \frac{1}{\pi} \lim_{\lambda \to \infty} f(x) \int_{\lambda \ln(1-\delta)}^{\lambda \ln(1+\delta)} \frac{\sin t}{t} dt$$

$$= \frac{f(x)}{\pi} \int_{-\infty}^{\infty} \frac{\sin t}{t} dt.$$

The integral in this final formula is well-known and has value π; from this the desired conclusion that $I = f(x)$ obviously follows. □

Readers have undoubtedly noticed that one of the key steps in the demonstration of each of the above results concerning Mellin transforms was inverting the order of integration. We shall assume that the functions involved are always sufficiently well-behaved to allow this interchange.

7.3 Transform pairs and the Parseval formulas

The Mellin Transform

$$\mathscr{F}(s) = M[f(x)] \equiv \int_0^{\infty} x^{s-1} f(x) dx \qquad a < \text{Re } s < b$$

and its inverse **(7.3–1)***

$$f(x) = M^{-1}[\mathscr{F}(s)] \equiv \frac{1}{2\pi i} \int_{\sigma - i\infty}^{\sigma + i\infty} x^{-s} \mathscr{F}(s) ds,$$

for which we have now established the necessary correspondence, form but one of a number of integral transform pairs that arise in various applications. Historically, the earliest was the (complex) Fourier transform

$$F(\omega) \equiv \frac{1}{\sqrt{2\pi}} \int_{-\infty}^{\infty} f(t)\, e^{i\omega t} dt$$

$$f(t) = \frac{1}{\sqrt{2\pi}} \int_{-\infty}^{\infty} F(\omega)\, e^{-i\omega t} d\omega, \qquad \textbf{(7.3–2)}$$

which appeared in J. B. J. Fourier's (1768–1830) fundamental work "Théorie analytique de la chaleur" in 1811 as well as in Cauchy's roughly contemporaneous

*It should not go unnoticed that the expression (7.3–1), as well as the transform representations (7.3–2)–(7.3–4) that follow, is analogous to the earlier expansion (2.5–1). The x^{-s} plays the role of a continuously indexed eigenfunction and integration has replaced summation. Continuous analogues of Parseval's equality (2.5–2) are contained in (7.3–7)–(7.3–12).

studies of wave propagation. Closely related to (7.3–2) are the Fourier cosine transform

$$F_c(\omega) \equiv \sqrt{\frac{2}{\pi}} \int_0^\infty f(t) \cos \omega t \, dt,$$

$$f(t) = \sqrt{\frac{2}{\pi}} \int_0^\infty F_c(\omega) \cos \omega t \, d\omega,$$

and sine transform

$$F_s(\omega) \equiv \sqrt{\frac{2}{\pi}} \int_0^\infty f(t) \sin \omega t \, dt,$$

$$f(t) = \sqrt{\frac{2}{\pi}} \int_0^\infty F_s(\omega) \sin \omega t \, d\omega.$$

Other pairs of importance include the Laplace transform

$$\mathscr{L}(s) \equiv \int_0^\infty f(t) \, e^{-st} dt \qquad Re\, s > a$$

$$f(t) = \frac{1}{2\pi i} \int_{\sigma - i\infty}^{\sigma + i\infty} \mathscr{L}(s) \, e^{st} \, ds, \tag{7.3–3}$$

the Hankel transform

$$\mathscr{H}_\nu(s) \equiv \int_0^\infty x f(x) \, J_\nu(xs) dx \qquad \nu \geq -\tfrac{1}{2}$$

$$f(x) = \int_0^\infty s \mathscr{H}_\nu(s) \, J_\nu(xs) ds,$$

where J_ν is the Bessel function of the first kind of order ν (which is discussed in Chapter 3), and the Stieltjes transform

$$S(s) = \int_0^\infty \frac{f(x)}{x + s} dx \qquad |\arg s| < \pi$$

$$f(x) = \frac{i}{2\pi} [S(xe^{i\pi}) - S(xe^{-i\pi})]. \tag{7.3–4}$$

The Fourier and Laplace transforms can be considered "special cases" of the Mellin transform. Indeed, if we change variables in (7.3–1) by setting $x = e^t$, then

$$\mathscr{F}(\sigma + i\omega) = \int_{-\infty}^\infty e^{\sigma t} f(e^t) \, e^{i\omega t} \, dt$$

and thus $F(\omega) \equiv \mathscr{F}(\sigma + i\omega)/\sqrt{2\pi}$ is the complex Fourier transform of $e^{\sigma t} f(e^t)$. In analogous fashion, starting with a function $f(x)$ that vanishes for $0 < x < 1$, and setting $\omega = i\beta$, we readily establish that $\mathcal{L}(\beta) \equiv \mathcal{F}(\sigma - \beta)$ is the Laplace transform of the causal function that vanishes for $t < 0$ but is equal to $e^{\sigma t}f(e^t)$ for $t > 0$. [The Mellin transform is therefore related to the *two-sided* Laplace transform; see Carrier et al. (1966).] We will be using these relationships several times in this chapter.

The connections between the Mellin and other integral transforms are less direct and considerably more complicated. As we will demonstrate in Section 7.5, if $\mathcal{F}(s)$ and $\mathcal{H}_\nu(s)$ are the Mellin and Hankel transforms, respectively, of a given function $f(x)$, then

$$s\mathcal{H}_\nu(s) = M^{-1}\left[\frac{2^s\Gamma(\frac{1}{2}\nu + \frac{1}{2} + \frac{1}{2}s)}{\Gamma(\frac{1}{2}\nu + \frac{1}{2} - \frac{1}{2}s)}\mathcal{F}(1 - s)\right]. \qquad \textbf{(7.3–5)}$$

The similar expression for the Stieltjes transform (7.3–4) is

$$S(s) = M^{-1}\left[\frac{\pi}{\sin \pi s}\mathcal{F}(s)\right]. \qquad \textbf{(7.3–6)}$$

Parseval formulas

There are a number of convolution-type relations for various integral transforms that turn out to have practical importance. The Fourier transform version of these formulas can be traced back to Lord Rayleigh (John William Strutt, 1842–1919), while the comparable expressions for other transforms were derived by more recent researchers. Nevertheless, since a special case of the result is equivalent to Parseval's identity (2.5–2) for Fourier series, the relations are usually termed *Generalized Parseval Formulas*. We begin by formally deriving the expressions appropriate for Mellin transforms.

Let $\mathcal{F}(s)$ and $\mathcal{G}(s)$ be the Mellin transforms of the given functions $f(x)$ and $g(x)$, respectively. Then for $a > 0$ and complex α, β,

$$\frac{1}{2\pi i}\int_{\sigma-i\infty}^{\sigma+i\infty} a^{-s}\,\mathcal{F}(s + \alpha)\,\mathcal{G}(1 - \alpha + \beta - s)ds$$

$$= \frac{1}{2\pi i}\int_{\sigma-i\infty}^{\sigma+i\infty} \mathcal{G}(1 - \alpha + \beta - s)a^{-s}\left[\int_0^\infty f(y)y^{s+\alpha-1}dy\right]ds$$

$$= \frac{1}{2\pi i}\int_{\sigma-i\infty}^{\sigma+i\infty} \mathcal{G}(1 - \alpha + \beta - s)\left[\int_0^\infty f(ax)x^{s+\alpha-1}\,a^\alpha dx\right]ds$$

$$= a^\alpha\int_0^\infty f(ax)\left[\frac{1}{2\pi i}\int_{\sigma-i\infty}^{\sigma+i\infty} \mathcal{G}(1 - \alpha + \beta - s)x^{s+\alpha-1}ds\right]dx$$

$$= a^\alpha\int_0^\infty f(ax)\left[\frac{1}{2\pi i}\int_{\Sigma-i\infty}^{\Sigma+i\infty} \mathcal{G}(S)\,x^{\beta-S}\,dS\right]dx.$$

Thus

$$\frac{1}{2\pi i}\int_{\sigma-i\infty}^{\sigma+i\infty} a^{-s}\,\mathscr{F}(s+\alpha)\,\mathscr{G}(1-\alpha+\beta-s)ds = a^{\alpha}\int_0^{\infty} f(ax)\,g(x)\,x^{\beta}\,dx. \quad \textbf{(7.3–7)}$$

In similar fashion, we find that

$$\frac{1}{2\pi i}\int_{\sigma-i\infty}^{\sigma+i\infty} \alpha^{-s}\,\mathscr{F}(s+a)\mathscr{G}(1+\alpha+\beta+s)ds$$

$$= \frac{1}{2\pi i}\int_{\sigma-i\infty}^{\sigma+i\infty} \mathscr{G}(1+\alpha+\beta+s)a^{-s}\left[\int_0^{\infty} f(y)y^{s+\alpha-1}dy\right]ds$$

$$= \frac{1}{2\pi i}\int_{\sigma-i\infty}^{\sigma+i\infty} \mathscr{G}(1+\alpha+\beta+s)\left[\int_0^{\infty} f(a/x)x^{-1-\alpha-s}a^{\alpha}dx\right]ds$$

$$= a^{\alpha}\int_0^{\infty} f(a/x)\left[\frac{1}{2\pi i}\int_{\Sigma-i\infty}^{\Sigma+i\infty}\mathscr{G}(S)\,x^{\beta-S}dS\right]dx,$$

and hence

$$\frac{1}{2\pi i}\int_{\sigma-i\infty}^{\sigma+i\infty} a^{-s}\mathscr{F}(s+\alpha)\mathscr{G}(1+\alpha+\beta+s)ds$$

$$= a^{\alpha}\int_0^{\infty} f(a/x)\,g(x)\,x^{\beta}dx. \quad \textbf{(7.3–8)}$$

The corresponding relations for Fourier and Laplace transforms follow directly from these Mellin transform Parseval formulas (7.3–7), (7.3–8). In particular, if $F(\omega)$ and $G(\omega)$ are the complex Fourier transforms of $f(t)$ and $g(t)$, respectively, then by virtue of the relationship between the Mellin and Fourier transforms discussed above, (7.3–7) gives rise to the identity

$$\int_{-\infty}^{\infty} F(\omega)\,G(\gamma-\omega)e^{-i\omega b}\,d\omega = \int_{-\infty}^{\infty} f(b+t)\,g(t)\,e^{i\gamma t}\,dt, \quad \textbf{(7.3–9)}$$

while (7.3–8) yields

$$\int_{-\infty}^{\infty} F(\omega)\,G(\gamma+\omega)e^{-i\omega b}\,d\omega = \int_{-\infty}^{\infty} f(b-t)\,g(t)\,e^{i\gamma t}\,dt \quad \textbf{(7.3–10)*}$$

(see Problem 7). The analogous results for Laplace transforms are

$$\frac{1}{2\pi i}\int_{\sigma-i\infty}^{\sigma+i\infty} \mathscr{L}_f(s)\,\mathscr{L}_g(r-s)e^{bs}\,ds = \int_{\max(0,-b)}^{\infty} f(b+t)\,g(t)e^{-rt}\,dt$$

and **(7.3–11)**

$$\frac{1}{2\pi i}\int_{\sigma-i\infty}^{\sigma+i\infty} \mathscr{L}_f(s)\,\mathscr{L}_g(r+s)e^{bs}\,ds = \int_0^{\max(0,b)} f(b-t)\,g(t)e^{-rt}\,dt.$$

The parameters b and γ are real throughout these expressions.

---

*This second of the Parseval formulas for Fourier transforms could alternatively have been derived from the first relation (7.3–9) using the reciprocity of the transform pair (7.3–2).

The Parseval relations, in the form (7.3–9), (7.3–10), can be profitably rewritten by taking advantage of the specific character of the definition (7.3–2). We note that if $g(t)$ has Fourier transform $G(\omega)$, then $\bar{g}(t)$ and $\bar{g}(-t)$ have Fourier transforms $\bar{G}(-\omega)$ and $\bar{G}(\omega)$, respectively. It follows then that (7.3–9) and (7.3–10) are actually not independent identities but are together equivalent to

$$\int_{-\infty}^{\infty} F(\omega)\,\bar{G}(\omega - \gamma)e^{-i\omega b}\,d\omega = \int_{-\infty}^{\infty} f(b + t)\,\bar{g}(t)\,e^{i\gamma t}\,dt.$$

A very important special case of this latter formula is obtained by associating g with f and setting $b = \gamma = 0$, whence

$$\int_{-\infty}^{\infty} |F(\omega)|^2\,d\omega = \int_{-\infty}^{\infty} |f(t)|^2\,dt. \tag{7.3–12}$$

Readers will recognize this result as the precise analogue of the Parseval equation (2.5–2) mentioned above and discussed earlier in Chapter 2.

7.4 Examples of some Mellin transforms

In this section we consider the Mellin transforms associated with a number of specific functions. For the most part we derive these transforms directly, using the definition (7.3–1). Our small list of functions can be expanded considerably, of course, by employing various helpful (and easily verified) properties* such as

Property 1 $\quad M[f(ax)] = a^{-s}\,\mathcal{F}(s) \qquad (a > 0)$

Property 2 $\quad M[F(x^{\lambda})] = \dfrac{1}{|\lambda|}\,\mathcal{F}(s/\lambda) \qquad (\lambda \text{ real})$

Property 3 $\quad M[x^{\alpha}f(x)] = \mathcal{F}(s + \alpha)$

Property 4 $\quad M[\ln x\, f(x)] = \dfrac{d}{ds}\,\mathcal{F}(s)$

Property 5 $\quad M\left[x\,\dfrac{df(x)}{dx}\right] = -s\,\mathcal{F}(s)$

Property 6 $\quad M\left[\displaystyle\int_0^x f(t)dt\right] = -\dfrac{1}{s}\,\mathcal{F}(s + 1).$

For more extensive lists of Mellin transforms see Oberhettinger (1974) or Erdélyi et al. (1954). Appendix 6 contains a number of examples of Fourier and Laplace transforms of elementary functions.

---

*Some of these properties have been implicit in our earlier analyses in this chapter.

While considering the following examples, readers should keep in mind that the Mellin transform of a well-behaved function generally turns out to be a regular analytic function of s at least in some strip $a < Re\ s < b$. The behavior of the function $f(x)$ in the vicinity of the origin $x = 0$ determines the smallest possible such constant a, while the behavior of f for large x fixes the largest possible b. Since the associated strip of analyticity may be important for manipulations with these Mellin transforms, we will indicate these strips (in some cases half-planes) wherever appropriate.

Algebraic functions

Example 1 If

$$f(x) = \begin{cases} 1 & 0 < x < 1 \\ 0 & x > 1, \end{cases}$$

then

$$M[f] = \int_0^1 x^{s-1}\, dx$$

$$= \frac{1}{s} \qquad \text{Re } s > 0.$$

□

Example 2 If

$$f(x) = \begin{cases} (1 - x)^{\nu-1} & 0 < x < 1 \\ 0 & x > 1 \end{cases}$$

with Re $\nu > 0$, then

$$M[f] = \int_0^1 x^{s-1}(1 - x)^{\nu-1}\, dx$$

$$= \frac{\Gamma(s)\Gamma(\nu)}{\Gamma(s + \nu)} = B(s,\nu) \qquad \text{Re } s > 0,$$

where $B(s,\nu)$ is the standard beta function (see Appendix 5). □

Example 3 If

$$f(x) = (1 + x)^{-\nu},$$

then

$$M[f] = \int_0^\infty x^{s-1}(1 + x)^{-\nu}\, dx$$

$$= \int_1^\infty (y - 1)^{s-1} y^{-\nu}\, dy$$

$$= \int_0^1 (1 - t)^{s-1} t^{\nu - s - 1} dt$$

$$= B(s, \nu - s) \qquad 0 < \text{Re } s < \text{Re } \nu.$$

The special case of this example when $\nu = 1$ is of interest; we then have

$$M[(1 + x)^{-1}] = \frac{\pi}{\sin \pi s} \qquad 0 < \text{Re } s < 1.$$

□

Exponential functions

Example 4 If

$$f(x) = e^{-x},$$

then

$$M[f] = \int_0^\infty x^{s-1} e^{-x} dx$$

$$= \Gamma(s) \qquad \text{Re } s > 0.$$

□

Example 5 If

$$f(x) = \begin{cases} (1 - x)^{\nu - 1} e^x & 0 < x < 1 \\ 0 & x > 1 \end{cases}$$

with Re $\nu > 0$, then

$$M[f] = \int_0^1 x^{s-1}(1 - x)^{\nu - 1} e^x dx$$

$$= B(s, \nu) M(s; \nu + s; 1) \qquad \text{Re } s > 0,$$

where $M(a;c;x)$ is the confluent hypergeometric function considered in Chapter 3. (Recall the integral representation (3.5–4).) □

Example 6 If

$$f(x) = \frac{1}{e^x - 1},$$

then

$$M[f] = \int_0^\infty \frac{x^{s-1}}{e^x - 1} dx$$

$$= \int_0^\infty x^{s-1} \left[\sum_{n=1}^{\infty} e^{-nx} \right] dx$$

$$= \sum_{n=1}^{\infty} n^{-s} \int_0^\infty y^{s-1} e^{-y} \, dy$$

$$= \zeta(s) \, \Gamma(s) \qquad \text{Re } s > 1,$$

where $\zeta(s)$ is called *Riemann's zeta function*. □

Logarithmic functions

Example 7 If

$$f(x) = \begin{cases} \ln x & 0 < x < 1 \\ 0 & x > 1, \end{cases}$$

then

$$M[f] = \frac{d}{ds} \left[\frac{1}{s} \right]$$

$$= -\frac{1}{s^2} \qquad \text{Re } s > 0,$$

by virtue of Property 4 applied to Example 1. □

Example 8 If

$$f(x) = \begin{cases} \ln(1 - x) & 0 < x < 1 \\ 0 & x > 1, \end{cases}$$

then

$$M[f] = \int_0^1 x^{s-1} \ln(1 - x) dx$$

$$= \frac{d}{d\nu} \int_0^1 x^{s-1} (1 - x)^{\nu-1} \, dx \Bigg|_{\nu=1}$$

$$= \frac{d}{d\nu} B(s, \nu) \Bigg|_{\nu=1}$$

$$= \frac{1}{s} [\psi(1) - \psi(s + 1)] \qquad \text{Re } s > -1,$$

where $\psi(z) \equiv \Gamma'(z)/\Gamma(z)$ is the psi function (see Appendix 5). □

Example 9 If

$$f(x) = \ln(1 + x),$$

then

$$M[f] = \int_0^\infty x^{s-1} \ln(1 + x)\, dx$$

$$= -\frac{d}{d\nu} \int_0^\infty x^{s-1} (1 + x)^{-\nu}\, dx \bigg|_{\nu=0}$$

$$= -\Gamma(s) \frac{d}{d\nu} \left[\frac{\Gamma(\nu - s)}{\Gamma(\nu)} \right]_{\nu=0}$$

$$= \Gamma(s)\Gamma(-s) \frac{\Gamma'(\nu)}{\Gamma^2(\nu)} \bigg|_{\nu=0}$$

$$= \frac{-\pi}{s \sin \pi s} \left[\frac{\nu\Gamma'(\nu + 1) - \Gamma(\nu + 1)}{\Gamma^2(\nu + 1)} \right]_{\nu=0}$$

$$= \frac{\pi}{s \sin \pi s} \qquad -1 < \text{Re } s < 0.$$

□

Trigonometric functions

Example 10 If

$$f(x) = e^{ix},$$

then

$$M[f] = \int_0^\infty x^{s-1} e^{ix}\, dx$$

$$= e^{is\pi/2} \int_0^{-i\infty} z^{s-1} e^{-z}\, dz$$

$$= e^{is\pi/2} \int_0^\infty z^{s-1} e^{-z}\, dz$$

$$= e^{is\pi/2}\, \Gamma(s) \qquad 0 < \text{Re } s < 1,$$

owing to the Cauchy-Goursat theorem for regular analytic functions and Example 4 above. In similar fashion we can show

$$M[e^{-ix}] = e^{-is\pi/2}\Gamma(s) \qquad 0 < \text{Re } s < 1$$

and hence,

$$M[\cos x] = \Gamma(s) \cos \tfrac{1}{2}\pi s \qquad 0 < \text{Re } s < 1,$$

while

$$M[\sin x] = \Gamma(s) \sin \tfrac{1}{2}\pi s \qquad -1 < \text{Re } s < 1.$$

(Note the larger strip of analyticity due to the simple zero of the sine function at the origin.)

Alternative forms for these last two Mellin transforms are often of value. They can be obtained by applying Legendre's duplication formula (see Appendix 5) to reexpress $\Gamma(s)$ in terms of $\Gamma(\tfrac{1}{2}s)$ and $\Gamma(\tfrac{1}{2}s + \tfrac{1}{2})$. Carrying out this procedure, we find

$$M[\cos x] = \sqrt{\pi}\, 2^{s-1} \frac{\Gamma(\tfrac{1}{2}s)}{\Gamma(\tfrac{1}{2} - \tfrac{1}{2}s)} \qquad 0 < \text{Re } s < 1$$

and

$$M[\sin x] = \sqrt{\pi}\, 2^{s-1} \frac{\Gamma(\tfrac{1}{2} + \tfrac{1}{2}s)}{\Gamma(1 - \tfrac{1}{2}s)} \qquad -1 < \text{Re } s < 1. \qquad \square$$

Bessel functions

Example 11 If

$$f(x) = x^{-\nu} J_\nu(x)$$

with Re $\nu > -\tfrac{1}{2}$, then, in view of the integral representation (3.6–6),

$$\begin{aligned}
M[f] &= \int_0^\infty x^{s-1} x^{-\nu} J_\nu(x)\, dx \\
&= \frac{2^{1-\nu}}{\sqrt{\pi}\,\Gamma(\nu + \tfrac{1}{2})} \int_0^\infty x^{s-1} \left[\int_0^1 \cos xt\, (1-t^2)^{\nu-\frac{1}{2}}\, dt\right] dx \\
&= \frac{2^{1-\nu}}{\sqrt{\pi}\,\Gamma(\nu + \tfrac{1}{2})} \int_0^1 (1-t^2)^{\nu-\frac{1}{2}} M[\cos xt]\, dt \\
&= \frac{2^{1-\nu}}{\sqrt{\pi}\,\Gamma(\nu + \tfrac{1}{2})} M[\cos x] \int_0^1 (1-t^2)^{\nu-\frac{1}{2}} t^{-s}\, dt \\
&= \frac{2^{s-\nu-1}}{\Gamma(\nu + \tfrac{1}{2})} \frac{\Gamma(\tfrac{1}{2}s)}{\Gamma(\tfrac{1}{2} - \tfrac{1}{2}s)} \int_0^1 (1-y)^{\nu-\frac{1}{2}} y^{-\frac{1}{2}s-\frac{1}{2}}\, dy \\
&= 2^{s-\nu-1} \frac{\Gamma(\tfrac{1}{2}s)}{\Gamma(\nu + 1 - \tfrac{1}{2}s)} \qquad 0 < \text{Re } s < \text{Re } \nu + 3/2. \qquad \square
\end{aligned}$$

Example 12 An analogous approach, employing the comparable integral representation for the modified Bessel function $K_\nu(x)$ given in Section 3.6, shows that

$$M[x^{-\nu}K_\nu(x)] = 2^{s-\nu-2}\,\Gamma(\tfrac{1}{2}s)\,\Gamma(\tfrac{1}{2}s-\nu) \qquad \text{Re } s > \max(0, 2\,\text{Re }\nu)$$

(see Problem 11). □

Hypergeometric functions

Example 13 If

$$f(x) = \begin{cases} (1-x)^{c-1}\,{}_2F_1\,(a,b;\,c;\,\gamma(1-x)) & 0 < x < 1 \\ 0 & x > 1 \end{cases}$$

with $\text{Re}(c - a - b) > 0$, $\text{Re } c > 0$, and $0 \le \gamma \le 1$, then

$$M[f] = B(c,s)\,{}_2F_1\,(a,b;\,c+s;\,\gamma) \qquad \text{Re } s > 0$$

(see Problem 13).

An interesting special case occurs when $a = \tfrac{1}{2} + p$, $b = \tfrac{1}{2} - p$, $c = \tfrac{1}{2}$ and $\gamma = \tfrac{1}{2}$. For these values we find

$$M[f] = B(\tfrac{1}{2},s)\,{}_2F_1\,(\tfrac{1}{2}+p,\,\tfrac{1}{2}-p;\,\tfrac{1}{2}+s;\,\tfrac{1}{2})$$

$$= \frac{2^{\frac{1}{2}-s}\,\pi\,\Gamma(s)}{\Gamma(\tfrac{1}{2}s+\tfrac{1}{2}-\tfrac{1}{2}p)\,\Gamma(\tfrac{1}{2}s+\tfrac{1}{2}+\tfrac{1}{2}p)}, \qquad \textbf{(7.4–1)}$$

where we have summed the hypergeometric function using the result of Problem 18, Chapter 3. In this case, by virtue of (3.4–2) (see also Problem 17 of Chapter 3),

$$f(x) = \begin{cases} \sqrt{2}\,(1-x^2)^{-\frac{1}{2}}\,T_p(x) & 0 < x < 1 \\ 0 & x > 1 \end{cases}$$

with $T_p(x)$ the Chebyshev function of the first kind, which we discussed at some length in Section 3.4.

In completely analogous fashion, we deduce for the special case $a = -p$, $b = 1+p$, $c = 1$, and $\gamma = \tfrac{1}{2}$ that if

$$f(x) = \begin{cases} P_p(x) & 0 < x < 1 \\ 0 & x > 1, \end{cases}$$

where $P_p(x)$ is the Legendre function of the first kind, then

$$M[f] = \frac{2^{-s}\sqrt{\pi}\,\Gamma(s)}{\Gamma(\tfrac{1}{2}s+\tfrac{1}{2}-\tfrac{1}{2}p)\,\Gamma(\tfrac{1}{2}s+1+\tfrac{1}{2}p)} \qquad \text{Re } s > 0 \qquad \textbf{(7.4–2)}$$

(recall the representation (3.4–1).) □

Example 14 If

$$f(x) = {}_2F_1\,(a,b;\,c;\,-x),$$

then

$$M[f] = \Gamma(s)\,\frac{\Gamma(a - s)\Gamma(b - s)\Gamma(c)}{\Gamma(a)\,\Gamma(b)\,\Gamma(c - s)} \qquad 0 < \text{Re}\, s < \min(\text{Re}\, a, \text{Re}\, b)$$

(see Problem 14). Note that when $b = c$ we obtain

$$M[{}_2F_1\,(a,c;\,c;\,-x)] = \frac{\Gamma(s)\Gamma(a - s)}{\Gamma(a)} \qquad 0 < \text{Re}\, s < \text{Re}\, a$$

which agrees with the result established earlier as Example 3. □

7.5 Further manipulations with Mellin transforms

In Section 7.3 we derived two convolution-type relations involving Mellin transforms. Owing to the form of the inverse expression (7.3–1), these Parseval formulas can also be rewritten as

$$M[x^\alpha \int_0^\infty f(xy)g(y)y^\beta dy] = \mathcal{F}(s + \alpha)\,\mathcal{G}(1 - \alpha + \beta - s) \qquad \textbf{(7.5–1)}$$

and

$$M[x^\alpha \int_0^\infty f(x/y)g(y)y^\beta dy] = \mathcal{F}(s + \alpha)\,\mathcal{G}(1 + \alpha + \beta + s) \qquad \textbf{(7.5–2)}$$

where $\mathcal{F}(s)$ and $\mathcal{G}(s)$, respectively, are the Mellin transforms of f and g. Now we want to exploit these identities and to deduce a number of useful additional results concerning special functions and integrals, the other transforms mentioned in Section 7.3, and some integral equations.

A gamma function identity

If we take $f(x) = g(x) = e^{-x}$ and assume that $\alpha + a$, $\beta + a + b - 1$ where a and b are real, (7.5–1) becomes, using Example 4 of Section 7.4,

$$M[x^a(1 + x)^{-a-b}]\Gamma(a + b) = \Gamma(a + s)\Gamma(b - s) \qquad -a < \text{Re}\, s < b.$$

(Note that we could have alternatively obtained this relation from Example 3 on page **205**.) Inverting and setting $x = 1$ then leads to

$$\frac{1}{2\pi}\int_{-\infty}^{\infty} \Gamma(a + \sigma + i\omega)\Gamma(b - \sigma - i\omega)d\omega = 2^{-a-b}\,\Gamma(a + b)$$

with $-a < \sigma < b$. When $b = a > 0$, we can choose $\sigma = 0$, and hence

$$\int_0^\infty |\Gamma(a + i\omega)|^2\, d\omega = \pi\, 2^{-2a}\,\Gamma(2a). \qquad \textbf{(7.5–3)}$$

Two convolution integrals

Applying Property 2 of Section 7.4 to Example 2 shows that the function

$$f(x) = \begin{cases} (1 - x^2)^{\nu-1} & 0 < x < 1 \\ 0 & x > 1 \end{cases}$$

has the Mellin transform

$$M[f] = \tfrac{1}{2}B(\tfrac{1}{2}s,\nu) \qquad \text{Re } s > 0.$$

Thus, substituting in (7.5–1) and simplifying, we find that

$$M[x^{1-\beta-2\nu} \int_0^x y^\beta (x^2 - y^2)^{\nu-1} g(y)dy] = \tfrac{1}{2}\mathcal{G}(s)B(\tfrac{1}{2} + \tfrac{1}{2}\beta - \tfrac{1}{2}s,\nu).$$

The special case $\beta = 0$, $\nu = \tfrac{1}{2}$ gives rise to

$$M\left[\int_0^x \frac{g(y)}{\sqrt{x^2 - y^2}}dy\right] = \tfrac{1}{2}\sqrt{\pi}\,\mathcal{G}(s)\,\frac{\Gamma(\tfrac{1}{2} - \tfrac{1}{2}s)}{\Gamma(1 - \tfrac{1}{2}s)}. \qquad \textbf{(7.5–4)}$$

In similar fashion we can establish

$$M\left[\int_0^\infty \frac{g(y)}{\sqrt{y^2 - x^2}}dy\right] = \tfrac{1}{2}\sqrt{\pi}\,\mathcal{G}(s)\,\frac{\mathcal{G}(\tfrac{1}{2}s)}{\Gamma(\tfrac{1}{2} + \tfrac{1}{2}s)}. \qquad \textbf{(7.5–5)}$$

(see Problem 15).

Another hypergeometric function representation

The Mellin transform of the Gauss hypergeometric function with negative argument was given earlier as Example 14 on page **211**. If we apply the inverse Mellin transform to both sides of that identity we obtain, as expected, the integral representation

$$\frac{\Gamma(a)\Gamma(b)}{\Gamma(c)}\,{}_2F_1(a,b;\,c;\,-x) = \frac{1}{2\pi i}\int_{\sigma-i\infty}^{\sigma-i\infty} x^{-s}\Gamma(s)\,\frac{\Gamma(a-s)\Gamma(b-s)}{\Gamma(c-s)}\,ds \qquad \textbf{(7.5–6)}$$

where $o < o < \min(\text{Re } a, \text{Re } b)$. (This relation, incidentally, was used by E. W. Barnes as a convenient point-of-departure for his 1908 exposition on these important special functions.) A companion representation for hypergeometric functions of positive argument, also involving gamma functions (and hence sometimes referred to as a Barnes integral), may be deduced from the Parseval formula (7.5–1).

We begin by letting

$$f(x) = \begin{cases} (1-x)^{-b} & 0 < x < 1 \\ 0 & x > 1, \end{cases}$$

$$g(x) = \begin{cases} (1-x)^{c-a-1} & 0 < x < 1 \\ 0 & x > 1, \end{cases}$$

where Re $b < 1$ and Re $c >$ Re $a > 0$, and observe that

$$\mathscr{F}(s) = B(s, 1 - b) \qquad \text{and} \qquad \mathscr{G}(s) = B(s, c - a)$$

for Re $s > 0$ again by virtue of Example 2 of Section 7.4. Substituting into (7.5–1) with $\alpha = 0$, $\beta = a - 1$ then leads to

$$M\left[\int_0^{\min(1,1/x)} (1-xy)^{-b}(1 - y)^{c-a-1}\, y^{a-1}dy\right] = B(s,1 - b)B(a - s,c - a)$$

for $0 <$ Re $s <$ Re a. In view of the Euler representation (3.3–6), we see that the definite integral in this last expression can be rewritten in terms of a hypergeometric function. Therefore, inverting finally gives rise, for $0 < x \leq 1$, to

$$\frac{\Gamma(a)}{\Gamma(1 - b)\Gamma(c)}\, {}_2F_1\,(a,b; c; x) = \frac{1}{2\pi i}\int_{\sigma - i\infty}^{\sigma + i\infty} x^{-s}\frac{\Gamma(s)\Gamma(a - s)}{\Gamma(1 - b + s)\Gamma(c - s)}ds, \quad \textbf{(7.5–7)}$$

where $0 < \sigma <$ Re a.

Hypergeometric functions as convolutions

The case $\gamma = 1$ of Example 13 is equivalent to

$$(1 - x)^{c-1}{}_2F_1\,(a,b; c; 1 - x) = \frac{1}{2\pi i}\int_{\sigma - i\infty}^{\sigma + i\infty} x^{-s}\Gamma(c)\frac{\Gamma(s)\Gamma(s + c - a - b)}{\Gamma(s + c - a)\Gamma(s + c - b)}ds$$

(7.5–8)

for $0 < x < 1$ (see Problem 13). This expression can be used to generate two interesting integral representations of hypergeometric functions valid for $c = 1,2$.

The first such relation uses the fact that the function

$$f(x) = \begin{cases} (1 - x^2)^{-1/2}\, T_p(x^{-1}) & 0 < x < 1 \\ 0 & x > 1, \end{cases}$$

where T_p is the Chebyshev function of the first kind, has the Mellin transform

$$M[f] = 2^{s-2}\frac{\Gamma(\tfrac{1}{2}s + \tfrac{1}{2}p)\,\Gamma(\tfrac{1}{2}s - \tfrac{1}{2}p)}{\Gamma(s)}$$

for Re $s > |\text{Re } p|$. (Sneddon (1972) gives this result when p is integral). Thus if we take f as given above, let

$$g(x) = \begin{cases} (1 - x^2)^{-1/2}\, T_q(x) & 0 < x < 1 \\ 0 & x > 1, \end{cases}$$

where T_q is another Chebyshev function, assume $\beta = -1$, and substitute in (7.5–2), we are led, in view of (7.4–1), to

$$M\left[x^{\alpha}\int_x^1 \frac{T_p(y/x)T_q(y)}{\sqrt{y^2-x^2}\sqrt{1-y^2}}\,dy\right]$$
$$= \tfrac{1}{4}\pi\,\frac{\Gamma(\tfrac{1}{2}s+\tfrac{1}{2}\alpha+\tfrac{1}{2}p)\Gamma(\tfrac{1}{2}s+\tfrac{1}{2}\alpha-\tfrac{1}{2}p)}{\Gamma(\tfrac{1}{2}s+\tfrac{1}{2}\alpha+\tfrac{1}{2}+\tfrac{1}{2}q)\Gamma(\tfrac{1}{2}\alpha+\tfrac{1}{2}-\tfrac{1}{2}q)}$$

with Re $s > |\text{Re}\, p|$. Setting $\alpha = p$ and making the proper associations, we can then use (7.5–8) to invert this expression, thus obtaining

$$x^p \int_x^1 \frac{T_p(y/x)T_q(y)}{\sqrt{y^2-x^2}\sqrt{1-y^2}}\,dy$$
$$= \tfrac{1}{2}\pi\,{}_2F_1\,(\tfrac{1}{2}-\tfrac{1}{2}p+\tfrac{1}{2}q,\tfrac{1}{2}-\tfrac{1}{2}p-\tfrac{1}{2}q;1;1-x^2) \qquad \textbf{(7.5–9)}$$

for $0 < x < 1$.

In order to derive the second representation, we apply the identity (7.5–2) to the functions

$$f(x) = \begin{cases} P_p(x^{-1}) & 0 < x < 1 \\ 0 & x > 1 \end{cases}$$

and

$$g(x) = \begin{cases} P_q(x) & 0 < x < 1 \\ 0 & x > 1, \end{cases}$$

where P_p, P_q are Legendre functions of the first kind. Since

$$M[f] = 2^{s-1}\,\frac{\Gamma(\tfrac{1}{2}s+\tfrac{1}{2}+\tfrac{1}{2}p)\,\Gamma(\tfrac{1}{2}s-\tfrac{1}{2}p)}{\sqrt{\pi}\,\Gamma(1+s)}$$

with Re $s > -\tfrac{1}{2} + |\text{Re}(p+\tfrac{1}{2})|$ in this case (see Sneddon (1972) or Oberhettinger (1974), for example) and $M[g]$ is given by (7.4–2), we find

$$M[x^{\alpha}\int_x^1 P_p(y/x)P_q(y)dy]$$
$$= \tfrac{1}{4}\,\frac{\Gamma(\tfrac{1}{2}s+\tfrac{1}{2}\alpha+\tfrac{1}{2}+\tfrac{1}{2}p)\,\Gamma(\tfrac{1}{2}s+\tfrac{1}{2}\alpha-\tfrac{1}{2}p)}{\Gamma(\tfrac{1}{2}s+\tfrac{1}{2}\alpha+3/2+\tfrac{1}{2}q)\,\Gamma(\tfrac{1}{2}s+\tfrac{1}{2}\alpha+1-\tfrac{1}{2}q)}\,.$$

Again setting $\alpha = p$ and making the proper associations, (7.5–8) now gives rise to

$$x^p \int_x^1 P_p(y/x)P_q(y)dy$$
$$= \tfrac{1}{2}(1-x^2)\,{}_2F_1\,(1-\tfrac{1}{2}p+\tfrac{1}{2}q,\tfrac{1}{2}-\tfrac{1}{2}p-\tfrac{1}{2}q;2;1-x^2) \qquad \textbf{(7.5–10)}$$

for $0 < x < 1$.

Integral equations with product or quotient kernels

Equations in which the unknown function occurs within an integration are generally difficult to solve (see Chapter 10). However, if the integral equation has the form

$$G(x) = \int_0^\infty K(xy)f(y)dy \qquad x > 0, \tag{7.5–11}$$

where the forcing function g and the kernel K (with the product xy as its argument) are known, the solution can be expressed, at least formally, by making use of our earlier Parseval identities for Mellin transforms. Indeed, denoting the transforms of f,g, and K by $\mathcal{F}$, $\mathcal{G}$, and $\mathcal{K}$, respectively, (7.5–1) shows that

$$\mathcal{G}(s) = \mathcal{K}(s)\,\mathcal{F}(1-s)$$

for this important class of integral equations. It obviously follows that

$$f(x) = M^{-1}[\mathcal{G}(1-s)/\mathcal{K}(1-s)]$$

or

$$f(x) = \int_0^\infty L(xy)g(y)dy \qquad x > 0, \tag{7.5–12}$$

where $L(x)$ is the function (if it exists) whose Mellin transform is $1/\mathcal{K}(1-s)$. If instead of (7.5–11), the integral equation of interest is

$$g(x) = \int_0^\infty K(y/x)f(y)dy \qquad x > 0, \tag{7.5–13}$$

where K is now a "quotient **kernel,**" then comparable reasoning using (7.5–2) leads to

$$f(x) = \int_0^\infty L(x/y)g(y)\,\frac{dy}{y^2} \qquad x > 0. \tag{7.5–14}$$

For those cases in which there does not exist a function that has $1/\mathcal{K}(1-s)$ as its Mellin transform, the above procedure must be modified appropriately. This can often be accomplished by redefining L as

$$L(x) \equiv M^{-1}\left[\frac{(-s)^{-n}}{\mathcal{K}(1-s)}\right],$$

where n is the smallest integer for which this definition is meaningful. It follows that the solution of (7.5–11) is now given by

$$f(x) = \left(x\,\frac{d}{dx}\right)^n \int_0^\infty L(xy)g(y)dy \qquad x > 0, \tag{7.5–15}$$

while the solution of (7.5–13) can be expressed as

$$f(x) = \left(x\,\frac{d}{dx}\right)^n \int_0^\infty L(x/y)g(y)\,\frac{dy}{y^2} \qquad x > 0 \tag{7.5–16}$$

(see Problem 16).

The formulas (7.5–12) and (7.5–14)–(7.5–16) provide integral representations for the solutions of integral equations having product or quotient kernels, a class of problems that occurs rather often in practical applications. We illustrate the utility of these expressions with several examples.

Example 1 If

$$g(x) = \sqrt{\frac{2}{\pi}} \int_0^\infty \cos xy\, f(y)dy,$$

then

$$\mathcal{K}(s) = \sqrt{\frac{2}{\pi}}\, \Gamma(s) \cos \tfrac{1}{2}\pi s$$

$$= 2^{s-\frac{1}{2}} \frac{\Gamma(\frac{1}{2}s)}{\Gamma(\frac{1}{2}-\frac{1}{2}s)},$$

so that

$$1/\mathcal{K}(1-s) = 2^{s-\frac{1}{2}} \frac{\Gamma(\frac{1}{2}s)}{\Gamma(\frac{1}{2}-\frac{1}{2}s)},$$

and hence by (7.5–12),

$$f(x) = \sqrt{\frac{2}{\pi}} \int_0^\infty \cos xy\, g(y)dy.$$

□

Example 2 Analogously, if

$$g(x) = \sqrt{\frac{2}{\pi}} \int_0^\infty \sin xy\, f(y)dy,$$

then

$$f(x) = \sqrt{\frac{2}{\pi}} \int_0^\infty \sin xy\, g(y)dy.$$

□

Example 3 If

$$g(x) = \int_0^\infty \sqrt{xy}\, J_\nu(xy) f(y)dy,$$

then

$$\mathcal{K}(s) = 2^{s-\frac{1}{2}} \frac{\Gamma(\frac{1}{2}\nu + \frac{1}{4} + \frac{1}{2}s)}{\Gamma(\frac{1}{2}\nu + \frac{3}{4} - \frac{1}{2}s)},$$

$$1/\mathcal{K}(1-s) = 2^{s-\frac{1}{2}} \frac{\Gamma(\frac{1}{2}\nu + \frac{1}{4} + \frac{1}{2}s)}{\Gamma(\frac{1}{2}\nu + \frac{3}{4} - \frac{1}{2}s)},$$

and thus

$$f(x) = \int_0^\infty \sqrt{xy}\, J_\nu(xy) g(y) dy. \qquad \square$$

The above are examples of integral equations with so-called *Fourier kernels*, that is, ones for which

$$\mathcal{K}(s)\,\mathcal{K}(1-s) = 1 \tag{7.5–17}$$

(see Problem 17). Since we had already encountered these kernels in our discussion of specific transforms in Section 7.3, we could have anticipated the symmetric forms of the solution expressions.

A somewhat different situation is illustrated by

Example 4* If

$$g(x) = \int_0^x \frac{f(y)}{\sqrt{x^2 - y^2}}\, dy,$$

or equivalently

$$x^2 g(x) = \int_0^x (x/y)\,[1 - (y/x)^2]^{-1/2}\, y f(y) dy,$$

then

$$\mathcal{K}(s) = \tfrac{1}{2} B(\tfrac{1}{2}s - \tfrac{1}{2}, \tfrac{1}{2})$$

and

$$1/\mathcal{K}(1-s) = \frac{2}{\sqrt{\pi}} \frac{\Gamma(\tfrac{1}{2} - \tfrac{1}{2}s)}{\Gamma(-\tfrac{1}{2}s)}.$$

Since this latter expression is not invertible, we use the alternative (7.5–16), defining L by

$$\begin{aligned} L(x) &\equiv M^{-1}\left[\frac{(-s)^{-1}}{\mathcal{K}(1-s)}\right] \\ &= \frac{1}{\sqrt{\pi}} M^{-1}\left[\frac{\Gamma(\tfrac{1}{2} - \tfrac{1}{2}s)}{\Gamma(1 - \tfrac{1}{2}s)}\right] \\ &= \begin{cases} 0 & 0 < x < 1 \\ \dfrac{2}{\pi}(x^2 - 1)^{-1/2} & x > 1 \end{cases} \end{aligned}$$

*This example is interesting from an historical point of view since it is related to an integral equation first solved (in rather different fashion, of course) by Niels Abel in 1823.

whence

$$xf(x) = x \frac{d}{dx} \int_0^x \frac{2}{\pi} [(x/y)^2 - 1]^{-1/2} y^2 g(y) \frac{dy}{y^2}$$

or

$$f(x) = \frac{2}{\pi} \frac{d}{dx} \int_0^x \frac{y\, g(y)}{\sqrt{x^2 - y^2}} dy.$$

□

A slightly more general case of Example 4 is treated in Problem 18. Additional examples appear in Sneddon (1972) and Titchmarsh (1959) and among the other problems at the end of this chapter.

Relationships with other transforms

If the Fourier cosine and sine transforms of a function f of a positive real variable are as defined in Section 7.3, then implicit in Examples 1 and 2 of the last subsection are the identities

$$M[F_c] = \sqrt{\frac{2}{\pi}}\, \Gamma(s) \cos \tfrac{1}{2}\pi s\; \mathcal{F}(1-s),$$

$$M[F_s] = \sqrt{\frac{2}{\pi}}\, \Gamma(s) \sin \tfrac{1}{2}\pi s\; \mathcal{F}(1-s),$$

where $\mathcal{F}(s)$ is the Mellin transform of f. In similar fashion, underlying Example 3 is the Hankel transform relationship

$$M[\mathcal{H}] = 2^{s-1} \frac{\Gamma(\tfrac{1}{2}\nu + \tfrac{1}{2}s)}{\Gamma(1 + \tfrac{1}{2}\nu - \tfrac{1}{2}s)} \mathcal{F}(2-s).$$

Comparable expressions relating the other transforms discussed in Section 7.3 to the Mellin transform can easily be derived from (7.5–1) and (7.5–2). Carrying out the necessary steps, we find

$$M[\mathcal{L}] = \Gamma(s)\, \mathcal{F}(1-s)$$

for the Laplace transform and

$$M[S] = \frac{\pi}{\sin \pi s} \mathcal{F}(s) \tag{7.5–18}$$

for the Stieltjes transform. Related results for several other transforms are given in Oberhettinger (1974) (see also Problem 21).

7.6 The inversion process

Up to now we have been concerned, for the most part, with the straightforward calculation of Mellin transforms and their manipulation. Resolving practical problems, however, typically involves the reverse of this process. Sometimes the determination of an inverse Mellin transform can be accomplished simply "by inspection" or by "table search." (Oberhettinger (1974) has a rather extensive tabulation of inverse transforms; the list of Erdélyi et al. (1954) is slightly smaller.) More frequently the inversion process must be carried out from first principles. We adduce two examples to illustrate the nature of such calculations.

Example 1 If

$$\mathscr{F}(s) = 1/(s + \alpha) \qquad \mathrm{Re}(s + \alpha) > 0,$$

we know, by virtue of Property 3 of Section 7.4 and the definition of the inverse Mellin transform, that

$$f(x) = \frac{x^\alpha}{2\pi i}\int_{\sigma - i\infty}^{\sigma + i\infty} \frac{x^{-s}}{s}\,ds \qquad \sigma > 0.$$

For $x > 1$, we use the Cauchy-Goursat theorem (of Chapter 5) to show that the value of the line integral in this expression is unchanged if we shift the vertical contour to the *right*. Thus

$$\begin{aligned} f(x) &= \frac{x^\alpha}{2\pi i}\lim_{\sigma\to\infty}\int_{\sigma - i\infty}^{\sigma + i\infty} \frac{x^{-s}}{s}\,ds \\ &= 0 \qquad\qquad x > 1. \end{aligned}$$

For $0 < x < 1$, we shift the contour to the *left*. In this case, the singularity of the integrand at the origin contributes, and the Cauchy Integral formula is applicable. It follows that

$$\begin{aligned} f(x) &= x^\alpha x^{-s}\Big|_{s=0} + \frac{x^\alpha}{2\pi i}\lim_{\sigma\to-\infty}\int_{\sigma - i\infty}^{\sigma + i\infty} \frac{x^{-s}}{s}\,ds \qquad 0 < x < 1. \\ &= x^\alpha \end{aligned}$$

Combining these two results yields the function to be expected in light of Example 1 of Section 7.4. □

The method employed in this simple example is an archetype for determining the inverse transform in more general situations, though the details usually become more complicated, as the next example suggests.

Example 2 If

$$\mathscr{F}(s) = \Gamma(s) \qquad \mathrm{Re}\, s > 0,$$

then

$$f(x) = \frac{1}{2\pi i}\int_{\sigma - i\infty}^{\sigma + i\infty} x^{-s}\Gamma(s)\,ds \qquad \sigma > 0. \tag{7.6–1}$$

In order to determine the effects of contour shifts in this case, we need to employ Stirling's formula for asymptotic approximation of the gamma function, viz.

$$\Gamma(z) = \sqrt{2\pi/z}\, e^{z\ln z - z}\,[1 + O(1/z)]$$

as $|z| \to \infty$ with $|\arg z| < \pi$ (see Appendix 5). It follows that

$$|x^{-s}\Gamma(\sigma + i\omega)|$$

$$= \sqrt{2\pi}(\sigma^2+\omega^2)^{-1/4} e^{\sigma[\frac{1}{2}\ln(\sigma^2+\omega^2) - \ln x - 1] - \omega \tan^{-1}(\omega/\sigma)}[1 + 0(1/\sqrt{\sigma^2 + \omega^2})] \quad \textbf{(7.6–2)}$$

and hence the contour in (7.6–1) should be shifted to the left for all values of x. In fact, noting from (7.6–2) that

$$\lim_{|\omega|\to\infty} |x^{-s}\,\Gamma(\sigma+i\omega)| = \lim_{|\omega|\to\infty} \sqrt{2\pi/|\omega|}\, e^{\sigma \ln|\omega| - |\omega|\pi/2}$$

$$= 0,$$

and again employing the Cauchy-Goursat theorem and its various implications, we see that we can rewrite (7.6–1) in terms of a series of integrals around small circular contours C_k enclosing the singularities of $\Gamma(s)$ at $s = -k$, $k = 0,1,2, \ldots$.

Carrying out the steps just indicated, we obtain

$$f(x) = \sum_{k=0}^{\infty} \frac{1}{2\pi i} \oint_{C_k} x^{-s}\Gamma(s)ds + \frac{1}{2\pi i} \lim_{\sigma\to-\infty} \int_{\sigma-i\infty}^{\sigma+i\infty} x^{-s}\Gamma(s)ds.$$

The last integral vanishes in the limit of large negative σ, by (7.6–2). The integral along the circular contour around the singularity at $s = -k$, on the other hand, has the value

$$\frac{1}{2\pi i}\oint_{C_k} x^{-s}\Gamma(s)ds = \frac{1}{2\pi i}\oint_{C_k} \frac{x^{-s}}{\Gamma(1-s)} \frac{\pi}{\sin \pi s} ds$$

$$= \frac{1}{2\pi i}\oint_{C_k} \frac{x^{-s}}{\Gamma(1-s)} \frac{(-1)^k \pi(s+k)}{\sin \pi(s+k)} \frac{ds}{s+k}$$

$$= \frac{x^{-k}(-1)^k}{\Gamma(1+k)}$$

$$= \frac{(-x)^k}{k!},$$

by virtue of the Cauchy Integral formula. As a consequence, we find

$$f(x) = e^{-x} \qquad x > 0,$$

which we could, of course, have anticipated from Example 4, Section 7.4. Our analysis is now complete. □

The inversion process for more complicated integrands is often no more difficult than that encountered in this second example. If the singularities are all simple poles, then the procedure is essentially identical*, although the contour may have to be shifted in both directions, as in Example 1, in order to accommodate the full range of x. We defer any further examples to the next section and to the problems at the end of this chapter.

7.7 Some problems in heat conduction

In Section 7.1 we initiated an examination of the mixed boundary-value problem for the steady-state temperature distribution $\phi(x,y)$ on a semi-infinite flat plate with partially insulated free edge. The equations of interest had the form (in polar coordinates)

$$\begin{aligned} \Delta\phi &= 0 \qquad r > 0, \quad 0 < \theta < \pi \\ \phi(r,0) &= f(r) \\ \frac{\partial\phi}{\partial\theta}(r,\pi) &= 0, \end{aligned} \tag{7.7–1}$$

and we had been led, by boundedness and linearity considerations, to try

$$\phi(r,\theta) = \int_\Gamma [A(\lambda)\cos\lambda\theta + B(\lambda)\sin\lambda\theta]r^{-\lambda}d\lambda,$$

as a representation for the solution. Here A and B were yet to be determined functions of λ and Γ was a contour along which Re λ was nonnegative. We noted that the boundary conditions would be satisfied if

$$B(\lambda) = A(\lambda)\tan\lambda\pi$$

where

$$\int_\Gamma A(\lambda)r^{-\lambda}d\lambda = f(r),$$

and thus the efficacy of this approach to the problem rested upon the possibility of inverting this integral equation for $A(\lambda)$.

If we select for our contour Γ the infinite vertical line Re λ = constant > 0, the above integral becomes nothing more than the inverse Mellin transform of $A(\lambda)$. It follows then that if $f(r)$ is well behaved, $A(\lambda)$ must be given by

$$A(\lambda) = \frac{1}{2\pi i}\int_0^\infty r^{\lambda-1}f(r)dr.$$

*Readers well-versed in the "residue calculus" for analytic functions may be able to eliminate some intermediate steps.

Carrying out this integration, therefore, allows us to complete the formal solution of the given boundary-value problem.

Example Consider the specific case of $f(r) = e^{-r}$. Then

$$A(\lambda) = \frac{\Gamma(\lambda)}{2\pi i},$$

$$B(\lambda) = \frac{\Gamma(\lambda)}{2\pi i} \tan \lambda\pi,$$

and

$$\phi(r,\theta) = \frac{1}{2\pi i}\int_{\sigma-i\infty}^{\sigma+i\infty} r^{-\lambda}\Gamma(\lambda)\,[\cos\lambda\theta + \tan\lambda\pi \sin\lambda\theta]\,d\lambda$$

with $\sigma > 0$. A more practical representation for the solution in this example can be obtained by applying the inversion procedure outlined in the previous section; we find

$$\begin{aligned}\phi(r,\theta) &= \sum_{k=0}^{\infty} \frac{(-r)^k}{\Gamma(k+1)} \cos k\theta + \frac{1}{\pi}\sum_{k=0}^{\infty} r^{k+1/2}\Gamma(-k-1/2)\sin(k+1/2)\theta \\ &= \operatorname{Re}\left[\sum_{k=0}^{\infty} \frac{(-re^{i\theta})^k}{\Gamma(k+1)}\right] - \sqrt{r}\,\operatorname{Im}\left[e^{i\theta/2}\sum_{k=0}^{\infty}\frac{(-re^{i\theta})^k}{\Gamma(k+3/2)}\right] \\ &= e^{-r\cos\theta}\cos(r\sin\theta) - 2\sqrt{r/\pi}\,\operatorname{Im}\,[e^{i\theta/2}M(1;3/2;-re^{i\theta})]\end{aligned}$$

where $M(a;c;x)$ is the confluent hypergeometric function introduced in Chapter 3. Alternatively, using the integral representation (3.5–4),

$$\phi(r,\theta) = e^{-r\cos\theta}\cos(r\sin\theta) + \sqrt{r/\pi}\int_0^1 e^{-rt\cos\theta}\,\frac{\sin(rt\sin\theta - 1/2\theta)}{\sqrt{1-t}}\,dt.$$

With this latter form of the solution, it is easy to verify that the boundary conditions are indeed satisfied. □

A problem with prescribed heat flux

An interesting variant of the mixed boundary-value problem just considered is provided by

$$\begin{aligned}\Delta\phi &= 0 \qquad r > 0, \quad 0 < \theta < \pi \\ \phi(r,0) &= 0 \qquad\qquad\qquad\qquad (7.7\text{–}2) \\ \frac{\partial\phi}{\partial\theta}(r,\pi) &= g(r).\end{aligned}$$

In the terminology of heat conduction, one-half of the free edge of a semi-infinite plate is kept at constant temperature (chosen to be zero merely for convenience);

along the other half of the edge the heat flux is specified. Reasoning as before, we are again led to try

$$\phi(r,\theta) = \int_\Gamma [A(\lambda) \cos \lambda\theta + B(\lambda) \sin \lambda\theta] r^{-\lambda} d\lambda \qquad \textbf{(7.7–3)}$$

as a possible solution. However, the new boundary conditions necessitate

$$A(\lambda) = 0$$

and

$$\int_\Gamma \lambda B(\lambda) \cos \lambda\pi \, r^{-\lambda} d\lambda = g(r).$$

The Mellin transform theory, therefore, yields

$$\lambda B(\lambda) \cos \lambda\pi = \frac{1}{2\pi i} \int_0^\infty r^{\lambda-1} g(r) \, dr$$

for the value of $B(\lambda)$ that must be used in (7.7–3).

Example As a specific case we take

$$g(r) = \frac{r}{(1+r)^2}.$$

Owing to Example 3 of Section 7.4 (page **205**), it follows that

$$\lambda B(\lambda) \cos \lambda\pi = \frac{\Gamma(\lambda + 1)\Gamma(1 - \lambda)}{2\pi i}$$

$$= \frac{-\lambda i}{2 \sin \lambda\pi} \qquad -1 < \text{Re}\, \lambda < 1,$$

and thus

$$B(\lambda) = \frac{-i}{\sin 2\lambda\pi}.$$

Substituting in (7.7–3) and explicitly carrying out the indicated inversion then gives

$$\phi(r,\theta) = -\sum_{k=0}^{\infty} r^{\pm \frac{1}{2}k} \sin k(\tfrac{1}{2}\phi + \pi),$$

where the plus sign is associated with $0 < r \leq 1$ and the minus sign with $r \geq 1$. Summing the series, we finally obtain

$$\phi(r,\theta) = -\text{Im} \sum_{k=0}^{\infty} [r^{\pm \frac{1}{2}} e^{i(\frac{1}{2}\theta + \pi)}]^k$$

$$= -\text{Im}\, [1 - r^{\pm \frac{1}{2}} e^{i(\frac{1}{2}\theta + \pi)}]^{-1}$$

$$= \frac{\sqrt{r} \sin \frac{1}{2}\theta}{1 + 2\sqrt{r} \cos \frac{1}{2}\theta + r}.$$

The mathematical problem leading to the special case of (7.7–2) just considered is a simplified model for a thermal defect in a conductive coating. The boundary-value problem whose solution $\Phi(u,v)$ is required is

$$\begin{aligned} \triangle\Phi &= 0 && -\infty < u < \infty,\ v > 0 \\ \Phi(u,0) &= 0 && |u| > a \\ \frac{\partial\Phi}{\partial v}(u,0) &= \text{constant} && |u| < a. \end{aligned}$$

Under the conformal transformation $z = a(1 + w)/(1 - w)$ where $z = re^{i\theta}$ and $w = u + iv$, this is equivalent to the problem we have solved. (See Karush and Young (1952) for additional discussion of this application.)

Problems with rotational symmetry

The mixed boundary-value problems discussed above have three-dimensional generalizations. For example, one analogue of (7.7–1) is the rotationally symmetric problem (expressed in cylindrical coordinates ρ,θ,z)

$$\begin{aligned} \triangle\phi &= 0 && \rho \geq 0,\ 0 \leq \theta \leq 2\pi,\ z > 0 \\ \phi(\rho,\theta,0) &= f(\rho) && \rho < a \\ \frac{\partial\phi}{\partial z}(\rho,\theta,0) &= 0 && \rho > a. \end{aligned} \qquad \textbf{(7.7–4)}$$

(Sneddon (1966) calls this the "first basic (axisymmetric) problem" and considers a number of physical situations in which this mixed problem arises.) Gratifyingly, an operational approach similar to that of Section 7.1 can be applied in this case.

The separated solutions of Laplace's equation in cylindrical coordinates involve Bessel functions in ρ, sinusoids in θ, and exponentials in z. The natural assumption of boundedness of the solution (needed again for uniqueness) removes the Bessel functions of the second kind and any increasing exponentials from consideration; the rotational symmetry in the problem, moreover, suggests that any dependence upon the angular variable θ is actually extrinsic. These observations lead us to try

$$\phi(\rho,\theta,z) = \phi(\rho,z) = \int_0^\infty A(\lambda)J_0(\lambda\rho)e^{-\lambda z}d\lambda,$$

where $A(\lambda)$ is to be determined, as a ration for the solution of (7.7–4). To fulfill the boundary conditions, the function $A(\lambda)$ must satisfy

$$\int_0^\infty A(\lambda)J_0(\lambda\rho)d\lambda = f(\rho) \qquad 0 \leq \rho < 1,$$

and **(7.7–5)***

$$\int_0^\infty A(\lambda)\lambda J_0(\lambda\rho)d\lambda = 0 \qquad \rho > 1.$$

*If the second of these equations were absent and the first equation held for all ρ, the problem would be equivalent to that considered as Example 3 in Section 7.5.

The coupled equations (7.7–5) constitute a set of *dual integral equations* for the unknown $A(\lambda)$. Although the general theory for such equations is quite complex (see Sneddon (1966) or Green (1969), for example), the specific case of interest here can be handled using the Mellin transform. However, we leave further investigation of these matters to the interested reader (Titchmarsh (1959) has the details), and merely indicate that the procedure relies in part on the generalized Parseval formulas (7.3–7), (7.3–8) and notions from basic analytic function theory.

References

Apostol, T. M. (1974): *Mathematical Analysis,* 2nd ed., Addison-Wesley, Reading, Mass.; Section 15.8.

Carrier, G. F., M. Krook, and C. E. Pearson (1966): *Functions of a Complex Variable,* McGraw-Hill, New York; Chapter 7.

Carslaw, H. S. and J. C. Jaeger (1959): *Conduction of Heat in Solids,* 2nd ed., Clarendon Press, Oxford.

Courant, R. and D. Hilbert (1953): *Methods of Mathematical Physics, Vol. I,* Interscience, New York; Section II.10.8.

Davies, B. (1978): *Integral Transforms and Their Applications*, Springer-Verlag, New York.

Erdélyi, A., et al. (1954): *Tables of Integral Transforms, Vol. 1*, McGraw-Hill, New York; Chapters VI, VII.

Green, C. D. (1969): *Integral Equation Methods*, Barnes & Noble, New York; Chapter 6.

Henrici, P. (1974): *Applied and Computational Complex Analysis, Vol. I*, Wiley-Interscience, New York; Chapter 5.

Karush, W. and G. Young (1952): Temperature Rise in a Heat-Producing Solid Behind a Surface Defect, *Journal of Applied Physics* 23(No. 11): 1191–1193.

Oberhettinger, F. (1974): *Tables of Mellin Transforms*, Springer-Verlag, New York.

Özisik, M. N. (1968): *Boundary Value Problems in Heat Conduction*, International Textbook, Scranton, Penn.

Sneddon, I. N. (1966): *Mixed Boundary Value Problems in Potential Theory*, North-Holland, Amsterdam; Chapter I, IV.

Sneddon, I. N. (1972): *The Use of Integral Transforms*, McGraw-Hill, New York.

Titchmarsh, E. C. (1959): *Introduction to the Theory of Fourier Integrals*, 2nd ed., Clarendon Press, Oxford.

Tranter, C. J. (1966): *Integral Transforms in Mathematical Physics*, 3rd ed., Methuen, London.

Problems

Sections 7.1, 7.2, 7.3

1. (a) Transform the mixed boundary-value problem of the Section 6.3 example [whose solution is given by equation (6.3–7)] to the canonical form (7.1–2).

   (b) Carry out the analogous conversion for the boundary-value problem that appears as Problem 12 of Chapter 6.

2. (a) Find the complex Fourier transforms $F(\omega)$ of

   (i) $$f(t) = \begin{cases} e^{-at} & t \geq 0 \\ 0 & t < 0 \end{cases}$$

   and

   (ii) $$f(t) = e^{-a|t|},$$

   where $a > 0$ and indicate clearly the regions of the complex ω-plane where each is well-defined and analytic.

   (b) Restrict ω to real values and compare the transforms obtained in part (a) as regards their behavior for large $|\omega|$, and hence their integrability.

★3. (a) Let $f(x)$ be a continuous function such that for some real constants a,b

   $$f(x) = O(e^{ax}) \text{ as } x \to \infty, \; f(x) = O(e^{bx}) \text{ as } x \to -\infty.$$

   Show that the complex Fourier transform

   $$F(s) = \frac{1}{\sqrt{2\pi}} \int_{-\infty}^{\infty} f(x)\, e^{isx} dx$$

   of f, viewed as a function of the complex variable s, is analytic in s for $a < \text{Im } s < b$.

   [Remark: The order relation $f(x) = O(g(x))$ as $x \to x_0$ implies that $f(x)/g(x)$ remains bounded as $x \to x_0$; see Chapter 11.]

   (b) Assume that the function f of part (a) has a continuous derivative that satisfies the same order relations for large $|x|$ as f itself does.

   In this case, verify that if $\int_{-\infty}^{\infty} |f(x)| dx < \infty$, $\int_{-\infty}^{\infty} |f'(x)| dx < \infty$,

   then $s\, F(s)$ is uniformly bounded for $a + \delta \leq \text{im } s \leq b - \delta$, $\delta > 0$ (arbitrary).

★4. State and prove results for the Mellin transform, which are the analogues of those in Problem 3.

5. (a) Calculate the Laplace transforms of sin αt and cos αt *for real* α, and verify your results by a careful application of the inversion process.

(b) Use the results of part (a) to determine the complex Fourier transforms of

$$f(t) = \begin{cases} \sin \alpha t & t > 0 \\ 0 & t < 0 \end{cases} \qquad g(t) = \begin{cases} \cos \alpha t & t > 0 \\ 0 & t < 0 \end{cases}.$$

How would you verify these latter results by inversion?

6. (a) Given a function f of a positive real variable, the *Weyl fractional integral of order* μ of f is defined as

$$W_\mu [f] \equiv \frac{1}{\Gamma(\mu)} \int_s^\infty f(x)\,(x-s)^{\mu-1}\,dx.$$

Show that

$$W_\mu[W_\nu(f)] = W_{\mu+\nu}[f].$$

(b) As an extension of the ordinary Stieltjes transform $S(s)$ (7.3–4) there exists a *generalized Stieltjes transform*

$$S_\nu(s) \equiv \int_0^\infty \frac{f(x)}{(x+s)^\nu}\,dx.$$

Verify that these two transforms are related according to

$$S(s) = \Gamma(\nu)\, W_{\nu-1}\,[S_\nu].$$

7. (a) Show that, owing to the relations between Mellin and Fourier transforms, the Parseval formulas (7.3–9) and (7.3–10) are a consequence of (7.3–7) and (7.3–8).

(b) Carry out a comparable derivation for the Laplace convolution-type expressions (7.3–11).

8. (a) Demonstrate that the analogue for the Fourier cosine integral $F_c(\omega)$ (defined in Section 7.3) of both of the Fourier integral Parseval formulas (7.3–9) and (7.3–10) is

$$\int_0^\infty F_c(\omega)\,[G_c(\gamma+\omega)e^{i\omega b} + G_c(\gamma-\omega)e^{-i\omega b}]d\omega$$

$$= \int_0^\infty [f(b+t)e^{i\gamma t} + f(b-t)e^{-i\gamma t}]g(t)dt.$$

(b) Derive the corresponding result for the Fourier sine integral $F_s(\omega)$.

Section 7.4

9. Verify Properties 1–6 of the Mellin transform (Section 7.4) under the assumption that the function specified generates a transform analytic in some infinite vertical strip.

10. Use the transform properties given in Section 7.4 and the examples of that section to establish the following:

(a) $$M\left[\frac{1}{e^x+1}\right] = \Gamma(s)\,(1-2^{1-s})\,\zeta(s) \qquad \text{Re } s > 0;$$

(b) $$M[\ln|1-x|] = \pi\,\frac{\cot \pi s}{s} \qquad -1 < \text{Re } s < 0.$$

11. Employ the integral representation

$$K_\nu(x) = \frac{\sqrt{\pi}(x/2)^\nu}{\Gamma(\nu+\tfrac{1}{2})}\int_1^\infty e^{-xt}\,(t^2-1)^{\nu-1/2}\,dt$$

(Re $\nu > -\tfrac{1}{2}$) to show that

$$M[K_\nu(x)] = 2^{s-2}\Gamma(\tfrac{1}{2}s+\tfrac{1}{2}\nu)\Gamma(\tfrac{1}{2}s-\tfrac{1}{2}\nu)$$

for Re $s > |\text{Re } \nu|$.

12. Apply the approach used in Example 11 on page **209** to demonstrate that

$$M[\sin x\, J_\nu(x)] = 2^{\nu-1}\,\frac{\Gamma(\tfrac{1}{2}-s)\Gamma(\tfrac{1}{2}+\tfrac{1}{2}\nu+\tfrac{1}{2}s)}{\Gamma(1+\nu-s)\Gamma(1-\tfrac{1}{2}\nu-\tfrac{1}{2}s)}$$

for $-1-\text{Re }\nu < \text{Re } s < \tfrac{1}{2}$ with Re $\nu > -\tfrac{1}{2}$.

13. (a) Use the series representation (3.3–4) for the hypergeometric function to show that if

$$f(x) = \begin{cases} (1-x)^{c-1}\,{}_2F_1\,(a,b;c;\gamma(1-x)) & 0<x<1 \\ 0 & x>1 \end{cases}$$

with Re$(c-a-b) > 0$, Re $c > 0$, and $0 \le \gamma \le 1$, then

$$M[f] = B(c,s)\,{}_2F_1\,(a,b;c+s;\gamma) \qquad \text{Re } s > 0.$$

(b) Verify that when $\gamma = 1$, the result in part (a) leads to

$$M[g] = \Gamma(c)\,\frac{\Gamma(s)\,\Gamma(a+b-c+s)}{\Gamma(a+s)\Gamma(b+s)} \qquad \text{Re } s > \text{Re}(c-a-b),$$

where

$$g(x) = \begin{cases} x^{a+b-c}(1-x)^{c-1}{}_2F_1(a,b;c;1-x) & 0<x<1 \\ 0 & x>1 \end{cases}$$

with Re$(c-a-b) > 0$, Re $c > 0$.

14. Employ Euler's integral formula (3.3–6) to establish that

$$M[{}_2F_1(a,b;c;-x)] = \Gamma(s)\,\frac{\Gamma(a-s)\Gamma(b-s)\Gamma(c)}{\Gamma(a)\Gamma(b)\Gamma(c-s)}$$

for $0 < \text{Re } s < \min(\text{Re } a, \text{Re } b)$

Section 7.5

15. Verify the Mellin transform (7.5–5).

★16. (a) Show that whenever $1/\mathcal{K}(1-s)$ is invertible, where $\mathcal{K}(s)$ is the Mellin transform of the kernel $K(x)$, the solution of the integral equation (7.5–13) may be expressed in the form (7.5–14).

(b) If $1/\mathcal{K}(1-s)$ is not invertible but

$$L(x) \equiv M^{-1}\left[\frac{(-s)^{-n}}{\mathcal{K}(1-s)}\right]$$

is well-defined for some positive integer n, validate the alternative solutions (7.5–15), (7.5–16).

17. Demonstrate that the condition (7.5–17) for a kernel $K(x)$ to be a *Fourier kernel* is equivalent to

$$\int_0^\infty K(xy)\left[\int_0^{ty} K(u)du\right]\frac{dy}{y} = \begin{cases} 1 & 0 < x < t \\ 0 & x > t. \end{cases}$$

18. By an approach analogous to that used in Example 4 on page **206**, verify that the solution of the integral equation

$$g(x) = \int_0^x \frac{f(y)}{(x^2 - y^2)^\alpha}\,dy \qquad (0 < \alpha < 1)$$

can be expressed as

$$f(x) = \frac{2}{\pi}\sin\alpha\pi\frac{d}{dx}\int_0^x \frac{y\,g(y)}{(x^2 - y^2)^{1-\alpha}}\,dy.$$

★19. Assume that the kernel $K(x)$ has the Mellin transform $\mathcal{K}(s)$.

(a) Employ the Parseval identities (7.5–1), (7.5–2) to establish that the integral equation

$$g(x) = \int_0^\infty K(x/y)\,f(y)dy \qquad x > 0$$

has the solution

$$f(x) = \frac{1}{x^2}\int_0^\infty L(y/x)\,g(y)dy, \qquad x > 0$$

where $L(x)$ is the inverse Mellin transform of $1?\mathcal{K}(1 - s)$ (assume it exists).

(b) If $1/\mathcal{K}(1 - s)$ is not invertible but $L(x) \equiv M^{-1}[s^{-n}/\mathcal{K}(1-s)]$ (where n is a positive integer) does exist, show that the solution of the given integral equation may now be expressed as

$$f(x) = \frac{1}{x^2}\left[x\frac{d}{dx}\right]^n\int_0^\infty L(y/x)\,g(y)dy, \qquad x > 0.$$

20. Apply the result of Problem 19 to the integral equation

$$\int_x^\infty (y - x)^{c-1}{}_2F_1\,(a,b;c;1 - x/y)\,f(y)dy = g(x), \qquad x > 0$$

to demonstrate that the solution has the representation

$$f(x) = \frac{x^{-a}}{\Gamma(c)\Gamma(-c)} \int_x^{\infty} (y - x)^{-c-1} {}_2F_1(-a, b - c; -c; 1 - x/y) y^a g(y) dy, \qquad x > 0$$

(recall (7.5–8) and Problem 13).

21. Given a function f of a positive real variable, the Weyl fractional integral $W_\mu[f]$ introduced in Problem 6 can be viewed as an integral transform.
    (a) Use the Parseval formula (7.5–2) to show that

$$M[W_\mu] = \frac{\Gamma(s)}{\Gamma(\mu + s)} \mathscr{F}(\mu + s)$$

    where $\mathscr{F}(s)$ is the Mellin transform of $f(x)$.
    (b) Employ the relation exhibited in part (b) of Problem 6 and (7.5–18) to verify that

$$M[S_\nu] = B(s, \nu - s)\, \mathscr{F}(1 - \nu + s)$$

    for the generalized Stieltjes transform.

Sections 7.6, 7.7

22. Establish (7.5–6) for $0 < x < 1$ by inversion. [Shift the contour to the left. Verification of (7.5–6) by inversion for $x > 1$ requires an identity relating hypergeometric functions of arguments $-x$ and $1/(-x)$.]
23. Under the assumptions that $0 < \text{Re } a < \text{Re } c$ and $\text{Re } b < 1$, employ the inversion procedure to verify (7.5–7) for $0 < x \leq 1$.
24. Consider the boundary-value problem

$$\Delta\phi = 0 \qquad r > 0, \qquad |\theta| < \alpha \qquad (\alpha < \pi)$$

$$\phi(r, \pm\alpha) = \begin{cases} 1 & r < a \\ 0 & r > a \end{cases}$$

which describes the steady-state temperature distribution in a wedge-shaped plate of angle 2α whose edges are maintained at unit temperature for a distance a (measured from the apex) and at zero temperature elsewhere. Conformal mapping techniques, as introduced in Chapter 6, can be used to generate the solution for this problem:

$$\phi(r, \theta) = \frac{1}{\pi} \tan^{-1}\left[\frac{2(ar)^\gamma \cos \gamma\theta}{r^{2\gamma} - a^{2\gamma}}\right]$$

where $\gamma = \pi/2\alpha$ and $0 \leq \tan^{-1} x \leq \pi$.
    (a) Employ a Mellin transform procedure to deduce an integral representation for the temperature distribution.
    (b) Obtain the expression for the solution given above by performing an inversion on this integral representation, similar to that discussed in Sections 7.6 and 7.7.

8

Green's functions

8.1 The natural occurrence of Green's functions

In earlier chapters we have encountered and used integral representations for various functions. Although our point of view at the time may not have been so directed, in most cases we could have considered each function as having arisen as the solution of some differential equation subject to specific auxiliary conditions. We will now adopt this approach and discuss a systematic method for generating integral representations for such functions: the *Green's function technique*.

As most readers are aware, the method is by no means modern; it has its origins in an 1828 monograph by the English mathematician George Green (1793–1841). However, the wide application of the technique to current problems owes much to the efforts of scores of other researchers and applied scientists. As a method for solving boundary- and initial-value problems*, the Green's function approach is a valuable alternative to the more familiar generalized Fourier series technique (recall the one-dimensional discussion in Section 2.5). In higher dimensions the latter technique, which involves expansion of the solution in a (generally infinite) series of eigenfunctions, is applicable only if the partial differential equation separates in a coordinate system consonant with the (prescribed) boundary conditions. The Green's function approach does not suffer from this defect. In addition, the technique works quite well in situations where other procedures fail: for example, with eigenvalue problems, on unbounded domains, which possess a continuous spectrum (that is, have non-discrete eigenvalues). When Green's function can be determined in closed form (and even sometimes when it cannot), the method of Green usually offers the best insight into the general overall behavior of the solution as well as special features that it may possess.

The Poisson kernel

Our first brush with an integral representation of the type we propose to investigate occurred already in Chapter 5. For points z within the simply-connected region

*It is important to appreciate that the Green's function technique is not restricted merely to those differential equations associated with equilibrium phenomena. The method is much more broadly applicable, as we will see, and constitutes an extremely useful tool in the study of problems in diffusion or wave motion, for example.

bounded by the closed contour C, the Cauchy integral formula

$$f(z) = \frac{1}{2\pi i}\oint_C \frac{f(\zeta)}{\zeta - z}\,d\zeta \tag{8.1–1}$$

can be viewed as an *integral representation* for the regular analytic function $f(z)$ in terms of its *boundary values* along C.

In the special case when C is the circle $|\zeta| = \rho$, this representation generates a noteworthy expression. First we observe that (8.1–1) has the value zero when $|z| > \rho$. Thus, if t is a complex number satisfying $|z| > \rho > |t|$, then

$$\begin{aligned} f(z) &= f(z) - 0 \\ &= \frac{1}{2\pi i}\oint_{|\zeta|=\rho} \frac{f(\zeta)}{\zeta - z}\,d\zeta - \frac{1}{2\pi i}\oint_{|\zeta|=\rho} \frac{f(\zeta)}{\zeta - t}\,d\zeta \\ &= \frac{1}{2\pi i}\oint_{|\zeta|=\rho} \frac{z-t}{(\zeta - z)\,(\zeta - t)} f(\zeta)d\zeta. \end{aligned}$$

If we now let $\zeta = \rho e^{i\phi}$, $z = re^{i\theta}$ and $t = (\rho^2/r)e^{i\theta}$, then $r < \rho$, and this relation becomes

$$\begin{aligned} f(re^{i\theta}) &= \frac{1}{2\pi}\int_0^{2\pi} \frac{r^2e^{i\theta} - \rho^2 e^{i\theta}}{(\rho e^{i\phi} - re^{i\theta})(\rho re^{i\theta} - \rho^2 e^{i\theta})} f(\rho e^{i\phi})\rho e^{i\phi} d\phi \\ &= \frac{1}{2\pi}\int_0^{2\pi} \frac{\rho^2 - r^2}{\rho^2 + r^2 - 2\rho r\cos(\phi - \theta)} f(\rho e^{i\phi})d\phi. \end{aligned}$$

It follows that the harmonic functions which are the real and imaginary parts of f must satisfy

$$\begin{Bmatrix} u(r,\theta) \\ v(r,\theta) \end{Bmatrix} = \frac{1}{2\pi}\int_0^{2\pi} \frac{\rho^2 - r^2}{\rho^2 + r^2 - 2\rho r\cos(\phi - \theta)} \begin{Bmatrix} u(\rho,\phi \\ v(\rho,\phi) \end{Bmatrix} d\phi, \qquad r < \rho. \tag{8.1–2}$$

The expressions (8.1–2) are known as *Poisson's integral formulas* for the circle. The first one is the special case, for two dimensions, of a more general n-dimensional result ($n \geq 2$)

$$u(P) = \int_{S\rho} K_n\,(P,Q)u(Q)dS_Q, \tag{8.1–3}$$

which determines the values of a harmonic function in the interior of an n-sphere from its boundary values on that sphere. In the representation (8.1–3) P denotes an arbitrary point whose distance r from some fixed point (the origin in (8.1–2) is less than ρ, S_ρ is the surface of the n-ball of radius ρ centered at the fixed point, Q is the integration point which ranges over the n-sphere S_ρ, dS_Q is the area element on S_ρ, and $K_n(P,Q)$ is the *Poisson kernel*

$$K_n(P,Q) \equiv \frac{\rho^2 - r^2}{\rho\omega_n r_{PQ}^n} \tag{8.1–4}$$

(see Figure 8.1).

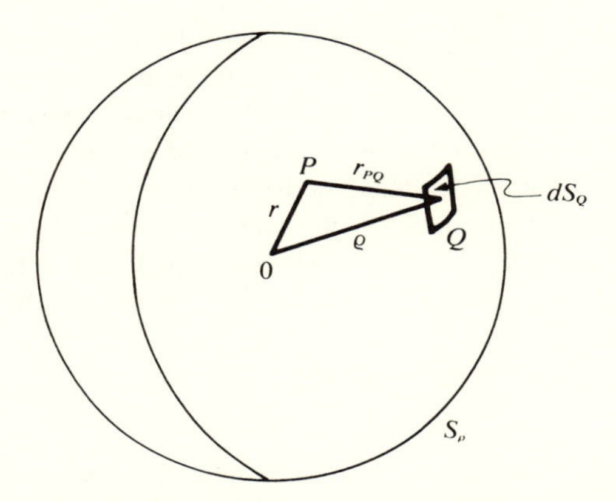

Figure 8.1 **The geometry for the Poisson integral formula in 3-dimensions**

For convenience here, we have designated the area of the *unit* n-sphere* by ω_n (see Problem 3), while r_{PQ} denotes the distance between the points P and Q. Since $K_n(P,Q) = 1/(\rho^{n-1}\omega_n)$ when $r = 0$, the important *mean-value theorem* for harmonic functions,

$$u(P_0) = \frac{1}{\rho^{n-1}\omega_n}\int_{S_\rho} u(Q)dS_Q,$$

is a consequence of (8.1–3). Viewed alternatively, the n-dimensional Poisson formula is a generalization of this mean-value theorem to points within S_ρ other than its center.

The Dirichlet problem for the n-ball

In our derivation of the Poisson integral formula (8.1–2) we started with a function f, which we assumed to be regular analytic in a region containing the disk $|z| \leq \rho$. Our result was a representation for each of the harmonic functions u,v (the real and imaginary parts of f) in terms of its values along the circumference of the disk. However, in keeping with the thrust of this chapter, we also require a theorem that proceeds in the opposite direction. In other words, we also need to discuss the

*Although it is somewhat nonstandard, we use the term *n-sphere* here and later to refer to the $(n - 1)$ – dimensional surface (boundary) of the n-ball.

general problem of finding a (actually, it is unique) function $u(z)$ that is harmonic for $|z| < \rho$ and "takes on" the boundary values $g(z)$ for $|z| \to \rho$, where g is a reasonably smooth, but otherwise arbitrary, given real function defined on $|z| = \rho$. Once we are assured of the existence of such a $u(z)$, we fully expect, from our discussion of (8.1–2), that the first of those formulas will provide a representation for it.

Excluding the representation question, this "inverse" problem is a special case of the *Dirichlet problem**, historically the oldest problem (of existence) in potential theory. (See Kellogg (1929) for an excellent review of the mathematics and the early history of the analytic approach to potential theory; Port and Stone (1978) give a nice account of the more modern probabilistic approach.) Thanks to the efforts of Riemann, Schwarz, Neumann, Hilbert, Poincaré, Fredholm and others, we know that the interior Dirichlet problem is uniquely soluble for smooth boundaries and continuous boundary-values (and in certain other cases as well; Garabedian (1964), for example, has the details). Moreover, for the n-sphere we have the following.

Theorem:

Let $g(Q)$ be a continuous function defined on the n-dimensional sphere S_ρ of radius ρ centered at some fixed point. Then the function $u(P)$ given by

$$u(P) \equiv \int_{S_\rho} K_n(P,Q)g(Q)dS_Q, \tag{8.1–5}$$

where $K_n(P,Q)$ is the Poisson kernel (8.1–4), is the unique function that

(a) is harmonic in the inside of S_ρ;

and (b) assumes the boundary-values $g(Q)$ in the sense that for every $Q \in S_\rho$

$$\lim_{P \to Q} u(P) = g(Q),$$

where P approaches Q from the inside of S_ρ.

The proof of this theorem is more intricate, and for our purposes less instructive, than the demonstration we gave of (the special case (8.1–2) of) its converse. Full details are available in Kellogg (1929) and Garabedian (1964).

We will require the relation (8.1–5) later. For the present, it suffices to remark that the behavior of the limit of $u(P)$ as P approaches the boundary S_ρ (in other words, the demonstration of conclusion (b) of the above theorem) depends on only three key features of the Poisson kernel (8.1–4):

---

*The general (interior) Dirichlet problem is the problem of finding a solution of Laplace's equation in a bounded region R, which takes on prescribed values on the boundary of R. Sometimes called the *first* boundary-value problem of potential theory, its more common designation honors P. G. Lejeune-Dirichlet (1805–1859) who laid the groundwork for a "variational" approach to its solution. The function $G(P,Q)$, which plays a role for the region R similar to the role that $K_n(P,Q)$ plays for the region inside S_ρ, is called *Green's function for R*.

(i) $K_n(P,Q) \geq 0$ for all P inside S_ρ, $Q \in S_\rho$;
(ii) For each $\delta > 0$ and $Q \in S_\rho$,
$\lim_{P \to S\rho} K_n(P,Q) = 0$ uniformly for $r_{PQ} \geq \delta$;
(iii) $\int_{S\rho} K_n(P,Q) dS_Q = 1$.

We omit the easy verifications. In a more general context, these three properties constitute the basic characteristics of "reproducing kernels," which we will meet more intimately in our discussion of generalized functions (the Dirac delta function and its associates) in the next chapter. In the situation at hand, however, we at least sense that the assumption of boundary-values for the Dirichlet problem is a local phenomenon.*

A diffusion example

One-dimensional nonsteady-state heat conduction provides another instructive illustration. If the medium of interest (say, an infinitesimally thin rod) is uniform and infinite in extent, then an appropriate "boundary-value" (actually initial-value) problem for the temperature distribution along the rod is

$$\frac{\partial^2 u}{\partial x^2} = k \frac{\partial u}{\partial t} \qquad -\infty < x < \infty, \ t > 0$$
$$u(x,0) = g(x) \qquad \textbf{(8.1–6)}$$

where k is a positive constant proportional to the product of the rod's specific heat and density and inversely proportional to its thermal conductivity. Separating variables and reasoning as in Section 7.1, we are led to

$$u(xt) = \int_{-\infty}^{\infty} A(\beta) e^{i\beta x - \beta^2 t/k} d\beta$$

as a possible representation for the solution of (8.1–6).

In order to satisfy the initial condition at $t = 0$, we must have

$$g(x) = \int_{-\infty}^{\infty} A(\beta) e^{i\beta x} d\beta \, .$$

This is easy to invert, however, using our knowledge of the Fourier transform pair (7.3–2). Using Fubini's theorem and completing the square, it then follows, that

$$u(x,t) = \int_{-\infty}^{\infty} \left[\frac{1}{2\pi} \int_{-\infty}^{\infty} g(y) e^{-i\beta y} dy \right] e^{i\beta x - \beta^2 t/k} d\beta$$
$$= \frac{1}{2\pi} \int_{-\infty}^{\infty} g(y) \left[\int_{-\infty}^{\infty} e^{i\beta(x-y) - \beta^2 t/k} d\beta \right] dy$$
$$= \frac{1}{2\pi} \int_{-\infty}^{\infty} g(y) \left\{ \int_{-\infty}^{\infty} e^{-\frac{t}{k}\left[\beta - i\frac{k(x-y)}{2t}\right]^2} d\beta \right\} e^{-\frac{k(x-y)^2}{4t}} dy.$$

*A similar state of affairs was encountered earlier when we verified the reciprocal nature of the Mellin transform pair; see the proof of Theorem 2, Section 7.2.

Since we can shift the contour and easily evaluate the inner integral to get $\sqrt{\pi k/t}$, we finally obtain

$$u(x,t) = \sqrt{\frac{k}{4\pi t}} \int_{-\infty}^{\infty} g(y) e^{-\frac{k(x-y)^2}{4t}} dy. \quad \textbf{(8.1–7)}$$

This last expression provides an integral representation for the solution of the given problem in terms of the "boundary values." Analogously to (8.1–5), the formula (8.1–7) can be generalized to more than one spatial dimension as

$$u(P) = \int_S K(P,Q) g(Q) dS_Q, \quad \textbf{(8.1–8)}$$

where P is the arbitrary point with space coordinate x and time coordinate t, S is the boundary surface with equation $t = 0$, Q is the integration point which ranges over S, dS_Q is the surface area element on S, and $K(P,Q)$ is the "diffusion kernel." In the present one-dimensional case

$$K(P,Q) = \left[\frac{k}{4\pi t}\right]^{1/2} \exp\left[-\frac{k(x-y)^2}{4t}\right].^* \quad \textbf{(8.1–9)}$$

Like the Poisson kernel, the diffusion kernel is also of reproducing type; it satisfies the three basic properties of reproducing kernels, and the assumption of the "boundary-values" again involves only local considerations.

A problem in acoustical propagation

We direct our final inquiry to the mathematics of sound propagation in fluids. The study of wave motion and acoustics is a natural outgrowth of the analysis of vibration of a taut string (or its higher-dimensional counterparts such as oscillating membranes or pulsating cylinders). However, often one is less concerned with the generator than with the nature of the resulting acoustical propagation, which is, in turn, determined by the characteristics of the surrounding medium. If this medium is a fluid, the analysis involves considerations not unlike those of Section 2.1.* (See also Section 6.1; standard references such as Morse and Ingard (1968) may also be helpful).

Rather than present a detailed study of the physical principles underlying acoustic wave motion in fluids, however, we content ourselves with noting that it is the density and elasticity of the fluid, which influence the propagation of sound in such media. If we (i) assume that the fluid, in the absence of sound, has uniform

---

*In n-dimensions the diffusion kernel is given by $\left(\frac{k}{4\pi t}\right)^{1/2 n} \exp\left[-\frac{kr^2{}_{PQ}}{4t}\right]$ where r_{PQ} is the spatial separation of the points P and Q.

*Remarkably, over 2000 years ago the Roman architect Vitruvius speculated about the possibility of a relationship between the nature of sound propagation in air and the propagation of ripples on the surface of water.

density ρ, pressure P and temperature T; (ii) neglect the effects of viscosity and heat conduction; and (iii) suppose that the acoustic pressure is small relative to P, then the resulting linearized equation for the velocity potential ϕ turns out to be (not unexpectedly) the scalar wave equation

$$\Delta\phi = \frac{1}{c^2}\frac{\partial^2\phi}{\partial t^2} . \tag{8.1–10}$$

By analogy with other wave phenomena, we call c the *speed of sound* in the fluid. It is related to the constitutive parameters of the medium by the equations

$$\frac{1}{c^2} = \kappa\rho = \frac{1}{\gamma}\left(\frac{\partial\rho}{\partial P}\right)_{T=\text{constant}},$$

in which κ denotes the (adiabatic) compressibility and $\gamma = C_p/C_V$ is the ratio of the specific heat at constant pressure to the specific heat at constant volume. The last expression is especially useful in the case of a so-called "perfect gas," for which $MP = \rho RT$ (where M is the molecular weight and R the universal gas constant), whence $c = \sqrt{\gamma RT/M}$.

An interesting and important class of problems arises when the sound is restricted in some way, as in the case of acoustic propagation along the interior of a hollow tube. If the wavelength of the sound is large compared to the diameter of the tube, the wave motion is predominantly one-dimensional. It is then an easy matter to analyze the equation subject to various subsidiary conditions and to derive classical results, such as the relationship between the wavelengths of propagating modes and the length of the tube. (See Problem 5; also see Rayleigh (1896) for an historical look at this problem.)

When the small-diameter approximation is inappropriate we must solve the full three-dimensional equation (8.1–10) subject to suitable auxiliary conditions, and this is an imposing task. However, there exist integral expressions, comparable to (8.1–5) and (8.1–8), which provide considerable insight into what constitutes a well-posed problem and the nature of its solution. Such an integral representation of the solution hinges upon the "Green's function" associated with the problem. Indeed, determining Green's function for the problem provides a methodical procedure for obtaining such representations. It is clear then that we need to know more about this method of Green; accordingly, we turn to a systematic discussion of that approach and postpone further study of acoustical propagation until Section 8.6.

8.2 One-dimensional Green's functions

Since we have already discussed Sturm-Liouville problems at some length (in Chapter 2), it will be helpful to begin our consideration of Green's functions in the setting of linear ordinary differential equations. We recall that

Definition 2.2.1 A linear differential operator L defined on an interval (a,b) is termed **self-adjoint** if there exists a function $F(u,v;x)$ with the property that, for any two (suitably smooth) functions $u(x)$, $v(x)$,

$$uLv - vLu = \frac{d}{dx}F(u,v;x).$$

Definition 2.2.2 Homogeneous linear boundary conditions are said to be **regular** for the operator L on (a,b) if, for all u,v that satisfy the conditions,

$$F(u,v;x)\Big|_a^b = 0\,.$$

The implication of these two results obviously is that if L is self-adjoint and the boundary conditions (B.C.) are regular, then

$$\int_a^b [uLv - vLu]\,dx = 0 \qquad \textbf{(8.2–1)}$$

for all u and v that satisfy the B.C.

Imagine, therefore, that we are interested in solving the Sturm-Liouville problem

$$\begin{aligned} &Lu = -f(x) \qquad a < x < b,\\ &\text{Homogeneous B.C.}\end{aligned} \qquad \textbf{(8.2–2)}$$

with linear self-adjoint operator L and regular boundary conditions. In view of (8.2–1), if we could somehow determine a two-variable function $v = v(x,x')$, which, as a function of x, satisfies the B.C. of (8.2–2) and, as a function of x', has the property that

$$u(x) = -\int_a^b u(x')Lv(x,x')dx' \qquad \textbf{(8.2–3)*}$$

for all u, then we would immediately have an integral representation for the solution of (8.2–2) in terms of this v, namely

$$u(x) = \int_a^b v(x,x')f(x')dx'. \qquad \textbf{(8.2–4)}$$

Green's function $G(x,x')$ is just such a v.

The second-order case

In this chapter we shall restrict our attention to equations in which the highest derivative is the second. Therefore, we assume that our linear differential operator has the canonical form (2.2–1)

---

*In this relation $-Lv$ is playing the role of a delta function. Those readers familiar with such functions can simplify, at least formally, some of the ensuing analysis. We will treat generalized functions in some detail in chapter 9.

$$Lu \equiv \frac{d}{dx}\left[p(x)\frac{du}{dx}\right] - q(x)u \qquad a < x < b$$

with dp/dx and $q(x)$ real-valued and continuous on $[a,b]$ and $p(x) > 0$. Green's function for this case is associated with the homogeneous version of the boundary-value problem (BVP) (8.2–2), wherein L is the above differential operator and the B.C. are given by (2.2–3). Indeed, if $\lambda = 0$ is *not* an eigenvalue of this homogeneous BVP, then

Definition:* The (regular or ordinary) **Green's function** for the given BVP is the two-variable function $G(x,x')$, $a \leq x,x' \leq b$, with the properties that, *as a function of x*,

(i) $L\,G(x,x') = 0 \qquad x \neq x'$;
(ii) $G(x,x')$ satisfies the homogeneous B.C.;
(iii) $G(x,x')$ is continuous for all $a \leq x \leq b$; and
(iv) $\partial G(x,x')/\partial x$ is continuous for $x \neq x'$ with

$$\frac{\partial G(x'+0,x')}{\partial x} - \frac{\partial G(x'-0,x')}{\partial x} = -\frac{1}{p(x')}$$

or, equivalently,

$$\frac{\partial G(x',x'+0)}{\partial x} - \frac{\partial G(x',x'-0)}{\partial x} = \frac{1}{p(x')}.$$

The principal result can now be stated as

Theorem 1:

If $\lambda = 0$ is not an eigenvalue of the homogeneous version of the BVP, then the solution of the Sturm-Liouville problem

$$\frac{d}{dx}\left[p(x)\frac{du}{dx}\right] - q(x)u = -f(x) \qquad a < x < b \qquad \textbf{(8.2–5)}$$

Homogeneous B.C.

may be represented in terms of the appropriate Green's function as

$$u(x) = \int_a^b G(x,x')f(x')dx'. \qquad \textbf{(8.2–6)}$$

Proof: We need to verify that $u(x)$, defined by (8.2–6), solves the given Sturm-Liouville problem. First we note that $u(x)$ satisfies the appropriate homogeneous B.C. since $G(x,x')$ has this property. Next we observe that from (8.2–6),

---

*For the comparable definition in the case of a higher-order linear differential operator, see Courant and Hilbert (1953) or Coddington and Levinson (1955).

$$u(x) = \int_a^x G(x,x')f(x')dx' + \int_x^b G(x,x')f(x')dx'$$

and thus, since G is continuous in x,

$$\frac{du}{dx} = G(x,x)f(x) + \int_a^x \frac{\partial G(x,x')}{\partial x} f(x')dx'$$

$$- G(x,x)f(x) + \int_x^b \frac{\partial G(x,x')}{\partial x} f(x')dx'$$

$$= \int_a^x \frac{\partial G(x,x')}{\partial x} f(x')dx' + \int_x^b \frac{\partial G(x,x')}{\partial x} f(x')dx'.$$

However, owing to the jump discontinuity in $\partial G/\partial x$,

$$\frac{d}{dx}\left[p(x)\frac{du}{dx}\right] = p(x)\frac{\partial G(x,x-0)}{\partial x} f(x) + \int_a^x \frac{\partial}{\partial x}\left[p(x)\frac{\partial G(x')}{\partial x}\right] f(x')dx'$$

$$-p(x)\frac{\partial G(x,x+0)}{\partial x} f(x) + \int_x^b \frac{\partial}{\partial x}\left[p(x)\frac{\partial G(x,x')}{\partial x} f(x')dx'\right.$$

$$= -f(x) + \int_a^x q(x)G(x,x')f(x')dx' + \int_x^b q(x)G(x,x')f(x')dx'$$

$$= -f(x) + q(x)u\ .$$

It follows that $u(x)$ indeed satisfies the given differential equation. Since we have already noted that $U(x)$ has the correct boundary behavior, the demonstration is complete. □

A number of observations are in order at this point.

1. The solution of the BVP (8.2–5) is *unique* by virtue of the linearity of the problem and the fact that $\lambda = 0$ is not an eigenvalue of the homogeneous version, that is, if $f \equiv 0$ in (8.2–5), then $u \equiv 0$. The relation (8.2–6) provides the *Green's function representation* of this unique solution.
2. Both the homogeneity and linearity of the boundary conditions are essential for the validity of Theorem 1; the regularity of the B.C. is actually extrinsic.
3. However, if the boundary conditions *are* regular, then Green's function has the useful symmetry property

$$G(x,x') = G(x',x). \qquad \textbf{(8.2–7)}$$

This result can be established most easily by defining

$$g(x,x',x'') \equiv G(x,x')LG(x,x'') - G(x,x'')LG(x,x')$$

where $a < x' < x'' < b$. Thence, using (2.2–2) and the properties of Green's function:

$$0 = \int_a^{x'} g(x,x',x'')dx + \int_{x'}^{x''} g(x,x',x'')dx + \int_{x''}^{b} g(x,x',x'')dx$$

$$= p(x)W[G(x,x'), G(x,x'')]\Big|_a^{x'} + p(x)W[G(x,x'),G(x,x'')]\Big|_{x'}^{x''}$$

$$+ p(x)W[G(x,x'),G(x,x'')]\Big|_{x''}^{b}$$

$$= p(x)W[G(x,x'),G(x,x'')]\Big|_a^{b} + G(x'',x') - G(x',x'').$$

The desired conclusion follows from this last relation by imposing the definition of regularity.

4. A BVP with boundary conditions that are *not* homogeneous can be treated by first subtracting off a function w (actually *any* sufficiently smooth function) that takes on the given B.C. In view of the linearity of the problem, the analogue of (8.2–6) in this case is

$$u(x) = w(x) + \int_a^b G(x,x')\,[f(x') + Lw(x')]dx'. \qquad \textbf{(8.2–8)}$$

(However some care should be exercised in the selection of w. See the remarks of Lebedev et al. (1965), pp. 105 ff., for example.)

5. The "singular" problems, which occur when the fundamental interval (a,b) is infinite, or $p(x)$ vanishes (or $q(x)$ becomes unbounded) at one or both of the endpoints, can often be handled, as in the case of general Sturmian theory (see Chapter 2), by modest extensions of our basic notions. We will consider several important examples of singular problems in the next section.

6. If the appropriate Green's function is known, the *eigenvalue* problem

$$Lu + \lambda ru = 0 \qquad a < x < b$$
Homogeneous B.C.

associated with (8.2–5) can be reexpressed as a "homogeneous Fredholm integral equation of the second kind." Indeed, using (8.2–6), we have

$$u(x) = \lambda \int_a^b G(x,x')r(x')u(x')dx'.$$

We will return to these considerations in Chapter 10.

Some elementary Green's functions

We now illustrate the basic systematic procedure for determining one-dimensional Green's functions. As an initial step, the general solution of the equation $LG(x,x') = 0$ is found for each of the regions $a < x < x'$ and $x' < x < b$. Since L is a second-order linear differential operator, this will lead to four arbitrary constants. The values of two of these constants may be obtained by imposing condition (ii) of the definition; the remaining two are then specified by the continuity (iii) of G and the jump condition (iv) for $\partial G/\partial x$. A few examples will help to fix these ideas.

Example 1 If the underlying homogeneous problem is

$$\frac{d^2u}{dx^2} = 0 \qquad 0 < x < 1$$

$$u(0) = 0 = u(1),$$

then

$$G(x,x') = \begin{cases} A + Bx & x < x' \\ C + Dx & x > x'. \end{cases}$$

The B.C. imply that $A = 0$ and $D = -C$, and thus

$$G(x,x') = \begin{cases} Bx & x < x'. \\ C(1-x) & x > x'. \end{cases}$$

Using this formula, we see that G will be continuous at $x = x'$ only if $C = Bx'/(1 - x')$, while the choice $B = (1 - x')$ is required by condition (iv). Therefore, we finally obtain for Green's function

$$G(x,x') = \begin{cases} x(1 - x') & x < x' \\ x'(1 - x) & x > x'. \end{cases}$$ □

In this first example, the boundary conditions were unmixed (recall definition 1 of Section 2.2). When this condition prevails, there is a general formalism that can be utilized:

Example 2 If

$$\frac{d}{dx}\left[p(x)\frac{du}{dx}\right] - q(x)u = 0 \qquad a < x < b$$

Homogeneous, Unmixed B.C.,

then

$$G(x,x') = \begin{cases} Au_1(x) & x < x' \\ Cu_2(x) & x > x', \end{cases}$$

where $u_1(x)$, $u_2(x)$ are solutions of the differential equation which satisfy the boundary condition at $x = a$ and at $x = b$, respectively. Applying the two remaining conditions gives rise to

$$A = C\frac{u_2(x')}{u_1(x')} \qquad \text{and} \qquad C = \frac{u_1(x')}{p(x')W(u_2,u_1)\,|_{x'}},$$

and thus the appropriate Green's function has the form

$$G(x,x') = \begin{cases} c\, u_1(x)u_2(x') & x < x' \\ c\, u_1(x')u_2(x) & x > x', \end{cases}$$

where

$$1/c = p(x')W(u_2,u_1)\Bigg|_{x'}.$$

We note that u_1 and u_2 are perforce linearly independent, otherwise $\lambda = 0$ would be an eigenvalue of the underlying homogeneous problem. The right-hand side of the last equation is therefore not 0, so the parameter c is well defined; it is actually a constant, independent of x' (see Problem 7, Chapter 2). □

Notice the symmetry of the above Green's functions. As we have observed earlier, this is a consequence of the regularity of the boundary conditions.

Example 3 If the underlying homogeneous problem is

$$\frac{d^2u}{dx^2} = 0 \qquad 0 < x < 1$$

$$u(0) = 0$$

$$\frac{du}{dx}(0) = 0,$$

then

$$G(x,x') = \begin{cases} 0 & x < x' \\ C + Dx & x > x', \end{cases}$$

where we have already made use of the "boundary" conditions. Continuity of G and the jump discontinuity in $\partial G/\partial x$ at $x = x'$ imply $C = -Dx'$ and $D = -1$, whence

$$G(x,x') = \begin{cases} 0 & x < x' \\ x' - x & x > x'. \end{cases}$$

□

Example 4 More generally, if

$$\frac{d}{dx}\left[p(x)\frac{du}{dx}\right] - q(x)u = 0 \qquad a < x < b$$

$$u(a) = 0$$

$$\frac{du}{dx}(a) = 0,$$

then

$$G(x,x') = \begin{cases} 0 & x < x' \\ Cu_1(x) + Du_2(x) & x > x', \end{cases}$$

where $u_1(x)$, $u_2(x)$ are two (arbitrary) linearly independent solutions of the differential equation. Applying conditions (iii) and (iv) in the definition of Green's function leads to

$$D = -C\frac{u_1(x')}{u_2(x')} \quad \text{and} \quad C = -\frac{u_2(x')}{p(x')W(u_2,u_1)\,|_{x'}}.$$

The final form for Green's function is thus

$$G(x,x') = \begin{cases} 0 & x < x' \\ c[u_2(x)u_1(x') - u_1(x)u_2(x')] & x > x', \end{cases}$$

where, as in Example 2, the constant c is given by

$$1/c = p(x')W(u;,u_1)\Big|_{x'}.$$

□

This last example is particularly instructive. It suggests that initial-value problems (IVP's) can be treated within the context of the Green's function methodology (the heat conduction example of the previous section provides more evidence). From a linear systems point of view this means that the impulse-response of the system (recall our discussion in Section 5.1; see also Kaplan (1962), for example) is nothing more than the Green's function associated with the appropriate IVP.

8.3 Extensions of the basic theory

Green's functions for singular problems

As in the case of general Sturmian theory, many "singular" cases can be treated with little additional effort.

Example 1 If the underlying homogeneous BVP is

$$\frac{d}{dx}\left[x\frac{du}{dx}\right] - (\nu^2/x)u = 0 \qquad (\nu \geq 0) \quad 0 < x < 1$$

$$u(0) \text{ finite}$$

$$u(1) = 0,$$

then

$$G(x,x') = \begin{cases} A + B\ln x & x < x' \\ C + D\ln x & x > x' \end{cases} \quad \nu = 0$$
$$\qquad\qquad \begin{cases} Ax^{\nu} + Bx^{-\nu} & x < x' \\ Cx^{\nu} + Dx^{-\nu} & x > x' \end{cases} \quad \nu \neq 0$$

Applying the now-familiar procedure eliminates all arbitrariness and leads to the Green's functions

$$G(x,x') = \begin{cases} -\ln x' & x < x' \\ -\ln x & x > x' \end{cases} \quad \nu = 0$$
$$\qquad\qquad -(xx')_{\nu}/(2\nu) + \begin{cases} (x/x')^{\nu}/(2\nu) & x < x' \\ (x'/x)^{\nu}/(2\nu) & x > x' \end{cases} \quad \nu \neq 0. \qquad \square$$

The same finiteness boundary condition suffices for the following example.

Example 2 If

$$\frac{d^2u}{dx^2} - k^2u = 0 \qquad (k > 0) \qquad -\infty < x < \infty$$

u bounded as $x \to \pm\infty$,

then

$$G(x,x') = \begin{cases} Ae^{kx} + Be^{-kx} & x < x' \\ Ce^{kx} + De^{-kx} & x > x'. \end{cases}$$

The boundary conditions imply that $B = C = 0$, whence

$$A = 1/(2k)e^{-kx'} \qquad \text{and} \qquad D = 1/(2k)e^{kx'}.$$

The final form for the appropriate Green's function is thus

$$G(x,x') = \begin{cases} 1/(2k)e^{k(x-x')} & x < x' \\ 1/(2k)e^{k(x'-x)} & x > x', \end{cases}$$

or, written compactly,

$$G(x,x') = \frac{1}{2k}e^{-k|x-x'|}. \qquad \square$$

If the sign in front of k^2 is changed in the above example, we obtain the one-dimensional reduced wave or Helmholtz equation on an infinite domain. In this case, a finiteness boundary condition is no longer relevant. On the other hand, if we imagine that the equation has arisen through the suppression of an assumed harmonic time dependence $e^{-i\omega t}$, then we can obtain an appropriate (and natural) B.C. by assuming that the solutions resemble the spatial portions of *outgoing* waves for large $|x|$. This so-called "radiation condition" was proposed by the theoretical physicist Arnold Sommerfeld (1868–1951) who was an active contributor in the wave-propagation field.

Example 3 If

$$\frac{d^2u}{dx^2} + k^2u = 0 \qquad (k>0) \qquad -\infty < x < \infty$$

$$\frac{du}{dx}(x) \mp iku(x) \to 0 \text{ as } x \to \pm\infty ,$$

then

$$G(x,x') = \begin{cases} Ae^{ikx} + Be^{-ikx} & x < x' \\ Ce^{ikx} + De^{-ikx} & x > x'. \end{cases}$$

From the boundary conditions we obtain $A = D = 0$, while the two remaining conditions imply

$$B = Ce^{2ikx'} \qquad \text{and} \qquad C = \frac{i}{2k}e^{-ikx'}.$$

Therefore, the appropriate Green's function for the one-dimensional Helmholtz equation with Sommerfeld radiation condition is

$$G(x,x') = \frac{i}{2k}e^{ik|x-x'|}. \tag{8.3–1}$$

□

Generalized one-dimensional Green's functions

The Green's function approach that we have detailed so far in this chapter no longer applies when $\lambda = 0$ is an eigenvalue of the underlying homogeneous boundary-value problem. In such cases, a solution of the *in*homogeneous BVP may not even exist, and if it does, it is obviously not unique.

In order to handle this situation, we need to introduce the concept of adjoint boundary conditions. For simplicity we restrict ourselves to the contest at hand in which the ordinary differential equations of interest are of order two. Thus the most general linear homogeneous boundary conditions we need to consider have the form (2.2–3), that is,

$$\begin{aligned} Au(a) + Bu'(a) + Cu(b) + Du'(b) &= 0 \\ Eu(a) + Fu'(a) + Gu(b) + Hu'(b) &= 0. \end{aligned} \tag{2.2–3'}$$

Definition 1 Given a set of general mixed homogeneous B.C. of the form (2.2–3) associated with a given operator L, the set of **adjoint** B.C. is that (unique) set of boundary conditions of the same general form such that

$$p(x)W(u,v)\Big|_a^b = 0$$

whenever u and v satisfy the given B.C. and the adjoint B.C., respectively.*

We may now state an important result that will be especially pertinent to the needed generalization of our Green's function approach:

Theorem 1:

The inhomogeneous BVP

$$Lu \equiv \frac{d}{dx}\left[p(x)\frac{du}{dx}\right] - q(x)u = -f(x) \qquad a < x < b \qquad \textbf{(8.3–2)}$$

Homogeneous B.C.

has a solution *if and only if*

$$\int_a^b f(x)v(x)dx = 0$$

for every solution of the homogeneous adjoint problem

$$Lv = 0 \qquad\qquad a < x < b$$

Homogeneous adjoint B.C. **(8.3–3)**

Proof: See Coddington and Levinson (1955), pp. 294 ff. or Ince (1944), pp. 213f. □

Theorem 1 is actually one portion of the famous *Fredholm Alternative* (which we shall discuss at length in Chapter 10) specialized to this differential equations setting. Part of the rest of the Alternative ensures that the homogeneous *adjoint* problem has precisely the same number of linearly independent solutions as the homogeneous version of the given BVP. Although this could in some cases be two (see Problem 9), we assume in this chapter that it is never more than one (see Stakgold (1979), pp. 215 ff. for a brief discussion of the more general situation). If $\lambda = 0$ is an eigenvalue of the underlying homogeneous BVP, therefore, there is a unique (up to a multiplicative factor) nontrivial solution of (8.3–3), and we denote it by $v_0(x)$.

Now we are ready to modify our original notion of Green's function to accommodate this generalized situation. We give the basic definition first and then make a number of remarks regarding certain key features.

*Some readers may feel uncomfortable with this definition. As a practical matter, however, it is at least as efficient for determining the adjoint B.C. as the alternative definition (see Problem 7). Moreover, it readily clarifies why regular B. C. are sometimes termed "self-adjoint" since in this special case the adjoint B. C. must obviously be identical to the given B. C.

Definition 2 **Let** $v_0(x)$ be the nontrivial solution of the homogeneous adjoint BVP (8.3–3) and denote by $r(x)$ the positive weighting function associated with the related eigenvalue problem (recall the earlier discussion of Chapter 2). The **generalized Green's function** for the BVP (8.3–2) is the two-variable function $G(x,x')$, $a \leq x,x' \leq b$, with the properties that, *as a function of* x,

(i) $LG(x,x') = r(x)\, v_0(x)v_0(x') \qquad x \neq x'$;
(ii) $G(x,x')$ satisfies the homogeneous B.C.;
(iii) $G(x,x')$ is continuous for all $a \leq x \leq b$; and
(iv) $\int_a^b r(x)v_0(x)G(x,x')dx = 0$ for all $a \leq x' \leq b$.

Some explanation is called for:

1. The form of the defining equation for the generalized Green's function is suggested by Theorem 1. The nontrivial function $v_0(x)$ cannot be orthogonal to itself, and thus we are assured that the BVP

$$Lu = r(x)v_0(x)v_0(x') \qquad a < x < b$$
$$\text{Homogeneous B.C.}$$

has no *smooth* solution.

2. Unlike the ordinary situation, in this generalized case the jump discontinuity in $\partial G/\partial x$ at $x = x'$ may not be specified arbitrarily; its value is determined automatically. Indeed, if we define

$$f(x,x') \equiv v_0(x)LG(x,x') - G(x,x')Lv_0(x)$$

where $a < x' < b$, then, on the one hand

$$\int_a^b f(x,x')dx = v_0(x') \int_a^b r(x)v_0^2(x)dx.$$

On the other hand, using (2.2–2) and noting the adjoint relationship between the B.C. for $v_0(x)$ and $G(x,x')$, respectively,

$$\int_a^b f(x,x')dx = \int_a^{x'} f(x,x')dx + \int_{x'}^b f(x,x')dx$$

$$= p(x)W[v_0(x),G(x,x')]\Big|_a^{x'} + p(x)W[v_0(x),G(x,x')]\Big|_{x'}^b$$

$$= -v_0(x')p(x')\left[\frac{\partial G(x'+0,x')}{\partial x} - \frac{\partial G(x'-0,x')}{\partial x}\right].$$

Equating these two results, we find

$$\left[\frac{\partial G(x'+0,x')}{\partial x} - \frac{\partial G(x'-0,x')}{\partial x}\right] = -\frac{1}{p(x')}\int_a^b r(x)\, v_0^2(x)dx.$$

If the nontrivial function $v_0(x)$ is given its customary normalization with respect to the weighting function $r(x)$, namely,

$$\int_a^b r(x)v_0^2(x)dx = 1,$$

which we will assume to be the case henceforth, then the jump discontinuity in $\partial G/\partial x$ turns out to be precisely what it was in the ordinary situation.

3. A new condition is therefore needed to replace (iv) of the earlier definition. The one selected has the desirable feature that it ensures the *symmetry* of the resulting generalized Green's function when the boundary conditions associated with the given BVP are regular (see Problem 10).

With the new Green's function in hand, we can generalize the representation result we first gave in the last section.

Theorem 2:

Assume that the homogeneous version of the BVP (8.3–2) has no more than one nontrivial solution. If a solution of the given inhomogeneous problem exists, it may be represented in terms of the appropriate Green's function as

$$U(x) = \int_a^b G(x,x')f(x')dx'. \quad \textbf{(8.3–4)}$$

When $\lambda = 0$ is not an eigenvalue of the homogeneous problem, the requisite Green's function is the ordinary one; otherwise $G(x,x')$ is the generalized Green's function. In this latter case, the general solution to (8.3–2) has the form

$$u(x) = Cu_0(x) + U(x)$$

where u_0 is the nontrivial solution of the homogeneous version of (8.3–2), C is an arbitrary constant, and U is given by (8.3–4).

Proof: Since the essence of the proof is similar to that of Theorem 1, Section 8.2, we leave the details to the interested reader (Problem 11). □

As before, certain "singular" problems and BVP's with inhomogeneous boundary conditions can be easily accommodated.

More examples of Green's functions

Example 1 If the underlying homogeneous problem is

$$\frac{d^2u}{dx^2} = 0 \qquad 0 < x < 1$$

$$\frac{du}{dx}(0) = 0 = \frac{du}{dx}(1),$$

then $\lambda = 0$ is an eigenvalue. The choice of $r(x) \equiv 1$ leads to $v_0(x) = 1$. Thence

$$G(x,x') = \tfrac{1}{2}x^2 + \begin{cases} A + Bx & x < x' \\ C + Dx & x > x'. \end{cases}$$

The B.C. imply that $B = 0$ and $D = -1$ and thus

$$G(x,x') = \tfrac{1}{2}x^2 + \begin{cases} A & x < x' \\ C - x & x > x'. \end{cases}$$

Using this new form for G, the continuity at $x = x'$ obviously leads to $A = C - x'$, while the choice $C = \tfrac{1}{2}x'^2 + \tfrac{1}{3}$ ensures the desired orthogonality to $v_0(x)$. Finally, we have

$$G(x,x') = \tfrac{1}{2}(x^2+x'^2) + \tfrac{1}{3} - \begin{cases} x' & x < x' \\ x & x > x' \end{cases}$$

for our generalized Green's function. □

Example 2 If

$$\frac{d}{dx}\left[(1-x^2)\frac{du}{dx}\right] = 0 \qquad -1 < x < 1$$

$$u(-1),\ u(1) \text{ finite},$$

then $\lambda = 0$ is again an eigenvalue. This time the choice $r(x) \equiv 1$* leads to $v_0(x) = 1/\sqrt{2}$, from which it follows that $G(x,x')$ has the form

$$G(x,x') = -\tfrac{1}{4}\ln(1-x^2) + \begin{cases} A + B\ln\dfrac{1+x}{1-x} & x < x' \\ C + D\ln\dfrac{1+x}{1-x} & x > x'. \end{cases}$$

The B.C. imply $B = -D = \tfrac{1}{4}$, while application of the two remaining conditions gives rise to

$$C = A + \tfrac{1}{2}\ln\frac{1+x'}{1-x'} \qquad \text{wherein } A = \ln 2 - \tfrac{1}{2} - \tfrac{1}{2}\ln(1+x').$$

The appropriate generalized Green's function for this problem thus is

$$G(x,x') = \ln 2 - \tfrac{1}{2} - \begin{cases} \tfrac{1}{2}\ln[(1-x)(1+x')] & x < x' \\ \tfrac{1}{2}\ln[(1-x')(1+x)] & x > x'. \end{cases}$$ □

Additional examples of one-dimensional Green's functions, both ordinary and generalized, are considered in the problems at the end of this chapter.

---

*$r(x) \equiv 1$ is not always the appropriate choice; recall the discussion of Section 2.5.

8.4 Green's functions and partial differential equations

The natural generalization to higher dimensions of the canonical second-order linear operator (2.2–1) is

$$Lu \equiv \nabla \cdot (p\nabla u) - qu \qquad \mathbf{x} \in R \tag{8.4–1}$$

where R is a region bounded by a sufficiently smooth surface S. We note that

$$uLv - vLu = \nabla \cdot [p(u\nabla v - v\nabla u)]$$

so that this operator is, in analogy with the concept in one-dimension, "self-adjoint." Making use of the Divergence Theorem (see (5.4–4) for the two-dimensional version), we find that in addition

$$\iint_R (uLv - vLu)dV = \int_S p\left(u\frac{\partial v}{\partial n} - v\frac{\partial u}{\partial n}\right)dS, \tag{8.4–2}$$

where the normal derivatives are assumed to be calculated in the *outward* direction relative to the bounding surface S.

Relation (8.4–2) is the analogue of (2.2–2); it is often called *Green's identity* and plays a fundamental role in our deliberations. For example, the identity suggests the following definition.

Definition 1 Homogeneous linear boundary conditions are said to be **regular** for the operator L given by (8.4–1) if, for all u,v that satisfy the conditions,

$$\int_S p\left(u\frac{\partial v}{\partial n} - v\frac{\partial u}{\partial n}\right)dS = 0.$$

The most common boundary conditions that occur in applied second-order linear BVP's have the form*

$$\alpha u + \beta\frac{\partial u}{\partial n} = \gamma$$

where $\alpha = \alpha(Q)$, $\beta = \beta(Q)$ and $\gamma = \gamma(Q)$ are given functions on S with $\alpha^2 + \beta^2 \neq 0$. If $\beta \equiv 0$, these are the Dirichlet B.C. mentioned earlier in Section 8.1. The case $\alpha \equiv 0$ corresponds to the so-called *Neumann B.C.* However, even in the general case, Green's identity shows that the homogeneous version ($\gamma \equiv 0$) of these B.C. is *regular*, a fortunate circumstance indeed.

---

*This is the generalization of "unmixed" one-dimensional B.C. Note that the present boundary conditions could be "mixed" in the sense in which this term was used in Section 7.1.

The ordinary Green's function

If p,q are nonnegative in R and $\alpha/\beta \geq 0$ on S, then the homogeneous version ($\gamma \equiv 0$) and $f \equiv 0$) of the BVP

$$\nabla \cdot (p \nabla u) - qu = -f \qquad \mathbf{x} \in R$$

$$\alpha u + \beta \frac{\partial u}{\partial n} = \gamma \text{ on } S \tag{8.4-3}$$

generally has only the trivial solution. (An exception is the case $q \equiv 0$, $\alpha \equiv 0$; see Problem 15.) In such a situation (or, more generally whenever $\lambda = 0$ is not an eigenvalue of the homogeneous BVP) an ordinary Green's function may be defined as follows:

Definition 2 The (regular or ordinary) **Green's function** appropriate for the BVP (8.4–3) is the two-variable function $G(P,Q)$ (P,Q in R) with the properties that, *as a function of P*,

(i) $\nabla \cdot [p(P) \nabla G(P,Q)] - q(P)G(P,Q) = 0 \qquad P \neq Q;$

(ii) $\alpha G(P,Q) + \beta \dfrac{\partial G(P,Q)}{\partial n_p} = 0 \qquad P \text{ on } S,\ Q \text{ in } R;$

(iii) $\displaystyle\lim_{\epsilon \to 0} \int_{S_\epsilon} \frac{\partial G(P,Q)}{\partial n_p}\, dS = -\frac{1}{p(Q)}$

where S_ϵ is the sphere of radius ϵ about the point Q.

We observe that this definition is the analogue of that given in Section 8.2 for the one-dimensional case. Conditions (i) and (ii) are, in essence, the same as before; the earlier conditions (iii) and (iv), however, have been combined. This latter step is necessary in this higher-dimensional situation since the desired "picking-out" property (see the additional discussion below is manifested as a "singularity" (albeit, integrable) in the function $G(P,Q)$ at the influence point Q. As before, Definition 2 generally leads to a unique* Green's function, symmetric in its arguments (see Problem 17), and we will illustrate this with some examples later.

As suggested earlier, the importance of Green's function inheres in its role in integral representations for the solutions of BVP's. In the case of Dirichlet B.C., the pertinent result for problems of the form (8.4–3) is

Theorem 1:

If $\lambda = 0$ is not an eigenvalue of the homogeneous version of the BVP, then the solution of

$$\nabla \cdot (p \nabla u) - qu = -f \qquad \mathbf{x} \in R$$

$$u \text{ given on } S$$

*The uniqueness question and, to an even greater extent, the existence question for Green's functions are fundamental to a complete theoretical understanding. A substantive discussion here, however, would lead us rather far afield, and thus interested readers should consult references such as Garabedian (1964) for desired details.

may be represented in terms of the appropriate ordinary Green's function as

$$u(P) = \iint_R G(P,Q)f(Q)dV - \int_S p(Q)u(Q)\frac{\partial G(P,Q)}{\partial n_Q}\,dS. \qquad \textbf{(8.4–4)}$$

Proof: *An indirect proof based upon the symmetry of $G(P,Q)$ is easiest.* We exlude from R a small n-ball V_ϵ of radius ϵ centered at P and evaluate

$$I(\epsilon) = \iint_{R-V_\epsilon} [u(Q)LG(P,Q) - G(P,Q)Lu(Q)]dV_Q$$

where L is the differential operator given above (acting on the Q variable). By symmetry $LG(P,Q) = 0$ for $Q \neq P$, and hence, on the one hand,

$$I(\epsilon) = \iint_{R-V_\epsilon} G(P,Q)f(Q)dV.$$

Alternatively, using (8.4–2),

$$I(\epsilon) = \int_S p(Q)\left[u(Q)\frac{\partial G(P,Q)}{\partial n_Q} - G(P,Q)\frac{\partial u(Q)}{\partial n_Q}\right]dS$$
$$-\int_{S_\epsilon} p(Q)\left[u(Q)\frac{\partial G(P,Q)}{\partial n_Q} - G(P,Q)\frac{\partial u(Q)}{\partial n_Q}\right]dS.$$

This last expression can be simplified considerably, however. Since $G(P,Q)$ vanishes for Q on S (by symmetry), the first integral is merely

$$I_1 = \int_S p(Q)u(Q)\frac{\partial G(P,Q)}{\partial n_Q}\,dS.$$

Moreover, in the limit of vanishingly small ϵ, the second integral becomes

$$I_2 = -\lim_{\epsilon\to 0}\int_{S_\epsilon} p(Q)u(Q)\frac{\partial G(P,Q)}{\partial n_Q}\,dS.$$

Again invoking symmetry, this is simply

$$I_2 = u(P)$$

by virtue of condition (iii)* of Definition 2. Combining these results yields the representation (8.4–4). Using a related argument, it is easy to verify that (8.4–4) takes on the appropriate boundary values (see Problem 16).

□

---

*Thus the net effect of this condition is to assure that the normal derivative of Green's function has the needed "reproducing" characteristic (recall our discussion of reproducing kernels in Section 8.1). For simplicity, we call this the "picking-out" property.

When the boundary conditions are of Neumann type, the above representation takes a slightly different form. In this case, the comparable expression is

$$u(P) = \iint_R G(P,Q)f(Q)dV + \int_S p(Q)G(P,Q)\frac{\partial u(Q)}{\partial n_Q}dS. \qquad \textbf{(8.4–5)}$$

For the general situation, the analogue of (8.4–4), (8.4–5) is

$$u(P) = \iint_R G(P,Q)f(Q)dV + \int_S p(Q)G(P,Q)\left[\alpha(Q)u(Q) + \frac{\partial u(Q)}{\partial n_Q}\right]dS, \qquad \textbf{(8.4–6)}$$

provided we suppose (without loss of generality) that $\beta \equiv 1$.

Generalized Green's functions

In a number of problems of distinct practical importance the functions p and q appearing in the operator L given by (8.4–1) are of opposite sign (witness the reduced wave or Helmholtz equation). In such cases the homogeneous version of the BVP (8.4–3) may have nontrivial solutions, and thus an ordinary Green's function approach may not be applicable. However, as in the one-dimensional situation, we can handle these cases by introducing a generalized Green's function. Assuming $\lambda = 0$ is a simple eigenvalue of the underlying homogeneous problem, the relevant definition takes the form:

Definition 3 Let $v_0(P)$ be the nontrivial solution of the BVP (8.4–3). The appropriate **generalized Green's function** is the two-variable function $G(P,Q)$, P,Q in R, with the properties that, *as a function of P*,

(i) $\nabla \cdot [p(P)\nabla G(P,Q)] - q(P)G(P,Q) = r(P)v_0(P)v_0(Q) \qquad P \neq Q;$

(ii) $\alpha G(P,Q) + \beta \dfrac{\partial G(P,Q)}{\partial n_p} = 0 \qquad P$ on S, Q in R;

(iii) $\displaystyle\lim_{\epsilon \to 0} \int_{S_\epsilon} \frac{\partial G(P,Q)}{\partial n_p}dS = -\frac{1}{p(Q)},$

where S_ϵ is the sphere of radius ϵ about the point Q; and

(iv) $\displaystyle\iint_R r(P)v_0(P)G(P,Q)dV = 0 \qquad Q$ in R.

In connection with this definition, the reader should especially note the following:

1. Since the equation of interest is self-adjoint and the B.C. are regular, the homogeneous BVP adjoint to (8.4–3) and the homogeneous version of (8.4–3) are identical (compare with Definition 2, Section 8.3). However, the question of how many nontrivial solutions actually exist to this homogeneous problem cannot be easily answered in general.

Nevertheless, $\lambda = 0$ often turns out to be a simple eigenvalue in most practical problems.*

2. Conditions (i) and (ii) are the natural generalizations of the corresponding conditions in Definition 2 of the previous section. On the other hand, condition (iii) is superfluous. If G is integrable in the neighborhood of the influence point and v_0 is normalized so that

$$\iint_R r(P)v_0^2(P)dS = 1,$$

then (like the one-dimensional situation) the discontinuity condition (iii) is satisfied automatically. We include the condition here, however, since it serves to emphasize the "picking-out" property of $G(P,Q)$.

3. As before, condition (iv) ensures the symmetry of the generalized Green's function (see Problem 17).

Just as in the one-dimensional case, the Fredholm Alternative governs the solvability of the BVP (8.4–3) when $\lambda = 0$ is an eigenvalue of the related homogeneous problem. If this eigenvalue is simple, a necessary and sufficient condition can be given which guarantees that a solution (obviously nonunique) to (8.4–3) exists. (See Problem 18.) When this condition is met, particular solutions for the inhomogeneous BVP's of the first (Dirichlet), second (Neumann) and third kinds may be represented by expressions of the same form as in the ordinary case, namely (8.4–4), (8.4–5), and (8.4–6), respectively. The G to be used in these formulas is, of course, the generalized Green's function.

8.5 Application of the Green's function approach in higher dimensions

As suggested in Section 8.1, the integral representations (8.2–6) and (8.3–4) in the one-dimensional case and (8.4–4), (8.4–5), and (8.4–6) for higher dimensions provide a valuable mechanism for analyzing and interpreting the solutions of various boundary-value problems. However, the utility of this approach depends upon our ability to determine Green's function associated with each problem. Although our earlier one-dimensional examples may suggest that this is generally a straightforward undertaking, such is not the case in higher dimensions. We illustrate with the simplest partial differential operator that has the form (8.4–1), namely the Laplacian.

From potential-theoretic considerations (or direct calculation) we know that Laplace's equation has the solutions

*For example, u = constant is the only solution to the homogeneous Neumann problem for Laplace's equation. The fact that the reduced wave equation $\Delta u + k^2u = 0$ (where k corresponds to the fundamental frequency for the region) has a unique solution that vanishes on the boundary provides another example.

$$\delta(\mathbf{x}) = \begin{cases} k \ln (1/r) & r \neq 0,\ k = \text{constant} \quad n = 2 \\ kr^{2-n} & \quad n > 2, \end{cases}$$

where r is the distance of the point $\mathbf{x}$ from some fixed point in the region of interest. In physical terms, such a "singular" solution corresponds to the electrostatic potential associated with a point charge of appropriate strength k, located at the fixed point. The total flux emanating from the point, that is, the integral of $\partial\delta/\partial n$ over the n-sphere surrounding the point) is proportional to the charge strength k. Comparing this observation with condition (iii) of Definitions 2,3 of Section 8.4 suggests that if the fixed point is identified with the influence point Q of Green's function, the above singular solutions, properly scaled possess the "picking-out" characteristic of the $G(P,Q)$ we seek.

However, for bounded regions, the singular solution by itself is not the entire story. In order to satisfy the other conditions of the definitions, a suitable smooth function must also be included. This leads to the convenient representation

$$G(P,Q) = g(P,Q) + \begin{cases} \dfrac{1}{2\pi} \ln(1/r_{PQ}) & n = 2 \\ \dfrac{1}{(n-2)\omega_n} r_{PQ}^{2-n} & n > 2 \end{cases} \qquad \textbf{(8.5–1)}$$

for Green's functions associated with Laplace's equation. In this expression ω_n is the area of the unit n-sphere and $g(P,Q)$ is a smooth solution that is, without any "singularity") of either Laplace's equation or a related Poisson equation, according as $G(P,Q)$ is either an ordinary or a generalized Green's function.

It is, of course, the determination of $g(P,Q)$ that renders the complete specification of the appropriate Green's function difficult for most regions. Only in a modest number of cases can $g(P,Q)$ be easily ascertained in closed form. One particular case, however, deserves special consideration because of its ubiquitous nature.

The Dirichlet problem for the half-plane*

We would like to determine the Green's function appropriate for the BVP

$$\Delta u(x,Y) = -f(x,y) \qquad -\infty < x < \infty, \quad y > 0$$
$$u(x,0) \text{ given.}$$

In view of the Dirichlet B.C., we can interpret the function $g(P,Q)$ in (8.5–1), electrostatically, as the potential (harmonic function without any singularities in $y \geq 0$) needed to nullify the effect of $1/(2\pi) \ln (1/r_{PQ})$ when P is on the x-axis. This

*In keeping with common practice, we designate the BVP for Poisson's (Laplace's) equation having Dirichlet B.C., the Dirichlet Problem.

is obviously $-1/(2\pi)\ \ln\,(1/r)_{PQ'}$, where Q' is the mirror image of Q in this axis. Hence, if $P = (x,Y)$, $Q = (x',y')$, then

$$
\begin{aligned}
G(P,Q) &= \frac{1}{2\pi}\left[\ln r_{PQ'} - \ln r_{PQ}\right] \\
&= \frac{1}{2\pi}\left[\ln \sqrt{(x-x')^2 + (y+y')^2} - \ln \sqrt{(x-x')^2 + (y-y')^2}\right] \\
&= \frac{1}{4\pi} \ln \left[\frac{(x-x')^2 + (y+y')^2}{(x-x')^2 + (y-y')^2}\right]. \qquad \textbf{(8.5–2)}
\end{aligned}
$$

The Dirichlet problem for other two-dimensional regions

The implications of the above example are twofold. If we rewrite Green's function (8.5–2) in complex variable notation where $P = z$ and $Q = \zeta$, then

$$G(P,Q) = -\frac{1}{2\pi} \ln \left|\frac{z-\zeta}{z-\zeta}\right| .$$

We recognize the function $F(z,\zeta) \equiv (z - \zeta)/(z - \zeta)$ inside the logarithm as the analytic transformation (actually a linear fractional transformation in this case) that maps the upper half-plane conformally onto the interior of the unit circle and takes the point ζ to the origin.

In light of our Section 6.3 discussion on conformal transformations of harmonic functions, this suggests a procedure for obtaining the solution to the Dirichlet problem for Poisson's equation in other simply-connected two-dimensional regions. If $F(z,\zeta)$ is the analytic function (engendered by the Riemann Mapping Theorem) that maps the given Region R one-to-one onto the interior of the unit circle and takes the point ζ to the origin, then (associating P with z and Q with ζ and denoting the boundary of R by S),

$$G(P,Q) = -\frac{1}{2\pi} \ln |F(z,\zeta)| \qquad \textbf{(8.5–3)}$$

is the appropriate Green's function for

$$
\begin{gathered}
\Delta u = -f \qquad P \in R \\
u \text{ given on } S.
\end{gathered}
$$

The expression (8.5–3) underscores, at least for the Laplace operator in two-dimensions, the intimate relationship between conformal mapping, discussed in Chapter 6, and Green's functions. On the other hand, the usefulness of the relation is constrained by our lack of knowledge of the conformal maps associated with complicated domains. (See Problems 19–21, however, for some interesting examples.)

The method of images

The second feature of the half-plane solution that warrants exploitation is the physically motivated approach to finding the appropriate $g(P,Q)$ for use in (8.5–1)*. Since the B.C. are of Dirichlet type and the operator L is the Laplacian, the bounding surface (curve) can be viewed as a perfect conductor that is held at zero potential. The function $g(P,Q)$, therefore, provides the necessary potential to "balance things off," and in this case it is merely the field associated with the *reflection* in the boundary of the point charge situated at the influence point Q.

It is virtually self-evident how this imaging technique can be generalized to more complicated domains. (The same procedure also works in higher dimensions owing to Kelvin's theorem; see Garabedian (1964), p. 84.) The basic restriction is that the region of interest is bounded by lines and circles in two dimensions, planes and spheres in three dimensions, and so forth. In two-dimensions, therefore, this gives an alternative procedure for finding the Green's function appropriate for the Laplace operator with Dirichlet B.C. in such domains as the circle, the convex wedge, the infinite-strip, the annulus, the semi-infinite strip, the triangle, and the rectangle, amongst others. The general form for the Green's function so obtained is

$$G(P,Q) = -\frac{1}{2\pi}\ln r_{PQ} + \frac{1}{22\pi}\sum_i \delta(Q_i)\ln(\sigma(Q_i)r_{PQ_i}) \qquad \textbf{(8.5–4)}$$

where the summation extends over all Q_i obtained by reflecting Q, and the Q_i's themselves, in the various boundaries. The coefficients $\delta(Q_i)$ and $\sigma(Q_i)$ are to be chosen so that $G(P,Q)$ has the proper boundary behavior.

The method of images obviously breaks down if any image falls back within the region in question at a point other than Q (recall the smoothness requirement on g). Moreover, if the number of images is infinite, some care generally needs to be taken in evaluating the summation appearing in (8.5–4). These caveats notwithstanding, the imaging technique remains the method of choice for determining Green's function in a number of important cases.

Before we leave the method of images, it is worth noting that variants of the approach are applicable with equations other than Laplace's. The technique can also be used to indicate the form of Green's function in the case of B.C. of Neumann type. We illustrate this latter observation with the following.

Example In the circle of radius ρ about the origin, the Dirichlet Green's function for Laplace's equation is, by imaging,

$$G(P,Q) = -\frac{1}{2\pi}\ln r_{PQ} + \frac{1}{2\pi}\ln\left(\frac{r' r_{PQ'}}{\rho}\right)$$

where $P = re^{i\phi}$, $Q = r'e^{if}$ and $Q' = (\rho^2/r')e^{i\phi}$ with $r,\ r' < \rho$. This suggests that

*Such an electrostatic analogy was the starting point for Green in his 1828 essay.

the Neumann Green's function for the Laplacian in this domain should have the form

$$G(P,Q) = -\frac{1}{2\pi}\ln r_{PQ} - \frac{1}{2\pi}\ln\left(\frac{r' r_{PQ'}}{\rho}\right) + h(P,Q).$$

Now $\lambda = 0$ is an eigenvalue for the underlying homogeneous BVP and has associated with it the single normalized eigenfunction $v_0 = 1/(\sqrt{\pi}\,\rho)$ (assuming unit weighting function $r(\mathbf{x})$). It follows then, invoking symmetry, that h can be chosen to be

$$h(P,q) = h(r,r') = \frac{r^2 + r'^2}{4\pi\rho^2} + C,$$

where C is a constant. Condition (iv) of Definition 3 of Section 8.4 determines C as $(1/\pi)(\ln\rho - 3/8)$, so that finally

$$G(P,Q) = -\frac{1}{2\pi}\left[\ln\left(\frac{r' r_{PQ'} r_{PQ}}{\rho^3}\right) - \frac{r^2 + r'^2}{2\rho^2} + \frac{3}{4}\right]. \tag{8.5–5}$$

□

The Helmholtz equation

Aside from Laplace's equation, virtually the only other specific partial differential equation of the type considered in Section 8.4 that has been studied at any length is the Helmholtz equation

$$\Delta u + k^2 u = 0. \tag{8.5–6}$$

We have already encountered the one-dimensional version of this equation in Example 3 of the first subsection of Section 8.3. In higher dimensions ($n \geq 2$), a singular solution of (8.5–6) with the necessary "picking-out" characteristic is

$$G_0(P,Q) = \tfrac{1}{4}\, i\left(\frac{2\pi}{k}\, r_{PQ}\right)^{1-\frac{1}{2}n} H^{(1)}_{\frac{1}{2}n-1}(kr_{PQ}), \tag{8.5–7}$$

where $H^{(1)}_\nu$ is the Hankel function of the first kind (see Appendix 2). This particular singular solution satisfies the n-dimensional version of the Sommerfeld radiation condition (see Problem 23). In keeping with terminology to be introduced in the next section, we can therefore term the expression (8.5–7) the *free-space Green's function* for the Helmholtz equation (see Problem 24) and observe that, in the limit $k \to 0$, it gives rise to the expected *free-space Green's function* for Laplace's equation (see (8.5–1)). Since the Bessel functions of order ½ are elementary, when $n = 3$ (8.5–7) is usually written in the more familiar form

$$G_0(P,Q) = \frac{e^{ikr_{PQ}}}{4\pi r_{PQ}}.$$

8.6 A Green's function approach to diffusion and wave-propagation problems

In the previous section, our attention was primarily focused on partial differential equations associated with equilibrium phenomena. We close this chapter with a brief consideration of the remaining practical problems that served to motivate our original interest in Green's functions.

Generalized diffusion

A time-varying operator closely related to that given by (8.4–1) is

$$Lu \equiv \nabla \cdot (p \nabla u) - qu - \frac{\partial u}{\partial t} \qquad \mathbf{x} \in R,\ t > 0, \tag{8.6–1}$$

where p and q are restricted to be functions of the spatial variable $\mathbf{x}$ only with $p > 0$. This new operator is the so-called *generalized diffusion operator*, which occurs in the description of a variety of diffusion processes. (See Morse and Feshbach (1953b) or Crank (1975), for example.) By convention, time is taken to be increasing since diffusion is an irreversible process; problems involving the operator (8.6–1) in which $t < 0$ are in general not well-posed.

In analogy with (8.4–3), practical generalized diffusion problems commonly give rise to BVP's/IVP's of the form

$$\begin{aligned} \nabla \cdot (p \nabla u) - qu - \frac{\partial u}{\partial t} &= -f(\mathbf{x},t) && \mathbf{x} \in R,\ t > 0 \\ \alpha u + \beta \frac{\partial u}{\partial n} &= \gamma(\mathbf{x},t) && \mathbf{x} \in S,\ t > 0 \\ u(\mathbf{x},0) &= h(\mathbf{x}) && \mathbf{x} \in R, \end{aligned} \tag{8.6–2}$$

where S, as usual, denotes the (suitably smooth) surface of the spatial region R and $\alpha = \alpha(\mathbf{x},t)$, $\beta = \beta(\mathbf{x},t)$ are such that $\alpha^2 + \beta^2 \neq 0$. As might be expected, a Green's function $G(P,Q;t)$ exists for (8.6–2). In the ordinary case $G(P,Q;t-\tau)$ satisfies, as a function of P and t

(i) $\nabla \cdot (p \nabla G) - qG - \dfrac{\partial G}{\partial t} = 0 \qquad P,Q \in R,\ t > \tau$

(ii) $\alpha G + \beta \dfrac{\partial G}{\partial n} = 0 \qquad P \text{ on } S,\ t > \tau$

(iii) $G(P,Q;0) = 0 \qquad P \neq Q$

(iv) $\displaystyle \lim_{\epsilon \to 0} \lim_{\delta \to 0} \int_0^{t-\delta} \int_{S_\epsilon} \frac{\partial G}{\partial n}\, ds\, d\tau = -\frac{!}{p(Q)},$

where the sphere S_ϵ is the boundary of the ball V_ϵ of radius ϵ about the point Q.

The last condition has the alternative form

$$\text{(iv}'\text{)} \lim_{\delta\to 0} \iint_{V_\epsilon} G(P,Q;\delta)dV = 1, \qquad (\epsilon > 0)$$

which emphasizes the relation of these Green's functions to reproducing kernels and, hence, to generalized function theory. (Recall the discussion of Section 8.1.)*

When the B.C. on S are of Dirichlet type, the solution of (8.6–2) can be represented in terms of the appropriate Green's function as

$$\begin{aligned} u(P,t) = & \int_0^t \iint_R G(P,Q;t-\tau)f(Q,\tau)dV\,d\tau \\ & - \int_0^t \int_S p(Q)u(Q,\tau)\frac{\partial G(P,Q;t-\tau)}{\partial n_Q}\,dS\,d\tau \\ & + \iint_R u(Q,0)G(P,Q;t)dV. \end{aligned} \qquad \textbf{(8.6–3)}$$

This is the obvious analogue of (8.4–4). Expressions comparable to (8.4–5), (8.4–6) exist for the case of more general B.C. on S as well (see Roach (1970) or Morse and Feshbach (1953a) for details).

In a number of interesting situations the functions p and q appearing in (8.6–2) are constant. When such is the case, q might as well be taken to be zero, since a simple exponential transformation relates the two cases. If we introduce k as the reciprocal of p, we then obtain the classical heat-conduction equation.

$$\Delta u - k\frac{\partial u}{\partial t} = -f, \qquad (k > 0). \qquad \textbf{(8.6–4)}$$

Like Laplace's equation, the homogeneous version of (8.6–4) has certain canonical solutions that are fundamental building blocks in constructing Green's functions (compare (8.5–1)). For (8.6–4) the "special solutions" are

$$\delta(\mathbf{x},t) = \left(\frac{k}{4\pi t}\right)^{\frac{1}{2}n} \exp\left[-\frac{kr^2}{4t}\right] \qquad (t > 0),$$

where $n \geq 1$ is the number of space dimensions and r is the distance of the point $\mathbf{x}$ from some fixed point in the region R. If, as before, we identify that fixed point with the influence point Q, these special solutions satisfy conditions (iii) and (iv) above, and thus they must be the *free-space Green's functions* for the heat equation.† It is therefore appropriate to term the expression (8.1–7) obtained earlier in Section 8.1 the "Green's function representation" of the solution of the conduction problem (8.1–6).

---

*This form of the condition also points up the physical interpretation of the Green's function as the impulse-response.

†From the standpoint of differential equations theory, these are *fundamental solutions* (see Coddington and Levinson (1955)). The term *free-space* refers to the fact that the solutions are appropriate to unbounded regions with no finite boundaries.

Scalar wave propagation

The analogue of (8.6–4) for wave phenomena is the inhomogeneous scalar wave equation

$$\Delta u - \frac{1}{c^2}\frac{\partial^2 u}{\partial t^2} = -f \qquad \textbf{(8.6–5)}$$

whose homogeneous version we met earlier as (8.1–10). This too has "special solutions" that we can recognize as free-space Green's functions. Without going into details (interested readers may consult any of a number of references), we merely note that in one and two dimensions, respectively, these fundamental solutions have the form

$$G_0(x,x';t-\tau) = \begin{cases} c/2 & |x-x'| < c(t-\tau) \\ 0 & |x-x'| > c(t-\tau) \end{cases} \qquad \textbf{(8.6–6)}$$

and

$$G_0(P,Q;t-\tau) = \begin{cases} \dfrac{c/(2\pi)}{\sqrt{c^2(t-\tau)^2 - r_{PQ}^2}} & r_{PQ} < c(t-\tau) \\ 0 & r_{PQ} > c(t-\tau)\,. \end{cases} \qquad \textbf{(8.6–7)}$$

In higher dimensions ($n \geq 3$), the free-space Green's functions for the scalar wave equation involve delta and/or other generalized functions,* and thus it is inappropriate to discuss them at this juncture.

When finite boundaries are present, the "singular" functions (8.6–6), (8.6–7) must be augmented appropriately with smooth solutions of the homogeneous wave equation. If Green's function $G(P,Q;t - \tau)$ for a given region R (with boundary S) can be determined, then the solution of a typical BVP/IVP for (8.6–5) in that region may be represented in terms of G. Comparable to our earlier results, we find

$$\begin{aligned} u(P,t) = {} & \int_0^t \iint_R G(P,Q;t - \tau) f(Q,\tau)\,dV d\tau - \int_0^t \iint_S \Bigg[u(Q,\tau)\frac{\partial G(P,Q;;t - \tau)}{\partial n_Q} \\ & - G(P,Q;t-\tau)\frac{\partial u(Q,\tau)}{\partial n_Q}\Bigg] dS d\tau - \frac{1}{c^2}\iint_R \Bigg[u(Q,0)\frac{\partial G(P,Q;t - \tau)}{\partial \tau}\Bigg|_{\tau=0} \\ & - G(P,Q;t)\frac{\partial u(Q,0)}{\partial t}\Bigg] dV. \end{aligned} \qquad \textbf{(8.6–8)}$$

It is worth noting that (8.6–8) leads to the classical d'Alembert solution of the homogeneous wave equation

$$u(x,t) = \frac{1}{2}\left[\frac{1}{c}\int_{x-ct}^{x+ct} \frac{\partial u(x',0)}{\partial t}\,dx' + u(x+ct,0) + u(x-ct,0)\right] \qquad \textbf{(8.6–9)}$$

when the one-dimensional free-space Green's function (8.6–6) is used.

---

*In three-dimensions $G_0(P,Q;t-\tau) = \delta[(r_{PQ}/c) - (t-\tau)]/(4\pi r_{PQ})$.

In steady-wave problems, wherein a harmonic time dependence can be suppressed, the homogeneous version of (8.6–5) reduces to the Helmholtz equation

$$\Delta u + k^2 u = 0.$$

This equation governs a variety of sound-propagation problems. For example*, the time-independent portion of the velocity potential associated with the steady-state acoustic oscillations produced in a semi-infinite hollow tube of radius a by a membrane at $z = 0$, which vibrates harmonically and in axisymmetric fashion satisfies

$$\Delta\phi + k^2\phi = \frac{1}{\rho}\frac{\partial}{\partial\rho}\left(\rho\frac{\partial\phi}{\partial\rho}\right) + \frac{\partial^2\phi}{\partial z^2} + k^2\phi = 0 \qquad 0 \le \rho < a,\, z > 0$$

$$\phi(0,z) \text{ finite} \qquad z > 0$$

$$\frac{\partial\phi(a,z)}{\partial\rho} = 0 \qquad z > 0$$

$$\frac{\partial\phi}{\partial z} - i\beta\phi \to 0 \qquad \text{as } z \to \infty, \qquad 0 \le \rho < a$$

$$\frac{\partial\phi(\rho,0)}{\partial z} = f(\rho) \qquad 0 \le \rho < a. \tag{8.6–10}$$

Here $k \equiv \omega/c$, as usual, where ω is the frequency of the suppressed harmonic time-dependence and c is the sound-speed in the fluid filling the tube. The second boundary-condition specifies that the tube walls are rigid, while the third reminds us that the sound waves should be outgoing for large z.

Unfortunately, although this mixed boundary-value problem is conceptually quite simple, it falls into the class of BVP's for which the associated Green's functions have no known closed-form representations. Therefore, we are obligated to employ generalized Fourier series techniques and express the Green's function in the form

$$G(P,Q) = \frac{i}{\pi a^2}\sum_{n=0}^{\infty}\frac{J_0(\lambda_n\rho/a)\,J_0(\lambda_n\rho'/a)}{\beta_n J_0^2(\lambda_n)}\begin{cases} e^{i\beta_n z'}\cos\beta_n z & z < z' \\ e^{i\beta_n z}\cos\beta_n z' & z > z' \end{cases}. \tag{8.6–11}$$ †

In this formula, J_0 is the familiar Bessel function of the first kind of order zero, while the λ_n ($n = 0,1,2,\ldots$) are the turning points of J_0 (that is, the solutions of $J_0'(\lambda_n) = 0$), and $\beta_n \equiv \sqrt{k^2 - (\lambda_n/a)^2}$.

The expression (8.6–11) for Green's function is a modal expansion, that is, an expansion of G in terms of the modes supported by the rigid tube of radius a. Because the lowest eigenvalue λ_0 is zero, $\beta_0 = k$; accordingly the fundamental mode is a plane wave propagating at the sound-speed c. Only a finite number of higher-order modes will also be propagating, and since the accepted convention is that Im $\beta \ge 0$, all others will be attenuated.

---

*Recall the earlier acoustically-oriented discussion of Section 8.1.

†In the next chapter we will learn how to determine characteristic Green's function expansions such as this. Interested readers can find this particular expression in Morse and Ingard (1968).

With Green's function in hand, (8.6–8) can be used to provide a representation for the solution of the BVP (8.6–10). We find

$$\phi(\rho,z) = \frac{2i}{a^2} \sum_{n=0}^{\infty} e^{i\beta_n z} \frac{J_0(\lambda_n \rho/a)}{\beta_n J_0^2(\lambda_n)} \int_0^a \rho' f(\rho') J_0(\lambda_n \rho'/a) d\rho' .$$

The velocity potential turns out to be

$$\phi(\rho,z) = \frac{iA}{k} e^{ikz}$$

in the special case of a rigid membrane with $f(\rho) \equiv A$.

Final Remarks

Convinced of the utility of a Green's function approach by the above examples, some readers will want to learn more about this technique. Although Kellogg (1929) provides some interesting historical insights in addition to a careful treatment of the Green's functions of potential theory, and both Friedman (1956) and Roach (1970) consider a number of instructive examples (though from rather different points of view), the substantial volumes of Morse and Feshbach (1953a, b) are still the classic source. The variety of physically interesting problems treated in detail there makes this latter work a valuable reference indeed. Those readers interested in additional wave-oriented applications should consult Morse and Ingard (1968) for the scalar case and Felsen and Marcuvitz (1973) for the challenging vector situation.

References

Crank, J. (1975): *The Mathematics of Diffusion*, 2nd ed., Clarendon Press, Oxford.

Coddington, E. A. and N. Levinson (1955): *Theory of Ordinary Differential Equations*, McGraw-Hill, New York; Chapter 11.

Courant, R. and D. Hilbert (1953): *Methods of Mathematical Physics, Vol. I*, Interscience, New York; Chapter V.

Dettman, J. W. (1969): *Mathematical Methods in Physics and Engineering*, 2nd ed., McGraw-Hill, New York; Chapters 5, 7.

Felsen, L. B. and N. Marcuvitz (1973): *Radiation and Scattering of Waves*, Prentice-Hall, Englewood Cliffs, New Jersey.

Friedman, B. (1956): *Principles and Techniques of Applied Mathematics*, Wiley, New York; Chapters 3, 4, 5.

Garabedian, P. R. (1964): *Partial Differential Equations*, Wiley, New York; Chapters 5, 7, 8, 10.

Green, G. (1928): *An Essay on the Application of Mathematical Analysis to the Theories of Electricity and Magnetism*, Nottingham.

Ince, E. L. (1944): *Ordinary Differential Equations*, Dover, New York; Longmans, London (1927): Chapters IX, XI.

Kaplan, W. (1962): *Operational Methods for Linear Systems*, Addison-Wesley, Reading, Mass.; Chapter 8.

Kellogg, O. D. (1929): *Foundations of Potential Theory*, Springer-Verlag, Berlin; also Dover, New York; Chapters VII, IX, XI, XII.

Lebedev, N. N., I. P. Skalskaya, and Y. S. Uflyand (1965): *Problems of Mathematical Physics*, Transl. of Russian edition, Prentice-Hall, Englewood Cliffs, N. J.; Chapters 2, 4, 5.

Morse, P. M. and H. Feshbach (1953a): *Methods of Theoretical Physics, Part I*, McGraw-Hill, New York; Chapter 7; (1953b): "Methods of Theoretical Physics, Part II"; Chapters 10, 11, 12.

Morse, P. M. and K. U. Ingard (1968): *Theoretical Acoustics*, McGraw-Hill, New York; Chapters 6, 7, 8, 9.

Port, S. C. and C. J. Stone (1978): *Brownian Motion and Classical Potential Theory,* Academic Press, New York.

Rayleigh, Lord (J. W. Strutt) (1896): *The Theory of Sound, Vol. II*, 2nd ed., Macmillan, London; reprinted (1945) by Dover, New York; Chapters XII, XVI, Appendix A.

Roach, G. F. (1970): *Green's Functions: Introductory Theory with Applications*, Van Nostrand Reinhold, London

Stakgold, I. (1979): *Green's Functions and Boundary Value Problems*, Wiley, New York; Chapters 0, 1, 3, 6, 8.

Problems

Section 8.1

1. Let $u(r,\theta)$ be the real part of a function f of the complex variable $z = re^{i\theta}$, regular analytic in some region containing the disk $|z| < R$. Deduce from the Poisson integral formula (8.1–2) that $f(z)$ has the integral representation

$$f(z) = ic + \frac{1}{2\pi}\int_0^{2\pi} \frac{\rho e^{i\phi} + re^{i\theta}}{\rho e^{i\phi} - re^{i\theta}}\, u(\rho,\phi)d\phi$$

for all z *with* $|z| = r < \rho < R$. Here c is a real constant.

2. Use the formula of Problem 1 to demonstrate in formal fashion the well-known fact that if $f(z)$ is regular analytic in the disk $|z-z_0| < R$, then it has the convergent power series expansion

$$f(z) = \sum_{n=0}^{\infty} c_n(z-z_0)^n$$

there.

★3. (a) Determine the volume V_n of the ball of radius ρ in n-dimensions ($n \geq 1$).

[Hint: Start by showing that V_n satisfies the recurrence relation

$$V_{n+1} = \rho V_n \int_{-1}^{1} (1-x^2)^{\frac{1}{2}n} dx.$$

The integral on the right side of this expression is related to a beta function.]

(b) Since the area of the n-sphere of radius ρ is the derivative, with respect to ρ, of the volume of the n-ball, verify that the area of the *unit* n-sphere is given by the formula

$$\omega_n = \frac{n\pi^{\frac{1}{2}n}}{\Gamma(\frac{1}{2}n + 1)} .$$

4. Show that the Poisson kernel (8.1–4) satisfies

$$\frac{\rho - r}{\rho\omega_n (\rho+r)^{n-1}} \leq K_n(P,Q) \leq \frac{\rho + r}{\rho\omega_n(\rho-r)^{n-1}} ,$$

and hence derive the n-dimensional *Harnack inequalities*

$$\left(1 - \frac{r}{\rho}\right)\left(1 + \frac{r}{\rho}\right)^{1-n} u(0) \leq u(P) \leq \left(1 + \frac{r}{\rho}\right)\left(1 - \frac{r}{\rho}\right)^{1-n} u(0),$$

which hold for any nonnegative harmonic function u.

5. Develop plane-wave solutions of the form $e^{i(\omega t \pm kx)}$ to the one-dimensional version of the scalar wave equation (8.1–9). Under the assumption that these functions can be used to describe the propagation of sound along the interior of small-diameter hollow tubes, determine the classical formulas that relate the wavelength $\lambda = 2\pi/k$ of propagating modes to the length of the tube. Consider closed, semiclosed (half-open), and open tubes. (Rayleigh (1926) developed "corrections" to the classical formulas applicable in the latter two cases.)

Sections 8.2, 8.3

6. Find the Green's function associated with the BVP

$$\frac{d^2u}{dx^2} = 0 \qquad 0 < x < 1$$

$$u(0) = 0$$

$$\frac{du}{dx}(1) - \alpha u(1) = 0 \qquad \alpha \neq 1.$$

★7. Verify that two sets of general mixed homogeneous boundary conditions of the form (2.2–3), one with coefficients $A, B, \ldots, H$ and the other with coefficients $A', B', \ldots, H'$, are adjoints of each other (with respect to a given operator L) if

$$p(b)\begin{pmatrix} A' & B' \\ E' & F' \end{pmatrix}\begin{pmatrix} B & F \\ -A & -E \end{pmatrix} = p(a)\begin{pmatrix} C' & D' \\ G' & H' \end{pmatrix}\begin{pmatrix} D & H \\ -C & -G \end{pmatrix}.$$

[Like Problem 3 of Chapter 2, this is not an easy exercise algebraically; Coddington and Levinson (1955), pp. 288ff. has the details.]

8. Let

$$\text{(I)}\ u(0) - u(1) = 0 \qquad \text{(II)}\ v(0) = 0$$

and

$$\frac{du}{dx}(1) = 0 \qquad \frac{dv}{dx}(0) - \frac{dv}{dx}(1) = 0$$

be two sets of B.C.'s associated with a canonical second-order operator L for which $p(x) = 1$.

(a) Use Definition 1 of Section 8.3 to show that set II is the unique set of adjoint B.C. for set I and vice versa.

(b) Alternatively, employ the identity established in Problem 7 to derive this adjoint relationship between sets I and II.

9. Construct a homogeneous boundary-value problem of the form

$$\frac{d}{dx}\left[p(x)\frac{du}{dx}\right] - q(x)u = 0$$

Homogeneous B.C.

for which the eigenvalue $\lambda = 0$ has multiplicity two.
[Hint: Periodic B.C. will suffice.]

★10. Assume that $\lambda = 0$ is a simple eigenvalue of the homogeneous version of the BVP (8.3–2). Show that the associated generalized Green's function $G(x,x')$ defined in Section 8.3 is symmetric in its arguments if and only if the boundary conditions are regular.
[Hint: Integrate $G(x,x')\,LG(x,x'') - G(x,x'')\,LG(x,x')$ over $[a,b]$].

11. Give a proof of Theorem 2 of Section 8.3.

12. Construct Green's function associated with the ordinary differential equation

$$\frac{d^2u}{dx^2} = 0 \qquad < x < 1$$

when the boundary conditions are

(a) $u(0) = u(1)$ and $\dfrac{du}{dx}(0) = \dfrac{du}{dx}(1)$;

(b) $\dfrac{du}{dx}(0) = u(1)$ and $\dfrac{du}{dx}(1) = 0$.

Why is there a lack of symmetry in the second Green's function?

13. (a) Determine explicitly under what conditions there exists a solution to the BVP

$$\frac{d^2u}{dx^2} = -f(x) \qquad a < x < b$$

$$\frac{du}{dx}(a) = A$$

$$\frac{du}{dx}(b) = B$$

with A,B real constants.

(b) In the case where a solution to the problem of part (a) exists, represent the most general solution in terms of an appropriate Green's function.

14. (a) Construct the one-dimensional Green's function that is appropriate for the BVP

$$\frac{d^2u}{dx^2} = -f(x) \qquad 0 < x < 1$$

$$u(0) = u(1)$$

$$\frac{du}{dx}(1) = 0$$

[See Problem 8].

(b) Under the assumption that the solution to the BVP given in part (a) exists, express this solution in terms of Green's function and explicitly verify that the representation so obtained has all the desired properties.

Section 8.4

15. (a) Show that if $p(\mathbf{x})$, $q(\mathbf{x})$ are nonnegative in R and $\alpha/\beta \geq 0$ on the bounding surface S, then the BVP

$$\nabla \cdot [p(\mathbf{x}) \nabla u] - q(\mathbf{x})u = 0 \qquad \mathbf{x} \in R$$

$$\alpha u + \beta \frac{\partial u}{\partial n} = 0 \text{ on } S$$

has only the trivial solution $u \equiv 0$, unless $q(\mathbf{x}) \equiv 0$ and $\alpha \equiv 0$.

(b) If $\alpha \equiv 0$ but $q \not\equiv 0$, verify that

$$\iint_R q(P)G(P,Q)dV = 1,$$

where $G(P,Q)$ is the associated ordinary Green's function for the problem.

16. (a) Verify that when $\lambda = 0$ is not an eigenvalue of the homogeneous version of the BVP

$$Lu \equiv \nabla \cdot (p\nabla u) = qu = -f \qquad \mathbf{x} \in R$$
$$u \text{ given on } S,$$

then the solution of this BVP may be represented in terms of the appropriate ordinary Green's function as

$$u(P) = w(P) + \iint_R G(P,Q)\,[f(Q) + Lw]dV,$$

where w is any sufficiently smooth function that takes on the given B.C.

(b) Infer from the result of part (a) that the representation (8.4–4) takes on the given B.C.

(c) Give the one-dimensional analogue of the representation (8.4–4) that is appropriate for the equation (8.2–5) with *non*homogeneous B.C.

★17. Adapt and generalize the one-dimensional arguments to demonstrate that, when $\lambda = 0$ is at most a simple eigenvalue of the homogeneous version of the BVP (8.4–3), the appropriate Green's fucntion given by Definitions 2 or 3 of Section 8.4 satisfies

$$G(P,Q) = G(Q,P).$$

18. (a) Let $\lambda = 0$ be an eigenvalue of the homogeneous version of the BVP (8.4–3) and let v_0 be its corresponding eigenfunction. Show that the Dirichlet problem (in which $\alpha \equiv 1$, $\beta \equiv 0$) has a solution only if the compatibility condition

$$\iint_R v_0 f\,dV = \int_S p\gamma \frac{\partial v_0}{\partial n}\,dS$$

is satisfied.

(b) Derive analogous conditions for the BVP's of the second (Neumann) and third (general) kinds.

Section 8.5

19. Use either conformal mapping or the method of images to derive a Green's function representation for the solution of the Dirichlet problem

$$\Delta u = -f \qquad P \in R$$
$$u \text{ given on } S$$

when

(a) R is the quadrant $x > 0$, $y > 0$;

(b) R is the disk $x^2 + y^2 < \rho^2$.

20. Determine Green's function for the Dirichlet problem

$$\Delta u(x,y) = -f(x,y) \qquad -\infty < x < \infty,\ 0 < y < a$$
$$u(x,0) = 0 = u(x,a).$$

21. Repeat Problem 20 for the case in which the region R is the positive half-space $x > 0$ bounded by the convex semicircle $x = \sqrt{\rho^2 - y^2}$ ($|y| \le \rho$) and the remaining portions of the $y-$axis ($|y| > \rho$).

★22. Verify that the function given by (8.5–5) is Green's function for the Neumann problem associated with Poisson's equation in the disk $x^2 + Y^2 < \rho^2$.

23. Verify that the expression (8.5–7) does indeed provide the solution of the Helmholtz equation (8.5–6) that satisfies the n-dimensional version of the Sommerfeld radiation condition $r^{\frac{1}{2}(n-1)}\,[\partial u/\partial r - iku] \to 0$ as $r \to \infty$.

24. Let $u(\mathbf{x})$ be a continuously differentiable function in n-space that satisfies the radiation condition $r^{\frac{1}{2}(n-1)}\,[\partial u/\partial r - iku] \to 0$ and also is such that $r^{\frac{1}{2}(n-3)}\, u \to 0$ as $r \to \infty$. Here r is the distance from a fixed point P to a variable point Q. Show that

$$\iint_{S_R} \left[G_0(P,Q)\,\frac{\partial u(Q)}{\partial n_Q} - u(Q)\,\frac{\partial G_0(P,Q)}{\partial n_Q} \right] dS \to 0$$

as $R \to \infty$, where $G_0\,(P,Q)$ is given by (8.5–7) and S_R is the n-sphere of radius R centered at the point P. This justifies the use of the terminology *free-space Green's function* for $G_0(P,Q)$.

Section 8.6

25. (a) Integrate over all space the free-space Green's function (with $\tau = 0$)

$$G(P,Q;t) = \left[\frac{k}{4\pi t}\right]^{\frac{1}{2}n} \exp\left[-\frac{kr_{PQ}^2}{4t}\right] \quad (t > 0)$$

for the classical heat equation (8.6–4) and show that

$$\iint G(P,Q;t)\, dV = 1.$$

(Recall Problem 3b.) Interpret this result physically.

(b) Integrate Green's function of part (a) over all time and relate the result to the free-space Green's function for Laplace's equation.

26. Derive the two-dimensional analogue of the d'Alembert solution (8.6–9) of the homogeneous wave equation by substituting the free-space Green's function (8.6–7) into the representation (8.6–8). Explain, in physical terms, the differences between the one- and two-dimensional results.

9

Generalized functions

9.1 Reproducing kernels and the delta function

Chapter 8 dealt with a unified methodology for obtaining integral representations for the solutions of a wide variety of boundary and initial value problems. The key construct in that approach was the Green's function. In this regard we recall, for example, that if we define the differential operator L by

$$L\,u \equiv \frac{d}{dx}\left[p(x)\frac{du}{dx}\right] - q(x)\,u, \qquad a < x < b$$

then the solution of the Sturm-Liouville problem

$$L\,u = -f(x) \qquad a < x < b$$

Homogeneous B.C.

is given in terms of the Green's function by the expression

$$u(x) = \int_a^b G(x,x')\,f(x')dx'. \qquad \textbf{(9.1–1)}$$

Let us apply the differential operator L to both sides of the relation (9.1–1). Formally interchanging the order of differentiation and integration, we obtain

$$\int_a^b L\,G(x,x')\,f(x')\,dx' = -f(x).$$

Then, boldly replacing $-L\ G(x,x')$ by the two-variable "function" $K(x,x')$, we arrive at

$$\int_a^b K(x,x')\,f(x')\,dx' = f(x). \qquad \textbf{(9.1–2)}$$

If equation (9.1–2) were generally valid, the "function" $K(x,x')$ could realistically be termed a *reproducing kernel*. Such a kernel would have the property that, given *any* (reasonable) function f, multiplication of f by the kernel and subsequent integration over the domain of interest gives us back the same function f. This is precisely the nature of the "function" that physicists call the (Dirac) delta function and that engineers sometimes term the unit impulse. In this chapter, among other efforts, we want to try to understand better this interesting and very useful "function."

The nature of a reproducing kernel

The notion of reproducing kernels has been encountered several times previously. Although we didn't single it out for attention as such, the so-called Dirichlet kernel

$$K(x,x';\lambda) = \frac{\sin\lambda\,(x - x')}{\pi(x - x')}, \tag{9.1–3}$$

which played an important role in the reciprocity relation for the Mellin and other transforms (see Theorem 2, Section 7.2), is, *in the limit of arbitrarily large values of the parameter* λ, a reproducing kernel. So also are the Fejér-Césaro kernel

$$K(x,x';\lambda) = \frac{\sin^2\lambda\,(x - x')}{\pi\,\lambda\,(x - x')^2}, \tag{9.1–4a}$$

the Cauchy kernel

$$K(x,x';\lambda) = \frac{(\lambda/\pi)}{1 + \lambda^2\,(x - x')^2}, \tag{9.1–4b}$$

the Weierstrass kernel

$$K(x,x';\lambda) = (\lambda/\sqrt{\pi})\,e^{-\lambda^2(x-x')^2}, \tag{9.1–4c}$$

and the Bessel kernel

$$K(x,x';\lambda) = \begin{cases} \lambda J_\lambda\,[\lambda + \lambda(x - x')] & x \geq x' \\ 0 & x < x'. \end{cases} \tag{9.1–4d}$$

(See Titchmarsh (1959) and Lamborn (1969)).

Clearly, however, if the above examples are indicative of the general case, a reproducing kernel is not a function in the usual mathematical sense. Indeed, with the exception of the Dirichlet kernel (9.1–3), we have for the other kernels

$$\lim_{\lambda\to\infty} K(x,x';\lambda) = \begin{cases} 0 & x \neq x' \\ \infty & x = x'. \end{cases}$$

Therefore, the total influence of these kernels, in the limit $\lambda \to \infty$, becomes concentrated at the single point $x = x'$. On the other hand, easy calculations show that in all cases

$$\int_{-\infty}^{\infty} K(x,x';\lambda)\,dx' = 1. \tag{9.1–5}$$

Thus, as the extremely "spiked" behavior of each of the $K(x,x';\lambda)$ in the large λ limit evolves, it does so in such a way that the unit value of the integral (9.1–5) is maintained.

These observations suggest that it is appropriate to talk about a reproducing kernel (or delta function) only in terms of its contribution under an integral sign and to "evaluate" this contribution by means of approximations provided by integrations well-defined in the conventional sense. Indeed, exploiting this notion, we can embed the reproducing kernel concept in a variety of satisfactory mathematical frameworks. One such approach involves the theory of distributions.* Rather than give a careful treatment of that rather deep subject, however, we will consider these "functions" in a somewhat less general context. This latter construction was first suggested by George Temple (1955) and has the appealing virtue of providing a sound foundation without obscuring the (physical) insights that motivated Dirac and his contemporaries.

9.2 From good functions to generalized functions

Delta functions in optics

Before continuing our theoretical discussion of reproducing kernels, which we shall henceforth call delta functions, we want to remind readers of their utility in various practical situations. For example, under modest assumptions the diffraction figure $F(u,v)$ formed (at infinity) by monochromatic light passing though an aperture A in an otherwise opaque planar screen can be characterized by the (double) Fourier transform

$$F(u,v) = \iint_A f(x,y)\, e^{2\pi i(ux+vy)}\, dx\, dy. \tag{9.2–1}$$

Here $f(x,y)$ is the component of the electric field vector in the aperture (with time dependence suppressed), and we have assumed that a simple lens system has been used to image the diffraction figure in some distant focal plane.

If $f(x,y)$ is of the form $g(x)\, h(y)$, then the composite Fourier transform in (9.2–1) separates into the product of the individual transforms of g and h. In particular, for a rectangular aperture with dimensions $2a$ and $2b$ under constant illumination we have

$$F(u,v) = \frac{\sin 2\pi\, au}{\pi u}\ \frac{\sin 2\pi\, bv}{\pi v}\,.$$

An important special case of this result occurs in the small-aperture limit. If, as a and b tend to zero, we adjust the light intensity so that the total output (product of intensity times the aperture area) remains constant, we obtain the equivalent of a *point source*. For this configuration

*Developed by Laurent Schwartz, see (1978); see also Gel'fand and Shilov (1964) or Zemanian (1965), for example.

$$\begin{aligned} F(u,v) &= \lim_{a,b\to 0} \frac{\sin 2\pi\, au}{2\pi\, au}\;\frac{\sin 2\pi\, bv}{2\pi\, bv} \\ &= 1, \end{aligned}$$

and we see that the diffraction figure associated with a point source has constant intensity throughout.

The point source can also be simulated by a circular aperture of vanishingly small radius. In this case we would let

$$g(r,\theta) \equiv f(x,y) = \begin{cases} \dfrac{1}{\pi a^2} & 0 \le r < a \quad\quad 0 \le \theta < 2\pi \\ 0 & \text{otherwise,} \end{cases}$$

and thence calculate

$$\begin{aligned} G(\rho,\phi) \equiv F(u,v) &= \frac{1}{\pi a^2}\int_0^a \int_0^{2\pi} e^{2\pi\rho r\,\cos\,(\theta-\phi)}\, r\, dr\, d\theta \\ &= \frac{J_1(2\pi a\rho)}{\pi a\rho}, \end{aligned}$$

where J_1 is the Bessel function of the first kind (Problem 3). As a tends to zero we again, as expected, obtain the constant intensity diffraction figure.

Both

$$f_{a,b}(x,y) = \begin{cases} \dfrac{1}{4ab} & |x| < a,\ |y| < b \\ 0 & \text{otherwise} \end{cases}$$

and

$$g_a(r,\theta) = \begin{cases} \dfrac{1}{\pi a^2} & |r| < a \\ 0 & \text{otherwise} \end{cases}$$

are, for small a, b, approximations to delta functions. So also are the kernels (9.1–4) as we have observed earlier. This underscores the fact that the delta function (and related "functions") can be approximated in a variety of ways, a feature that we will now exploit.

Good functions/fairly good functions

We begin by defining two classes of well-behaved real-valued functions. Our terminology will follow that of Lighthill (1959).

Definition 1 An infinitely differentiable function f of the real variable x is termed a **good function** if $\lim_{|x|\to\infty} x^n d^k f/dx^k = 0$ for all n and nonnegative

k; it is called a **fairly good function** if $\lim_{|x|\to\infty} x^n d^k f/dx^k = 0$ merely for some particular (perhaps negative) n and all nonnegative k.

As examples, we note that e^{-x^2} is a good function, whereas any polynomial is a fairly good function.

The following properties can be established for good/fairly good functions:

Property 1
The derivative of a good (fairly good) function is good (fairly good).

Property 2
The sum of two good (fairly good) functions is good (fairly good).

Property 3
The product of a good (fairly good) function with a fairly good function is good (fairly good).

Property 4
The indefinite integral of a good (fairly good) function is fairly good.

Property 5
Good functions are integrable on (∞,∞); fairly good functions need not be.

Property 6
The Fourier transform of a good function is good.

The proofs of the first four properties are straightforward; for the latter two see Problem 4.

The good functions are going to be the building blocks in our construction of the delta and other generalized functions. However, first we need a little more terminology.

Definition 2 A sequence $\{f_n(x)\}$ of good functions is said to be **regular** if

$$\lim_{n\to\infty} \int_{-\infty}^{\infty} f_n(x)\phi(x)dx \qquad \textbf{(9.2–2)}$$

exists for *every* choice of good function $\phi(x)$.

Definition 3 Two regular sequences $\{f_n\}$, $\{g_n\}$ of good functions are termed *equivalent* if for each good function $\phi(x)$

$$\lim_{n\to\infty} \int_{-\infty}^{\infty} f_n(x)\phi(x)dx = \lim_{n\to\infty} \int_{-\infty}^{\infty} g_n(x)\phi(x)dx.$$

Using these definitions it is easy to show that the sequences $\{e^{-n^\alpha} x^2\}$ are all regular for arbitrary real α. If we normalize with the factor $\sqrt{n^\alpha/\pi}$, then these sequences for which $\alpha > 0$ are all equivalent, as are those for which $\alpha < 0$.

Generalized functions

Now we are ready to give a mathematically sound definition of the delta function and those other "functions" that are closely related to it.

> **Definition 4** A **generalized function (g.f.)** $f(x)$ is defined by a regular sequence $\{f_n(x)\}$ of good functions. Two g.f.'s are said to be equal if the corresponding regular sequences are equivalent.

In view of this definition, more than one regular sequence can be used to define (essentially) the same generalized function. However, since for regular sequences the expressions (9.2–2) are well-defined, we can give unambiguous meaning to the integral

$$\int_{-\infty}^{\infty} f(x)\phi(x)dx$$

of the product of a g.f. and a food function by defining it as

$$\lim_{n\to\infty} \int_{-\infty}^{\infty} f_n(x)\phi(x)dx$$

where $\{f_n(x)\}$ is any one of the applicable regular sequences.

There are some cases in which the regular sequences converge in the usual (pointwise) sense. As an example we have $\{e^{-n\alpha\, x^2}\}$ for negative α. In such situations, the above definition merely provides an alternative way of looking at familiar real-valued functions (see Problem 6). More generally, however, Definition 4 leads us out of the classical setting. As a case-in-point we consider the regular sequences $\{\sqrt{n^\alpha/\pi}\, e^{-n\alpha\, x^2}\}$ for $\alpha > 0$. In view of our earlier definition (9.1–4c) of the Weierstrass kernel, we suspect that these sequences define a bona fide generalized function which is proportional to the delta function. To be precise, we can use the mean-value theorem to show that for good functions $\phi(x)$,

$$\left| \phi(x) - \sqrt{\frac{n^\alpha}{\pi}} \int_{-\infty}^{\infty} e^{-n\alpha(x-x')^2}\, \phi(x')dx' \right|$$

$$= \sqrt{\frac{n^\alpha}{\pi}} \left| \int_{-\infty}^{\infty} e^{-n\alpha(x-x')^2}\, [\phi(x) - \phi(x')]\, dx' \right|$$

$$\leq \sqrt{\frac{n^\alpha}{\pi}} \int_{-\infty}^{\infty} e^{-n\alpha(x-x')^2}\, |x-x'| \max |\phi'(x)| dx'$$

$$= \frac{\max |\phi'(x)|}{\sqrt{n^\alpha \pi}} .$$

Since this last expression tends to zero as $n \to \infty$, it follows that for positive α, $\{\sqrt{n^\alpha/\pi}\, e^{-n\alpha x^2}\} = \delta(x)$.

9.3 Generalized functions: special cases and properties

Basic properties

There are a number of properties, some natural, others perhaps unexpected, which are fundamental to the calculus of generalized functions. For example,

Property 1
If two regular sequences $\{f_n\}$ and $\{g_n\}$ of good functions serve to define two generalized functions f and g, then

$$\{f_n + g_n\} = f + g.$$

Property 2
If $\{f_n\} \equiv f$, then the sequence of derivative functions $\{f_n'\}$ is regular and

$$\{f_n'\} \equiv f'.$$

Property 3
If $\{f_n\} \equiv f$ and ψ is a fairly good function, then the sequence $\{\psi f_n\}$ is regular and

$$\{\psi f_n\} \equiv \psi f.$$

Property 4
If $\{f_n\} \equiv f$ and, for each n, F_n is the Fourier transform of f_n, then the sequence $\{F_n\}$ is regular and

$$\{F_n\} \equiv \text{Fourier transform of } f.$$

In truth, some of the above results probably should have been called definitions. Terming them properties, however, emphasizes their consistency with results common to the more familiar theory of "ordinary" functions.

Two observations are worth making at this point. First, from Property 2 we have

$$\begin{aligned}
\int_{-\infty}^{\infty} f'(x)\phi(x)dx &= \lim_{n\to\infty} \int_{-\infty}^{\infty} f_n'(x)\phi(x)dx \\
&= \lim_{n\to\infty} \left[f_n(x)\phi(x)\Big|_{\infty}^{\infty} - \int_{-\infty}^{\infty} f_n(x)\phi'(x)dx \right] \\
&= -\int_{-\infty}^{\infty} f(x)\phi'(x)dx
\end{aligned}$$

for arbitrary good $\phi(x)$, and thus *integration by parts* is valid in the generalized function context.* From this, such rules for differentiation as the product rule, for example, follow easily. Second, some important and useful transform pair relationships between generalized functions can be rigorously established using Property 4. For instance, since formally

$$\frac{1}{\sqrt{2\pi}} \int_{-\infty}^{\infty} e^{i\omega x}\, \delta(x) dx = \frac{1}{\sqrt{2\pi}},$$

it is reasonable to conjecture that $1/\sqrt{2\pi}$ and $\delta(x)$ constitute a Fourier transform pair in the generalized sense. Indeed, this expectation is given further credence by the calculation (again formal) of the inverse Fourier transform of $1/\sqrt{2\pi}$ as

$$\begin{aligned}
\frac{1}{\sqrt{2\pi}} \int_{-\infty}^{\infty} \frac{1}{\sqrt{2\pi}} e^{-i\omega x}\, d\omega &= \frac{1}{2\pi} \lim_{\lambda\to\infty} \int_{-\infty}^{\infty} e^{-i\omega x}\, e^{-|\omega|/\lambda}\, d\omega \\
&= \frac{1}{2\pi} \lim_{\lambda\to\infty} \left[\int_{-\infty}^{0} e^{-\omega(ix - 1/\lambda)}\, d\omega + \int_{0}^{\infty} e^{-\omega(ix+1/\lambda)}\, d\omega \right] \\
&= \frac{1}{2\pi} \lim_{\lambda\to\infty} \left[\frac{1}{1/\lambda - ix} + \frac{1}{1/\lambda + ix} \right] \\
&= \lim_{\lambda\to\infty} \frac{\lambda/\pi}{1+\lambda^2 x^2}\,.
\end{aligned}$$

We recognize this last result as the limiting case of the Cauchy kernel (9.1–4b), which exhibits reproducing kernel behavior, as we noted earlier. With the help of Property 4 above, we can now precisely establish this pairwise transform correspondence between $1/\sqrt{2\pi}$ and $\delta(x)$ (see Problem 8).

The delta function in disguise

The second observation above shows that one way of arriving at the delta function is as the Fourier transform of a constant function. This important generalized function also turns up in considerations relating to the Heaviside unit step function

$$h(x,x') \equiv \begin{cases} 0 & x < x' \\ 1 & x > x'. \end{cases}$$

Indeed,

$$\int_{-\infty}^{\infty} \frac{d\, h(x,x')}{dx'}\, \phi(x')dx' = h(x,x')\, \phi(x') \Bigg|_{-\infty}^{\infty} - \int_{-\infty}^{\infty} h(x,x')\, \phi'(x')dx'$$

---

*Note that this implies $\int_{-\infty}^{\infty} \delta'(x - x')\phi(x')dx' = \int_{-\infty}^{\infty} \frac{d\delta(x - x')}{dx'} \phi(x')dx' = \phi'(x)$ for arbitrary good functions $\phi(x)$.

$$= -\int_{-\infty}^{\infty} \phi'(x')dx'$$

$$= -\phi(x)$$

for arbitrary good $\phi(x)$. It follows that

$$\frac{d\,h(x,x')}{dx'} = -\delta(x - x')$$

or, in view of the symmetry of the Heaviside function,

$$\frac{d\,h(x,x')}{dx} = \delta(x - x'). \qquad \textbf{(9.3–1)}$$

A number of alternative expressions are consequences of the relation (9.3–1). For example, the familiar signum function has the definition

$$\text{sgn}\,(x - x') \equiv \begin{cases} -1 & x < x' \\ 1 & x > x', \end{cases}$$

so that

$$\text{sgn}\,(x - x') = h(x,x') - h(x',x).$$

Thus

$$\frac{1}{2}\frac{d\,\text{sgn}(x - x')}{dx} = \frac{1}{2}\left[\frac{d\,h(x,x')}{dx} - \frac{d\,h(x',x)}{dx}\right]$$

$$= \frac{1}{2}\left[\delta(x - x') + \delta(x' - x)\right]$$

$$= \delta(x - x').$$

It is not always so easy to recognize the delta function in its other guises. The following theorem provides a way that is particularly useful in practice. Its proof, employing a standard construction, is also instructive.

Theorem 1:

If $f(x)$ is a generalized function for which $xf(x) = 0$, then $f(x)$ is a constant times $\delta(x)$.

Proof: We compute the integral of the product of $f(x)$ and an arbitrary good function $\phi(x)$ as follows*:

$$\int_{-\infty}^{\infty} f(x - x')\phi(x')dx' = \int_{-\infty}^{\infty} f(x - x')\left\{\phi(x)\,e^{-(x-x')^2}\right.$$

$$\left. + (x - x')\left[\frac{\phi(x') - \phi(x)\,e^{-(x-x')^2}}{x-x'}\right]\right\} dx'$$

*Actually, $e^{-(x-x')^2}$ may be replaced in this construction by any other good function $F(x - x')$ that satisfies $F(0) = 1$.

$$= \phi(x) \int_{-\infty}^{\infty} f(x - x')\, e^{-(x-x')^2}\, dx'$$

$$+ \int_{-\infty}^{\infty} (x - x') f(x - x')\, \Phi(x,x')\, dx' .$$

We note that the function $\Phi(x,x')$ is, for each x, a good function of x'. As a consequence, the second integral vanishes. Since the first integral is merely a constant, independent of x, the demonstration is complete. □

There is a companion result to Theorem 1 regarding constant generalized functions.

Theorem 2:

If $f(x)$ is a generalized function and $f'(x) = 0$, then $f(x)$ is a constant.

Proof: The appropriate construction in this case takes the form

$$\int_{-\infty}^{\infty} f(x)\, \phi(x) dx = \int_{-\infty}^{\infty} f(x) \left[\frac{e^{-x^2}}{\sqrt{\pi}} \int_{-\infty}^{\infty} \phi(t) dt + \phi(x) \right.$$

$$\left. - \frac{e^{-x^2}}{\sqrt{\pi}} \int_{-\infty}^{\infty} \phi(t) dt \right] dx$$

$$= \left[\int_{-\infty}^{\infty} \frac{e^{-x^2}}{\sqrt{\pi}} f(x) dx \right] \left[\int_{-\infty}^{\infty} \phi(t) dt \right] + \int_{-\infty}^{\infty} f(x)\, \Phi(x) dx .$$

We note that for arbitrary good $\phi(x)$

$$\Phi(x) \equiv \phi(x) - \frac{e^{-x^2}}{\sqrt{\pi}} \int_{-\infty}^{\infty} \phi(t) dt$$

is a good function. More importantly, the indefinite integral of $\Phi(x)$ is likewise a good function. (Why?) Integrating the second integral above by parts, we obtain

$$\int_{-\infty}^{\infty} f(x)\, \Phi(x) dx = f(x) \int_{-\infty}^{x} \Phi(x') dx' \Bigg|_{-\infty}^{\infty} - \int_{-\infty}^{\infty} f'(x) \int_{-\infty}^{x} \Phi(x') dx'\, dx$$

$$= 0,$$

from which the desired result ensues. □

Ordinary functions as generalized functions

We have already observed that in many cases rather familiar ordinary real-valued functions can be given alternative definitions as generalized functions. A sufficient condition for this to occur is provided by the following.

Theorem:

Let $f(x)$ be a real-valued function of x, well-defined in the ordinary sense. If $(1 + x^2)^{-m} f(x)$ is integrable for some positive integer m, then there exists a regular sequence $\{f_n\}$ of good functions such that

$$\{f_n\} = f,$$

and thus f is also a bona fide generalized function.

Proof: See Lighthill (1959), for example. □

Note especially that this theorem says nothing about the ordinary functions being good, or even fairly good, and thus the collection of generalized functions is expanded enormously. The logarithmic function, and hence the function $f = 1/x$ (and its derivatives), as well as the Heaviside and signum functions, are all generalized functions.

9.4 The problem of the electrified disk

Having discussed some of the fundamental notions of generalized function theory, we now look at a problem in electrostatics to which we can apply these ideas. Although the problem is admittedly classical, it is representative of a collection of potential theoretic applications that can be best understood by employing a generalized function methodology. The underlying concept, which we intend to consider more closely, is the idea of the *Hadamard Finite Part* of an "improper" integral.

The charged disk

In Chapter 7 we analyzed various mixed boundary value problems involving Laplace's equation. One of the simplest physical situations in electrostatics that leads to such a mixed BVP is that of a charged thin circular disk whose electrostatic potential we wish to calculate. If we assume the disk has unit radius and position it in the plane $z = 0$, with its geometrical center at the origin, then the problem reduces to finding a single potential function $V(\rho,\theta,z)$, in $z > 0$, that satisfies

$$\begin{aligned} \Delta V &= 0 && \rho > 0,\ 0 \le \theta \le 2\pi,\ z > 0 \\ V(\rho,\theta,0) &= f(\rho,\theta) && \rho < 1 \\ \frac{\partial V}{\partial z}(\rho,\theta,0) &= 0 && \rho > 1\ , \end{aligned}$$

and vanishes as $\rho^2 + z^2 \to \infty$ with $z > 0$. Whenever the prescribed potential on the disk is rotationally symmetric (axisymmetric), then $f(\rho,\theta) = f(\rho)$, a function of ρ alone, and the mixed BVP assumes the form given earlier as (7.7–4). In this case, as observed in Section 7.7, the desired potential function may be represented as

$$V(\rho,\theta,z) = V(\rho,z) = \int_0^\infty A(\lambda)\, J_0(\lambda\rho)\, e^{-\lambda z}\, d\lambda, \qquad \textbf{(9.4–1)}$$

where J_0 is the zero-order Bessel function of the first kind and the kernel function $A(\lambda)$ satisfies the dual integral equations (7.7–5)

$$\int_0^\infty A(\lambda)\, J_0(\lambda\rho)\, d\lambda = f(\rho) \qquad 0 \le \rho < 1,$$

$$\int_0^\infty A(\lambda)\, \lambda J_0(\lambda\rho)\, d\lambda = 0 \qquad \rho > 1. \tag{9.4–2}$$

The Italian mathematician Eugenio Beltrami (1835–1900) first showed that the coupled integral equations (9.4–2) could be readily solved. Since Sneddon (1966), among others, has the details, we merely note that in Beltrami's approach the kernel function $A(\lambda)$ is represented as the finite cosine transform

$$A(\lambda) = \int_0^1 F(s) \cos \lambda s\, ds \tag{9.4–3}$$

of an auxiliary function $F(s)$, which itself is given in terms of the specified potential $f(\rho)$ on the disk as

$$F(s) = \frac{2}{\pi}\frac{d}{ds}\int_0^s \frac{t\, f(t)\, dt}{\sqrt{s^2 - t^2}}\,. \tag{9.4–4}$$

If $f(\rho)$ is continuously differentiable, we can easily verify that the relations (9.4–3), (9.4–4) do indeed provide a kernel function that satisfies the dual integral equations (9.4–2). On the other hand, if $f(\rho)$ is not this smooth, and in particular if $f(\rho)$ has a singularity at the origin $\rho = 0$, we run into some difficulty with a straightforward application of the classical approach. The source of our difficulty in these cases is our inability to use integration by parts in the expression (9.4–4) before we perform the indicated differentiation with respect to the variable s.

Jacques Hadamard (1865–1963) was perhaps the first to note that a comparable dilemma occurs in a variety of other physical problems of a potential- and wave-theoretic nature (see Hadamard (1952), for example). These are all problems in which a classical approach leads to representations of supposedly determinable physical quantities as "improper" integrals, and Hadamard developed an operational methodology of handling such situations. From our point of view, it is important that the formal approach of Hadamard can be naturally embedded in the theory of generalized functions, and this we will want to discuss rather fully in the next section. However, a few words about the essence of Hadamard's technique are appropriate now. We shall return to the problem of the electrified disk near the end of Section 9.5.

The Hadamard finite part, briefly

Assume that we are interested in the "value" of the definite integral

$$\int_0^a \frac{f(t)}{(a - t)^{3/2}}\, dt$$

for some well-behaved, say differentiable, function f. Consider instead the definite integral

$$\int_0^x \frac{f(t)}{(a-t)^{3/2}}\,dt,$$

which has meaning as long as $x < a$. Integrating by parts, we find

$$\int_0^x \frac{f(t)}{(a-t)^{3/2}}\,dt = 2\left.\frac{f(t)}{(a-t)^{1/2}}\right|_0^x - 2\int_0^x \frac{f'(t)}{(a-t)^{1/2}}\,dt$$

or

$$\int_0^x \frac{f(t)}{(a-t)^{3/2}}\,dt - 2\frac{f(x)}{(a-x)^{1/2}} = -2\frac{f(0)}{a^{1/2}} - 2\int_0^x \frac{f'(t)}{(a-t)^{1/2}}\,dt.$$

We observe that the right-hand side, and hence the left-hand side, of this last expression approaches a perfectly well-defined limit when x approaches a. Hadamard called this limiting value the *finite part* of the original "improper" definite integral. Thus, using the letters HFP to designate this finite part, we have

$$\begin{aligned}\text{HFP}\int_0^a \frac{f(t)}{(a-t)^{3/2}}\,dt &\equiv \lim_{x\to a}\left\{\int_0^x \frac{f(t)}{(a-t)^{3/2}}\,dt - 2\frac{f(x)}{(a-x)^{1/2}}\right\} \\ &= -2\frac{f(0)}{a^{1/2}} - 2\int_0^a \frac{f'(t)}{(a-t)^{1/2}}\,dt.\end{aligned} \tag{9.4–5}$$

It should be noted, however, that this particular form of the definition (9.4–5) is only one of a number of possibilities. For example, in cases where f' is not integrable over $(0,a)$ the second expression is obviously meaningless and an alternative form must be used. One such alternative is

$$\text{HFP}\int_0^a \frac{f(t)}{(a-t)^{3/2}}\,dt = -2\frac{f(a)}{a^{1/2}} + \int_0^a \frac{f(t)-f(a)}{(a-t)^{3/2}}\,dt \tag{9.4–6}$$

(as can be easily verified), and it is this relation that we will have occasion to employ subsequently when we return to the problem of the electrified disk.*

The formula (9.4–6) is certainly valid whenever $f(t)$ is differentiable in the neighborhood of $t = a$ and integrable elsewhere. Hadamard (1952) showed that if f has additional smoothness, then by removing these "fractional infinities at a" in an analogous manner, a satisfactory definition can also be provided for

$$\text{HFP}\int_0^a \frac{f(t)}{(a-t)^{n+1/2}}\,dt$$

---

*Note that

$$\frac{d}{da}\int_0^a \frac{f(t)}{(a-t)^{1/2}}\,dt = -\frac{1}{2}\,\text{HFP}\int_0^a \frac{f(t)}{(a-t)^{3/2}}\,dt\,.$$

where $n \geq 1$ is integral. In similar fashion the case of

$$\text{HFP} \int_0^a \frac{f(t)}{(a - t)^{n+\mu}} dt$$

for integral $n \geq 1$ and $0 < \mu < 1$, or even the situation wherein the integrand contains $\ln(a-t)$ as an additional factor, can be handled. The details of these extensions form a part of our considerations in the next section.

9.5 The finite parts of improper integrals

We recall from (9.3–1) that

$$\frac{dh(x,x')}{dx} = \delta(x - x') .$$

It follows that for all $\beta > 0$ and $x \leq a$

$$\frac{d}{dx}[(a - x)^{\beta} h(x,x')] = -\beta(a - x)^{\beta-1} h(x,x') + (a - x)^{\beta}\delta(x - x')$$

$$= -\beta(a-x)^{\beta-1} h(x,x') + \begin{cases} (a-x')^{\beta}\delta(x-x') & x' < a \\ 0 & x' = a. \end{cases}$$

In like manner, if $\beta > n - 1 \geq 0$,

$$\frac{d^n}{dx^n}[(a - x)^{\beta} h(x,x')] = \frac{(-1)^n \Gamma(\beta + 1)}{\Gamma(\beta + 1 - n)}(a - x)^{\beta-n} h(x,x')$$

$$+ \begin{cases} \sum_{i=0}^{n-1} \frac{(-1)^i \Gamma(\beta + 1)}{\Gamma(\beta + 1 - i)}(a - x')^{\beta-i}\,\delta^{(n-1-i)}(x - x') & x' < a \\ 0 & x' = a. \end{cases} \quad \textbf{(9.5–1)}$$

Since (9.5–1) provides a consistent definition for nonintegral values of $\beta < n-1$, we will use this latter expression to define other generalized functions for such β (see Lighthill 1959)).

With the help of these new generalized functions, we are able to "evaluate" improper integrals such as

$$I = \int_0^a \frac{f(x)}{(a - x)^{n+\mu}} dx \quad \textbf{(9.5–2)}$$

where n is integral and $0 < \mu < 1$. We begin by assuming that $f(x)$ is a good function and rewriting (9.5–2) as the generalized integral

$$I = \int_{-\infty}^{\infty} [(a - x)^{-n-\mu} h(x,0) - (a - x)^{-n-\mu} h(x,a)] f(x) dx .$$

Substituting from (9.5–1), with $\beta = -\mu$, we then obtain

$$I = \frac{\Gamma(1 - \mu - n)}{\Gamma(1 - \mu)}(-1)^n \int_{-\infty}^{\infty} \frac{d^n}{dx^n}[(a - x)^{-\mu} h(x,0) - (a - x)^{-\mu} h(x,a)] f(x) dx$$

$$- \sum_{i=0}^{n-1} \frac{(-1)^{n-i} \Gamma(1 - u - n)}{\Gamma(1 - \mu - i)} a^{-\mu-i} \int_{-\infty}^{\infty} \delta^{(n-1-i)}(x) f(x) dx,$$

which, using integration by parts, can be reexpressed as

$$I = \frac{\Gamma(1 - \mu - n)}{\Gamma(1 - \mu)} \int_{-\infty}^{\infty} [(a - x)^{-\mu} h(x,0) - (a - x)^{-\mu} h(x,a)] f^{(n)}(x) dx$$

$$+ \sum_{i=0}^{n-1} \frac{\Gamma(1 - \mu - n)}{\Gamma(1 - \mu - i)} a^{-\mu-i} f^{(n-1-i)}(0).$$

The integral appearing in this last expression is well-defined in the ordinary sense. Indeed, the combination of terms is precisely what would have been derived for the finite part of the integral I if we had carried out the steps in the manner outlined in the previous section (see Problem 12). The result is that now we are led to the extended definition

$$\text{HFP} \int_0^a \frac{f(x)}{(a - x)^{n+\mu}} dx \equiv \frac{\Gamma(1 - \mu - n)}{\Gamma(1 - \mu)} \int_0^a \frac{f^{(n)}(x)}{(a - x)^{\mu}} dx + \sum_{i=0}^{n-1} \frac{\Gamma(1 - \mu - n)}{\Gamma(1 - \mu - i)} \frac{f^{(n-1-i)}(0)}{a^{\mu-i}} \tag{9.5–3}$$

for the Hadamard finite part of an improper integral, which is valid for positive integers n, $0 < \mu < 1$, and sufficiently smooth functions $f(x)$.

As suggested earlier, there are alternative forms for this definition. We can derive one such expression by replacing $f(x)$ within the integral in (9.5–3) by

$$g(x) \equiv f(x) - \sum_{i=0}^{n-1} \frac{(x - a)^i f^{(i)}(a)}{\Gamma(i + 1)} . \tag{9.5–4}$$

(This is possible since $g^{(n)}(x) = f^{(n)}(x)$.) When we perform the indicated substitution and integrate by parts, we get

$$\text{HFP} \int_0^a \frac{f(x)}{(a - x)^{n+\mu}} dx = \int_0^a \frac{g(x)}{(a - x)^{n+\mu}} dx$$

$$+ \sum_{k=0}^{n-1} \frac{\Gamma(1 - \mu - n)}{\Gamma(1 - \mu - k)} \frac{g^{(n-1-k)}(x)}{(a - x)^{k+\mu}} \Bigg|_0^a$$

$$+ \sum_{i=0}^{n-1} \frac{\Gamma(1 - \mu - n)}{\Gamma(1 - \mu - i)} \frac{f^{(n-1-i)}(0)}{a^{\mu+i}}$$

$$= \int_0^a \frac{g(x)}{(a - x)^{n+\mu}} dx + \sum_{k=0}^{n-1} \frac{\Gamma(1 - \mu - n)}{\Gamma(1 - \mu - k)\, a^{k+\mu}}$$

$$\sum_{i=n-1-k}^{n-1} \frac{(-a)^{i-n+k+1} f^{(i)}(a)}{\Gamma(i - n + k + 2)} .$$

In arriving at this last expression, we have taken account of the fact that

$$\lim_{x \to a} \frac{g^{(n-1-k)}(x)}{(a-x)^{k+\mu}} = 0 \qquad k = 0, 1, \ldots, n-1.$$

We can simplify further if we use a combinatorial identity (see Problem 13). Employing the identity, we are finally led to

$$\text{HFP} \int_0^a \frac{f(x)}{(a-x)^{n+\mu}} dx = \int_0^a \frac{g(x)}{(a-x)^{n+\mu}} dx + \sum_{i=0}^{n-1} \frac{(-1)^{i+1} f^{(i)}(a)}{(\mu + n - 1 - i) i! \, a^{\mu+n-1-i}} \tag{9.5–5}$$

as the desired alternative formula.

The expression (9.5–5) for the Hadamard finite part has several advantages over the earlier formula (9.5–3). Foremost amongst these is a relaxed smoothness requirement on the function $f(x)$. For example, $f(x)$ need not be a good function, nor even infinitely differentiable in some interval that includes $(0,a)$; continuity throughout $(0,a)$ and, say, n-times differentiability in the neighborhood of $x=a$ suffices. This will be significant in our later considerations.

Before discussing some examples, we observe that the Hadamard finite part of an integral with a logarithmic factor in the integrand can be evaluated using parametric differentiation. That is to say,

$$\text{HFP} \int_0^a f(x)\,(a-x)^\alpha \ln(a-x) dx = \frac{d}{d\alpha} \text{HFP} \int_0^a f(x)\,(a-x)^\alpha \, dx \,.$$

The special case of the Hadamard finite part (with or without logarithmic factors) when α is a negative integer, however, requires the notion of the Cauchy Principle Value and is outside our current interests (see Lighthill (1959), for example, for details).

Two illustrative examples

Example 1 Using formula (9.5–5),

$$\begin{aligned} \text{HFP} \int_0^1 \frac{x\,dx}{(1-x)^{3/2}} &= \int_0^1 \frac{x-1}{(1-x)^{3/2}} dx - 2 \\ &= -\int_0^1 \frac{dx}{(1-x)^{1/2}} - 2 \\ &= -4. \end{aligned}$$

(Accordingly, the Hadamard finite part of an improper integral with a positive integrand may turn out to be negative.) □

Example 2 From (9.5–3),

$$\text{HFP}\int_0^1 \frac{(1+x)^{3/2}}{(1-x)^{5/2}}\,dx = \frac{3}{4}\frac{\Gamma(-3/2)}{\Gamma(1/2)}\int_0^1 (1+x)^{-1/2}(1-x)^{-1/2}\,dx$$
$$+\frac{3}{2}\frac{\Gamma(-3/2)}{\Gamma(1/2)} + \frac{\Gamma(-3/2)}{\Gamma(-1/2)}$$
$$= \int_0^1 \frac{dx}{(1-x^2)^{1/2}} + 4/3$$
$$= \pi/2 + 4/3.$$

□

An important special case

Example 1 above is more interesting than might at first be imagined. If we disregard the improper character of the integral for the moment, the expression is essentially the beta function $B(2, -1/2)$ (see Appendix 5). But

$$B(2, -1/2) = \frac{\Gamma(2)\,\Gamma(-1/2)}{\Gamma(1/2)}$$
$$= -4,$$

and we recognize this last value as precisely the real number we calculated as the Hadamard finite part of the given integral in the first place. This suggests that for certain improper integrals, the corresponding beta function and the Hadamard finite part are one and the same. Such is indeed the case as the following (formal) analysis demonstrates.

Consider the (possibly) improper integral

$$I \equiv \int_0^1 x^{\alpha-1}(1-x)^{\beta-1}\,dx,$$

where the real parameters α, β are restricted only by the fact that they cannot take on nonpositive integral values. For some $0 < \epsilon < 1$, form

$$I = \int_0^{\epsilon} x^{\alpha-1}(1-x)^{\beta-1}\,dx + \int_{\epsilon}^1 x^{\alpha-1}(1-x)^{\beta-1}\,dx$$
$$\equiv I_1 + I_2.$$

Now let $x = \epsilon(1-t)$, so that

$$I_1 = \epsilon^{\alpha}\int_0^1 (1-t)^{\alpha-1}(1-\epsilon+\epsilon t)^{\beta-1}\,dt.$$

If $\alpha > 0$, this can be expressed in terms of hypergeometric functions as

$$I_1 = \frac{\epsilon^\alpha (1-\epsilon)^{\beta-1}}{\alpha}\, {}_2F_1\left(1,1-\beta;\alpha+1;\frac{-\epsilon}{1-\epsilon}\right)$$

$$= \frac{\epsilon^\alpha}{\alpha}\, {}_2F_1(\alpha,1-\beta;\alpha+1;\epsilon),$$

by Euler's formula (3.3–6) and a well-known transformation relation (see Abramowitz and Stegun (1965), for example). On the other hand, if $\alpha < 0$, then setting $\alpha = 1 - n - \mu$ where $n \geq 1$ is integral and $0 < \mu < 1$, and using (9.5–5),

$$\text{HFP}(I_1) = \epsilon^\alpha \int_0^1 (1-t)^{-n-\mu}\left[(1-\epsilon+\epsilon t)^{\beta-1} - \sum_{i=0}^{n-1} \frac{(t-1)^i \epsilon^i}{i!}\frac{\Gamma(\beta)}{\Gamma(\beta-1)}\right] dt$$

$$+ \epsilon^\alpha \sum_{i=0}^{n-1} \frac{(-1)^{i+1}\epsilon^i}{(\mu+n-1-i)i!}\frac{\Gamma(\beta)}{\Gamma(\beta-i)}$$

$$= \epsilon^\alpha \sum_{i=n}^{\infty} \frac{(-1)^i \epsilon^i}{i!}\frac{\Gamma(\beta)}{\Gamma(\beta-i)}\int_0^1 (1-t)^{i-n-\mu}\, dt$$

$$+ \epsilon^\alpha \sum_{i=0}^{n-1} \frac{(-1)^{i+1}\epsilon^i}{(\mu+n-1-i)i!}\frac{\Gamma(\beta)}{\Gamma(\beta-i)}$$

$$= \epsilon^\alpha \sum_{i=0}^{\infty} \frac{(-1)^{i+1}\epsilon^i}{(\mu+n-1-i)i!}\frac{\Gamma(\beta)}{\Gamma(\beta-i)}$$

$$= \frac{\epsilon^\alpha}{\Gamma(1-\beta)}\sum_{i=0}^{\infty}\frac{\Gamma(\alpha+i)}{\Gamma(\alpha+1+i)}\frac{\Gamma(1-\beta+i)}{i!}\epsilon^i$$

$$= \frac{\epsilon^\alpha}{\alpha}\, {}_2F_1(\alpha,1-\beta;\alpha+1;\epsilon). \tag{9.5–6}$$

We note the similarity between the expressions for I_1 when $\alpha > 0$ and HFP (I_1) when $\alpha < 0$.

In completely analogous fashion

$$I_2 = \int_0^{1-\epsilon} (1-x)^{\alpha-1} x^{\beta-1}\, dx$$

$$= \frac{(1-\epsilon)^\beta}{\beta}\, {}_2F_1(\beta,1-\alpha;\beta+1;1-\epsilon) \tag{9.5–7}$$

for $\beta > 0$, with the identical result following for the Hadamard finite part of I_2 if $\beta < 0$. Combining (9.5–6) and (9.5–7) and making use of another well-known transformation formula for hypergeometric functions, we find then that if either α or β is negative (but nonintegral),

$$\text{HFP}(I) = \frac{\epsilon^\alpha}{\alpha}\, {}_2F_1(\alpha,1-\beta;\alpha+1;\epsilon)$$

$$+ \frac{(1-\epsilon)^\beta}{\beta}\, {}_2F_1(\beta,1-\alpha;\beta+1;1-\epsilon)$$

$$= \frac{\epsilon^{\alpha}}{\alpha}\left[\frac{\Gamma(1+\alpha)\Gamma(\beta)}{\Gamma(\alpha+\beta)}\, {}_2F_1\,(\alpha, 1-\beta; 1-\beta; 1-\epsilon)\right.$$

$$\left. - \frac{(1-\epsilon)^{\beta}\,\epsilon^{-\alpha}\,\alpha}{\beta}\, {}_2F_1\,(\beta, 1-\alpha; \beta+1; 1-\epsilon)\right]$$

$$+ \frac{(1-\epsilon)^{\beta}}{\beta}\, {}_2F_1\,(\beta, 1-\alpha; \beta+1; 1-\epsilon)$$

$$= \frac{\Gamma(\alpha)\Gamma(\beta)}{\Gamma(\alpha+\beta)}\,.$$

Thus we have the desired result that

$$\mathrm{HFP}\int_0^1 x^{\alpha-1}\,(1-x)^{\beta-1}\,dx = B(\alpha,\beta) \qquad \textbf{(9.5–8)}$$

whenever α,β are real parameters neither of which is equal to $0,\ -1,\ -2, \ldots$.

The electrified disk with power potentials

Let us return to the problem of calculating the general potential function (9.4–1) associated with the charged disk of Section 9.4. We would like to determine the qualitative effect of assuming that the potential $f(\rho)$ specified on the disk has the form

$$f(\rho) = C\rho^{\alpha}, \qquad \textbf{(9.5–9)}$$

where C and α are constants. The relevant equations are (9.4–4), (9.4–3), and the expression

$$\sigma(\rho) = \frac{1}{2\pi}\int_0^{\infty} A(\lambda)\lambda\, J_0(\lambda\rho)d\lambda \qquad 0 \leq \rho < 1$$

for the surface density of charge on the disk.

If $\alpha \geq 0$ in (9.5–9), the classical approach for determining $F(s)$ from (9.4–4) suffices (the cases $\alpha = 0,1$ are considered in Problem 11 at the end of the chapter). For negative α, however, a different procedure must be utilized. Interestingly, for $-2 < \alpha < 0$ the Hadamard finite-part methodology can be employed (see Problem 16). We find

$$F(s) = -\frac{2s}{\pi}\,\mathrm{HFP}\int_0^s \frac{t\,f(t)}{(s^2-t^2)^{3/2}}\,dt \qquad \textbf{(9.5–10)}$$

$$= -\frac{2s}{\pi}\,\mathrm{HFP}\int_0^s \frac{C\,t^{\alpha+1}}{(s^2-t^2)^{3/2}}\,dt$$

$$= -\frac{2s^{\alpha}}{\pi}\,\mathrm{HFP}\int_0^1 C\,x^{\alpha+1}\,(1-x)^{-3/2}\,(1+x)^{-3/2}\,dx\,.$$

This last expression can be evaluated using (9.5–8) and rearranging (or equivalently, employing Euler's formula). Carrying out the appropriate steps leads to

$$F(s) = \frac{4s^\alpha}{\sqrt{\pi}} C \frac{\Gamma(\alpha + 2)}{\Gamma(\alpha + 3/2)} {}_2F_1(\alpha + 2, 3/2; \alpha + 3/2; -1)$$
$$= \frac{2s^\alpha}{\sqrt{\pi}} C \frac{\Gamma(1/2\alpha + 1)}{\Gamma(1/2\alpha + 1/2)}.$$

(Problem 18, Chapter 3 provides the sum of the hypergeometric series with argument -1).

The above considerations show that potential functions $f(\rho)$ of the form ρ^α with $\alpha > -2$ give rise to auxiliary functions $F(s)$ of the same form. To compute the kernel function $A(\lambda)$, we then need to perform the integration

$$I_\alpha = \int_0^1 s^\alpha \cos \lambda s \, ds.$$

For $\alpha > -1$, this integral is well-defined and can be expressed in terms of confluent hypergeometric functions as

$$I_a = \frac{1}{2(\alpha + 1)} [M(\alpha + 1; \alpha + 2; i\lambda) + M(\alpha + 1; \alpha + 2; -i\lambda)]. \quad \textbf{(9.5–11)}$$

Unfortunately, for general α the relationship (9.5–11) cannot be simplified significantly. Even in the case $\alpha = -1/2$ we have

$$I_{-1/2} = \sqrt{2\pi/\lambda}\, \mathfrak{C}(\sqrt{2\lambda/\pi})$$

where $\mathfrak{C}(z)$ is the Fresnel cosine integral. And in order to calculate the charge density $\sigma(\rho)$ we would then have to integrate the product of this expression and $\lambda J_0(\lambda\rho)$, a nontrivial undertaking to say the least.

The approach of Hadamard comes to our rescue again, however. We proceed in alternative fashion by combining (9.4–3) with the expression for the charge distribution $\sigma(\rho)$ and interchanging the order of integration. The inner integral is of the form

$$\int_0^\infty \lambda \cos \lambda s \, J_0(\lambda\rho) \, d\lambda,$$

which can be evaluated in the context of generalized function theory as $-s(s^2 - \rho^2)^{-3/2}$ for $0 \le \rho < s$ and 0 otherwise (see Problem 12, Chapter 7). This means that $\sigma(\rho)$ is proportional to the finite part of the improper integral

$$I = -\int_\rho^1 s^{\alpha+1} (s^2 - \rho^2)^{-3/2} \, ds.$$

Furthermore, I can be rewritten as

$$I = -1/2(1 - \rho^2)^{-1/2} \int_0^1 (1 - x)^{-3/2} [1 - (1 - \rho^2)x]^{1/2\alpha} \, dx,$$

and hence using (9.5–3), (3.3–6), and a well-known transformation relation,

$$\begin{aligned}
\text{HFP}\,(I) &= -\tfrac{1}{2}\,\alpha(1-\rho^2)^{1/2}\int_0^1 (1-x)^{-1/2}\,[1-(1-\rho^2)x]^{1/2\alpha-1}\,dx \\
&\quad + (1-\rho^2)^{-1/2} \\
&= -\alpha(1-\rho^2)^{1/2}\,{}_2F_1\left(1,1-\int^{1/2\alpha};\tfrac{3}{2};\,1-\rho^2\right) + (1-\rho^2)^{-1/2} \qquad \textbf{(9.5–12)} \\
&= \frac{\alpha}{1-\alpha}(1-\rho^2)^{1/2}\,{}_2F_1\left(1,1-\int^{1/2\alpha};\tfrac{3}{2}-\int^{1/2\alpha};\rho^2\right) \\
&\quad -\tfrac{1}{2}\,\alpha\sqrt{\pi}\,\rho^{\alpha-1}\,\frac{\Gamma\left(\tfrac{1}{2}-\int^{1/2\alpha}\right)}{\Gamma\left(1-\int^{1/2\alpha}\right)} + (1-\rho^2)^{-1/2}.
\end{aligned}$$

The expressions (9.5–12) in terms of hypergeometric functions allow us to investigate the effect of different values of α on the resulting distribution of charge $\sigma(\rho)$.* We note that near the edge of the disk, the charge density becomes infinite like $(1-\rho^2)^{-1/2}$, independent of α. At the center of the disk, on the other hand, the growth is proportional to $\rho^{\alpha-1}$ if $\alpha < 1$, $\ln\rho$ if $\alpha = 1$, and a constant if $\alpha > 1$. Since

$$Q = 2\pi\int_0^1 \sigma(\rho)\,\rho\,d\rho$$

represents the total charge on the disk, it follows as a final result that this total charge is finite, provided $\alpha > -1$.

9.6 Eigenfunction expansions

Before we leave the subject of generalized functions, we add a few remarks about yet another kind of representation for these functions that is often of considerable utility in applied problems. We have in mind generalized Fourier series expansions for delta functions and Green's functions, which are particularly useful when explicit closed-form representations are not easily obtainable. (Recall the expression (8.6–13)).

For definiteness, we assume that the ordinary differential operator L is defined as we did in Chapters 2 and 8 and consider the Sturm-Liouville eigenvalue problem

$$Ly + \lambda ry \equiv (p(x)y')' + (\lambda r(x) - q(x))y = 0 \qquad a < x < b$$
$$\text{Homogeneous Regular B.C.} \qquad \textbf{(9.6–1)}$$

---

*It can be easily verified that the relations (9.5–12) lead to the same values of $\sigma(\rho)$ in the cases $\alpha = 0,1$ as the more classical approach carried out in Problem 11.

We recall that under modest assumptions on p, q, and r, which we presume to hold, such a problem has a countable infinity of discrete simple eigenvalues λ_n (all of which are real) and a corresponding complete set of eigenfunctions $y_i(x)$, which we may assume to be orthonormal with respect to the weight function $r(x)$. Moreover, we know that any reasonably nice function can be represented as a (mean) convergent series of these eigenfunctions.

For *fixed* $\lambda \neq \lambda_n$, let us designate the ordinary Green's function associated with the problem (9.6–1) as $G(x,x';\lambda)$. Since

$$Ly_n + \lambda_n r\, y_n = 0,$$

and hence

$$Ly_n + \lambda\, r\, y_n = -(\lambda_n - \lambda) r y_n\,,$$

it then follows from Theorem 1 of Section 8.2 that

$$y_n(x) = (\lambda_n - \lambda) \int_a^b G(x,x';\lambda)\, r(x') y_n(x') dx'$$

for each n. Thus, if we expand $G(x,x';\lambda)$ as $\sum_n \alpha_n y_n(x')$, we can immediately calculate the coefficients α_n to obtain

$$G(x,x';\lambda) = \sum_{n=0}^{\infty} \frac{y_n(x) y_n(x')}{\lambda_n - \lambda}\,. \qquad \textbf{(9.6–2)}$$

The expression (9.6–2) is the so-called *eigenfunction expansion for the Green's function*. It represents the Green's function for the general boundary-value problem in terms of the eigenfunctions (and eigenvalues) of the related Sturm-Liouville problem. Note especially that this representation clearly exhibits the expected symmetry of the Green's function.

Expansions for the delta function

In order to obtain eigenfunction expansions for the delta function that are comparable to (9.6–2), we can proceed in two basic ways. If we recognize that in the context of generalized function theory the differential equation for the Green's function $G(x,x';\lambda)$ can be rewritten as

$$LG(x,x';\lambda) + \lambda r(x)\, G(x,x';\lambda) = -\delta(x - x') \qquad a < x < b,$$

then we may conclude

$$\begin{aligned}\delta(x - x') &= -(L + \lambda r) \sum_n \frac{y_n(x) y_n(x')}{\lambda_n - \lambda} \\ &= -\sum_n \frac{y_n(x')}{\lambda_n - \lambda} [Ly_n(x) + \lambda r(x) y_n(x)] \\ &= \sum_n r(x) y_n(x) y_n(x'),\end{aligned}$$

and thus

$$\frac{\delta(x - x')}{r(x)} = \sum_{n=0}^{\infty} y_n(x)y_n(x'). \tag{9.6–3)*}$$

Alternatively, viewing $G(x,x';\lambda)$ as an analytic function of the complex variable λ and denoting by C a closed contour encircling all the singularities of G, we find

$$\begin{aligned}
-\frac{1}{2\pi i}\oint_C G(x,x';\lambda)d\lambda &= \frac{1}{2\pi i}\oint_C \sum_n \frac{y_n(x)y_n(x')}{\lambda - \lambda_n}d\lambda \\
&= \sum_n y_n(x)y_n(x')\frac{1}{2\pi i}\oint_C \frac{d\lambda}{\lambda - \lambda_n} \\
&= \sum_n y_n(x)y_n(x') \\
&= \frac{\delta(x - x')}{r(x)}.
\end{aligned} \tag{9.6–4}$$

The relations (9.6–3) and (9.6–4) provide distinct approaches to obtaining formal eigenfunction expansions for the delta function. If an appropriate set of eigenfunctions is known, then (9.6–3) can be obtained straightaway. On the other hand, if a closed-form expression for the Green's function $G(x,x';\lambda)$ is available, then the contour integral (9.6–4) is a viable alternative. Once again we note the multiplicity of ways of representing the delta function, a fact that is useful in many applications.

Some one-dimensional examples

Example 1 A most elementary case occurs when $p(x) = 1$, $q(x) = 0$, $r(x) = 1$ and the boundary conditions in (9.6–1) are $y(0) = 0 = y(1)$. The normalized eigenfunctions are obviously

$$y_n(x) = \sqrt{2}\sin n\pi x,$$

and thus

$$\delta(x - x') = 2\sum_{n=1}^{\infty} \sin n\pi x \sin n\pi x'. \tag{9.6–5}$$

This classical Fourier series expansion for the delta function can also be readily derived using (9.6–4). The relevant Green's function is

$$G(x,x';\lambda) = \begin{cases} \dfrac{\sin\sqrt{\lambda}\,x \sin\sqrt{\lambda}\,(1 - x')}{\sqrt{\lambda}\sin\sqrt{\lambda}} & x < x' \\[2ex] \dfrac{\sin\sqrt{\lambda}\,x' \sin\sqrt{\lambda}\,(1 - x)}{\sqrt{\lambda}\sin\sqrt{\lambda}} & x > x' \end{cases}$$

*Since $\delta(x - x') = 0$ for $x \neq x'$, the apparent nonsymmetry in this formal representation is actually illusory.

(recall Example 2, Section 8.2). Since the only singularities of $G(x,x';\lambda)$ are simple poles at the points $\lambda_n = n^2\pi^2$, $n=1,2,\ldots$, it follows that

$$-\frac{1}{2\pi i}\oint_C G(x,x';\lambda)d\lambda = -\sum_{n=1}^{\infty}\frac{\sin n\pi x \sin n\pi(1-x')}{n\pi(\cos n\pi)/(2n\pi)},$$

which simplifies to (9.6–5). □

Example 2 When $p(x) = 1$, $q(x) = 0$, and $r(x) = 1$ as in Example 1, but the interval of interest is $(-\infty,\infty)$, an appropriate set of boundary conditions is given by the "radiation condition" of Sommerfeld.

$$y'(x) \mp i\sqrt{\lambda}\, y(x) \to 0 \text{ as } x \to \pm\infty.$$

In this case the Green's function has the form (8.3–1)

$$G(x,x';\lambda) = \frac{i}{2\sqrt{\lambda}}e^{i\sqrt{\lambda}|x-x'|}.$$

The origin $\lambda = 0$ and the point at ∞ are branch-point singularities. Choosing the positive real-axis as the branch cut connecting these singularities, we are led to

$$\begin{aligned}
-\frac{1}{2\pi i}\oint_C G(x,x';\lambda)d\lambda &= -\frac{1}{4\pi}\left[\int_{\infty}^{0}\frac{\cos\sqrt{\lambda}(x-x')}{\sqrt{\lambda}}d\lambda + \int_0^{\infty}\frac{\cos\sqrt{\lambda}(x-x')}{-\sqrt{\lambda}}d\lambda\right]\\
&= \frac{1}{2\pi}\int_0^{\infty}\frac{\cos\sqrt{\lambda}(x-x')}{\sqrt{\lambda}}d\lambda\\
&= \frac{1}{\pi}\int_0^{\infty}\cos\xi(x-x')d\xi.
\end{aligned}$$

Thus we find that

$$\delta(x-x') = \frac{1}{\pi}\int_0^{\infty}\cos\xi(x-x')d\xi = \frac{1}{2\pi}\int_{-\infty}^{\infty}e^{i\xi(x-x')}d\xi.$$ □

Example 3 As a final one-dimensional example, consider the boundary-value problem with Bessel's equation

$$(xy')' + \lambda xy = 0 \qquad 0<x<R$$
$$u(0) \text{ finite}$$
$$u(R) = 0.$$

For this situation we obtain for $x < x'$ the Bessel function expansion

$$G(x,x';\lambda) = \frac{1}{2}\pi\frac{J_0(\sqrt{\lambda}\,x)}{J_0(\sqrt{\lambda}\,R)}[J_0(\sqrt{\lambda}\,x')Y_0(\sqrt{\lambda}\,R) - J_0(\sqrt{\lambda}\,R)Y_0(\sqrt{\lambda}\,x')],$$

with a comparable relation (having the roles of x and x' reversed) valid for $x > x'$. It is easy to verify that the origin $\lambda = 0$ is a regular point of $G(x,x';\lambda)$ and the (simple) real zeros of $J_0(\sqrt{\lambda}\,R)$ provide the only singularities. In view of this we can readily calculate

$$-\frac{1}{2\pi i}\oint_C G(x,x';\lambda)d\lambda = -\frac{1}{2}\pi\sum_{n=1}^{\infty}\frac{J_0(\sqrt{\lambda_n}\,x)J_0(\sqrt{\lambda_n}\,x')Y_0(\sqrt{\lambda_n}\,R)}{\dfrac{d}{d\lambda}[J_0(\sqrt{\lambda}\,R)]|_{\lambda=\lambda_n}}$$

$$= \pi\sum_{n=1}^{\infty}\frac{\sqrt{\lambda_n}}{R}\frac{J_0(\sqrt{\lambda_n}\,x)J_0(\sqrt{\lambda_n}\,x')Y_0(\sqrt{\lambda_n}\,R)}{J_1(\sqrt{\lambda_n}\,R)},$$

Making use of the Wronskian relation for Bessel functions, we finally derive the formula

$$\frac{\delta(x-x')}{x} = \sum_{n=1}^{\infty}\left[\frac{\sqrt{2}\,J_0(\sqrt{\lambda_n}\,x)}{R\,J_1(\sqrt{\lambda_n}\,R)}\right]\left[\frac{\sqrt{2}\,J_0(\sqrt{\lambda_n}\,x')}{R\,J_1(\sqrt{\lambda_n}\,R)}\right], \tag{9.6–6}$$

where the λ_n are given in terms of the roots of $J_0(\sqrt{\lambda}\,R) = 0$. □

It is interesting that as $R \to \infty$, the right-hand side of (9.6–6) can be formally interpreted as a Riemann integral, albeit improper. In this limit we are led to the representation

$$\frac{\delta(x-x')}{x} = \int_0^{\infty}\xi\,J_0(\xi x)J_0(\xi x')d\xi.$$

Alternatively, we can deduce this expression by considering an appropriate boundary-value problem in the semi-infinite domain $(0,\infty)$ (see Problems 19, 20).

A two-dimensional example

The earlier analysis in this section can be extended to higher dimensions (references such as Friedman (1956) have the details). A single illustration should suffice to suggest the nature of the generalization.

Example 1 Consider the reduced scalar wave equation in polar coordinates

$$\Delta u + \lambda u \equiv \frac{1}{\rho}\frac{\partial}{\partial\rho}\left(\rho\frac{\partial u}{\partial\rho}\right) + \frac{1}{\rho^2}\frac{\partial^2 u}{\partial\theta^2} + \lambda u = 0 \qquad 0<\rho<R,\ 0\le\theta\le 2\pi,$$

with the boundary conditions

$$u(0,\theta) \text{ finite}$$
$$u(R,\theta) = 0.$$

If we separate variables (as we did in the example of Section 3.1), we obtain a doubly-indexed set of eigenfunctions

$$y_{m,n}(\rho,\theta) = \begin{cases} 1/\sqrt{\pi}\,\dfrac{J_0(\sqrt{\lambda_{mo}}\,\rho)}{R\,J_0'(\sqrt{\lambda_{mo}}\,R)} & n=0 \\ \sqrt{2/\pi}\,\dfrac{J_n(\sqrt{\lambda_{mn}}\,\rho)}{R\,J_n'(\sqrt{\lambda_{mn}}\,R)}\,\begin{matrix}\cos\\ \sin\end{matrix}\,n\theta & n = 1,2,\ldots, \end{cases}$$

where the eigenvalues λ_{mn} are given by the positive roots of

$$J_n(\sqrt{\lambda_{mn}}\,R) = 0.$$

In close analogy with (9.6–3), we then find

$$\frac{\delta(\rho - \rho')}{\rho}\,\delta(\theta - \theta') = \sum_{n=1}^{\infty} \frac{J_0(\sqrt{\lambda_{mo}}\,\rho)\,J_0(\sqrt{\lambda_{mo}}\,\rho')}{\pi R^2[J_0'(\sqrt{\lambda_{mo}}\,R)]^2} + \sum_{m=1}^{\infty}\sum_{n=1}^{\infty} \frac{2J_n(\sqrt{\lambda_{mn}}\,\rho)\,J_n(\sqrt{\lambda_{mn}}\,\pi')\cos n(\theta - \theta')}{\pi R^2[J_n'(\sqrt{\lambda_{mn}}\,R)]^2}.$$

Alternatively, we can directly determine the Green's function associated with the given boundary-value problem as

$$G(\rho,\theta,\rho',\theta';\lambda) = \frac{1}{4}\frac{J_0(\sqrt{\lambda}\,\rho)}{J_0(\sqrt{\lambda}\,R)}\,[J_0(\sqrt{\lambda}\,\rho')\,Y_0(\sqrt{\lambda}\,R) - J_0(\sqrt{\lambda}\,R)\,Y_0(\sqrt{\lambda}\,\rho')]$$
$$+ \frac{1}{2}\sum_{n=1}^{\infty}\frac{J_n(\sqrt{\lambda}\,\rho)}{J_n(\sqrt{\lambda}\,R)}\,[J_n(\sqrt{\lambda}\rho')Y_n(\sqrt{\lambda}\,R) \qquad \textbf{(9.6–8)}$$
$$- J_n(\sqrt{\lambda}R)\,Y_n(\sqrt{\lambda}\,\rho')]\cos n(\theta - \theta')$$

for $\rho < \rho'$. (Note that this is a problem, like (8.6–12), lacking a convenient closed-form representation for the Green's function.) Observing that the only singularities of G as a function of λ occur when $J_n(\sqrt{\lambda}\,R) = 0$, $n = 1,2,\ldots$, we can use the contour integral approach embodied in (the appropriate analogue of) (9.6–4) to deduce the representation (9.6–7). □

References

Abramowitz, M. and I. A. Stegun (1965): *Handbook of Mathematical Functions*, Dover, New York; NBS Applied Math Series 55, U.S. Dept. of Commerce, Washington, D.C. (1964); Chapter 15.

Arsac, J. (1966): *Fourier Transforms and the Theory of Distributions*, Transl. of French edition, Prentice-Hall, Englewood Cliffs, New Jersey.

Cristescu, R. and G. Marinescu (1973): *Applications of the Theory of Distributions*, Transl. of Romanian edition, Wiley, New York.

Friedman, B. (1956): *Principles and Techniques of Applied Mathematics*, Wiley, New York; Chapter 5.

Gel'fand, I. M. and G. E. Shilov (1964): *Generalized Functions, Vol. I*, Transl. of Russian edition, Academic Press, New York.

Hadamard, J. (1952): *Lectures on Cauchy's Problem in Linear Partial Differential Equations*, Reprint of the 1923 Yale University Press edition, Dover, New York; Book III.

Hille, E. (1972): "Introduction to General Theory of Reproducing Kernels," *Rocky Mountain J. Math.* 2: 321–368.

Lamborn, B. N. A. (1969): "An Expression for the Dirac Delta Function," *SIAM Rev.* 11:603.

Lighthill, M. J. (1959): *Introduction to Fourier Analysis and Generalized Functions*, University Press, Cambridge.

Schwartz, L. (1978): *Théorie des distributions*, Nouvelle ed., Hermann, Paris.

Sneddon, I. N. (1966): *Mixed Boundary Value Problems in Potential Theory*, North-Holland, Amsterdam; Chapter III.

Temple, G. (1955): "The Theory of Generalized Functions," *Proc. Roy. Soc. London*, A 228:175–190.

Titchmarsh, E. C. (1959): *Introduction to the Theory of Fourier Integrals*, 2nd edition, Clarendon Press, Oxford.

Zemanian, A. H. (1965): *Distribution Theory and Transform Analysis*, McGraw-Hill, New York.

Problems

Sections 9.1, 9.2

1. Reproducing kernels or delta functions can be approximated by sequences of well-behaved kernels $K(x,x';\lambda)$, depending upon a parameter λ, such as given by equations (9.1–4 a, b, c).

   (a) Show that for these kernels

   $$\int_{-\infty}^{\infty} K(x,x';\lambda)\,dx' = 1$$

   for all λ.

   (b) In similar fashion demonstrate that

   $$\lim_{\lambda\to\infty} \int_{|x-x'|\geq\delta} K(x,x';\lambda)\,dx' = 0$$

   and hence

   $$\lim_{\lambda\to\infty} \int_{|x-x'|\leq\delta} K(x,x';\lambda)\,dx' = 1$$

   for these kernels. Here δ is an arbitrary positive number.

2. Assume that $K(x,x';\lambda)$ is a nonnegative, continuous kernel depending upon a parameter λ and satisfying the relations given in part (b) of Problem 1. Prove that for all bounded and continuous functions $f(x)$,

$$\lim_{\lambda\to\infty}\int_{-\infty}^{\infty} K(x,x';\lambda)\, f(x')\, dx' = f(x).$$

3. A *point source* in optics can be considered as an appropriate limit of a constant aperture source of vanishingly small dimensions. If

$$g(r,\theta) \equiv f(x,y) = \begin{cases} \dfrac{1}{\pi a^2} & |r| < a \\ 0 & |r| \geq a, \end{cases}$$

show that the diffraction figure $F(u,v)$ given by (9.2–1) satisfies

$$G(\rho,\theta) \equiv F(u,v) = \frac{J_1(2\pi a\rho)}{\pi a\rho},$$

where J_1 is the Bessel function of the first kind, and thus

$$\lim_{a\to 0} \quad G(\rho,\phi) = 1.$$

[The relations in Section 3.6 should prove helpful in carrying out the necessary integrations.]

4. Demonstrate the validity of Properties 5,6 of Section 9.2.

5. (a) Carefully establish the fact that the sequences $\{\sqrt{n^\alpha/\pi}\, e^{-n^\alpha x^2}\}$ are
(b) Prove the comparable result for the case $\alpha < 0$.

6. Verify that for negative α, $\{e^{-n^\alpha x^2}\} \equiv 1$.

Section 9.3

7. Demonstrate the consistency of Properties 2 and 4 of Section 9.3.

8. Carefully establish that $1/\sqrt{2\pi}$ and the delta function $\delta(x)$ constitute a Fourier transform *pair* in the generalized sense.

9. Show that $f(x) = \ln|x|$ is a well-defined generalized function by constructing a regular sequence $\{f_n(x)\}$ of good functions such that $\{f_n\} = f$.

Sections 9.4, 9.5

10. Verify that whenever the specified potential $f(\rho)$ is continuously differentiable, the kernel function $A(\lambda)$ defined by equations (9.4–3), (9.4–4) provides a solution of the coupled integral equations (9.4–2).

11. Consider a charged disk as described in Section 9.4, under the assumption that the prescribed potential on the disk is rotationally symmetric. Employ a classical approach to obtain a closed-form expression for

$$\left.\frac{\partial V}{\partial z}\right|_{z=0},$$

where $V(\rho,z)$ is the electrostatic potential, in the cases where

(a) $f(\rho) = 1$; (b) $f(\rho) = \rho$.

[The quantity $\partial V/\partial z|_{z=0}$ is proportional to the surface density of charge on the disk.]

12. Carry out the procedure suggested in the last subsection of Section 9.4 on the improper integral

$$I = \int_0^a \frac{f(x)}{(a - x)^{n+\mu}}\, dx,$$

where $f(x)$ is a good function; and thereby provide an alternate derivation of the formula (9.5–3) for the Hadamard finite part of I.

13. (a) Use the identity established in Problem 15 of Chapter 3 to verify that for positive integer n and $a \neq 0, -1, -2, \ldots,$

$$\sum_{k=0}^{n-1} \frac{\Gamma(a + k)}{\Gamma(k + 1)} = \frac{\Gamma(a + n)}{a\,\Gamma(n)}.$$

(b) Employ this result to rewrite

$$\sum_{k=0}^{n-1} \frac{\Gamma(1 - \mu - n)}{\Gamma(1 - \mu - k)a^{k+\mu}} \sum_{i=n-1-k}^{n-1} \frac{(-a)^{i-n+k+1} f^{(i)}(a)}{\Gamma(i - n + k + 2)}$$

as

$$\sum_{i=0}^{n-1} \frac{(-1)^{i+1} f^{(i)}(a)}{(\mu + n - 1 - i)i!\, a^{\mu+n-1-i}},$$

thereby completing the derivation of the alternative formula (9.5–5) for the Hadamard finite part.

14. (a) Determine

$$\text{HFP}\int_0^1 \frac{x^{-5/2}}{1 + x}\, dx$$

via the formula (9.5–3).

(b) Verify that the same result as in (a) is obtained from the alternative expression (9.5–5).

15. Show directly that

$$\text{HFP}\int_0^1 x^{-5/2}(1 - x)^{-1/2}\, dx = 0,$$

in keeping with the result expected from (9.5–8).

16. Demonstrate that if $g(t)$ is integrable over the interval (o,s), then the expression (9.5–10) is valid, that is,

$$\frac{d}{ds}\int_0^s \frac{g(t)}{\sqrt{s^2 - t^2}}\, dt = -s\, \text{HFP}\int_0^s \frac{g(t)}{(s^2 - t^2)^{3/2}}\, dt.$$

Section 9.6

17. By changing the boundary conditions in Example 1 of Section 9.6 to $y'(0) = 0 = y'(1)$, derive a cosine series representation for the delta function analogous to the sine series (9.6–5). Verify that both of the approaches outlined in the text yield the same result.

18. (a) Determine the Green's function $G(x,x';\lambda)$ appropriate for the boundary-value problem

$$y'' + \lambda y = 0 \qquad 0 < x < \infty$$
$$y'(0) = 0$$
$$y'(x) - i\sqrt{\lambda}\, y(x) \to 0 \quad \text{as } x \to \infty.$$

(b) Use (9.6–4) to obtain the expression

$$\delta(x - x') = 2/\pi \int_0^\infty \sin \xi x \sin \xi x' \, d\xi \, .$$

★19. Consider the expression (9.6–6). Show that in the limit $R \to \infty$, the series expansion in that relationship can be formally interpreted as a Riemann sum. (Hint: Where are the zeros located?) In this way, derive the representation

$$\frac{\delta(x - x')}{x} = \int_0^\infty \xi J_0(\xi x) J_0(\xi x') \, d\xi \, .$$

20. Formulate a one-dimensional boundary-value problem of the type (9.6–1) in the domain $(0,\infty)$ from which the representation for the delta function obtained in Problem 19 can be alternatively deduced using the procedure embodied in (9.6–4). [Hint: Recall the nature of the "radiation condition" in two-dimensions (Problem 24, Chapter 8).]

10

Linear integral equations

10.1 A problem in beam vibrations

In the course of our earlier discussion of Green's functions we noted that certain boundary-value problems for differential equations can be reformulated as integral equations. As an example, we observed that if the appropriate Green's function is known, the Sturm-Liouville eigenvalue problem

$$(pu')' + (\lambda r - q)u = 0 \qquad a < x < b$$

Homogeneous regular B.C.

can be reexpressed in the form

$$u(x) = \lambda \int_a^b G(x,x')\, r(x')\, u(x')\, dx' .$$

Now we want to explore these matters further and to develop a fuller appreciation for the value of an integral equation approach in various practical problems. In addition, since models involving integral equations arise naturally in numerous applications where there are no alternative methods, the theory of linear integral equations is indispensable to the applied mathematician. Therefore, we will need to consider the basics of that theory in some detail.

A problem in the oscillation of thin beams is a good place to begin our analysis. Let us consider a "rigid" thin beam of length L attached to a wall as depicted in Figure 10.1. For convenience we choose the coordinate system so that

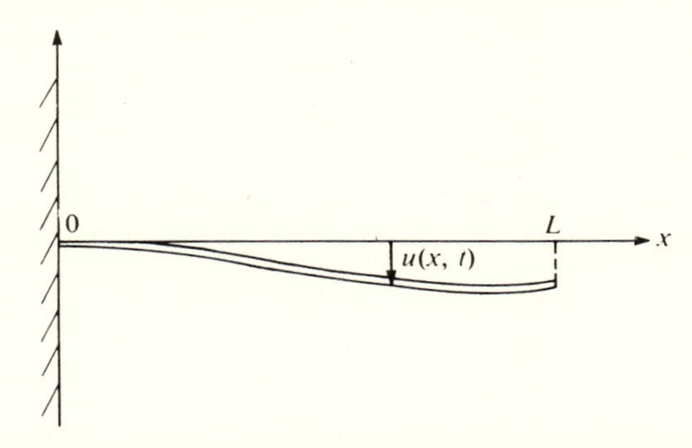

Figure 10.1 **A "rigid" thin beam of length L undergoing small deflections $u(x, t)$ from equilibrium**

the x-axis is parallel to the undeflected beam with the z-axis upward, and we presume that any beam motion that occurs is due to a continuously distributed load (force) $w(x,t)$ acting solely in the negative z-direction. If we designate the deflection of the beam from its equilibrium position by $u(x,t)$, the elementary theory of bending leads, for small deflections, to the differential relation

$$E\,I(x)\,\frac{\partial^2\, u(x,t)}{\partial x^2}\quad M(x,t). \tag{10.1–1}$$

In this expression E is the modulus of elasticity of the material of the beam and $I(x)$ is the moment of inertia of the cross-sectional area of the beam at the point x about the so-called neutral axis of that cross-section.* The quantity $M(x,t)$ in (10.1–1) is the total bending moment produced at the point x by all the forces acting on the beam. We assume that M is positive when counterclockwise, that is, $M > 0$ if it acts to bend the beam towards the positive z-axis.

For transverse loading, the bending moment $M(x,t)$ is related to the applied load by the equation

$$\frac{\partial^2 M}{\partial x^2} = -w.$$

If we decompose $w(x,t)$ into an external applied component $F(x,t)$ and an internal inertial component $\rho\partial^2 u/\partial t^2$, where $\rho = \rho\,(x)$ is the (linear) mass density of the beam at the point x, we then obtain the partial differential equation

$$\frac{\partial^2}{\partial x^2}\left[E\,I(x)\,\frac{\partial^2 u(x,t)}{\partial x^2}\right] + \rho(x)\,\frac{\partial^2 u(x,t)}{\partial t^2} = -F(x,t). \tag{10.1–2}$$

Under the assumption that the end of the beam at $x = 0$ is fixed while the end at $x = L$ is free, appropriate boundary and initial conditions for (10.1–2) take the form

$$\begin{aligned} u(0,t) &= 0 & \frac{\partial^2 u(L,t)}{\partial x^2} &= 0, \\ \frac{\partial u(0,t)}{\partial x} &= 0 & \frac{\partial}{\partial x}\left[E\,I\,\frac{\partial^2 u}{\partial x^2}\right]_{x=L} &= 0, \\ u(x,0) &= g(x) & \frac{\partial u(x,0)}{\partial t} &= h(x). \end{aligned} \tag{10.1–3}$$

The two conditions at $x = L$ reflect our assumption that no moment or shearing force is applied at the free end.

The boundary/initial-value problem consisting of (10.1–2), (10.1–3) can be analyzed in a number of ways. For the case of harmonic oscillations in which

$$\begin{aligned} F(x,t) &= f(x)\sin(\omega t + \theta) \\ u(x,t) &= \phi(x)\sin(\omega t + \theta), \end{aligned}$$

*In the case of a homogeneous beam, symmetric about the x-axis, the neutral axis is perpendicular to the xz-plane and passes through the center of mass of the cross-section in question. See Skudrzyk (1968), for example.

(10.1–2) reduces to

$$(E\,I\,\phi'')'' - \rho\omega^2\phi = -f. \qquad \textbf{(10.1–4)}$$

This equation is in self-adjoint form (see Problem 1), and a generalization of the Sturmian theory given in Chapter 2 for second-order operators can be applied. The special situation in which the beam is homogeneous and uniform, so that I and ρ are constant, has been studied extensively by this method. It turns out that the eigenfunctions (normal modes) for the homogeneous problem in this case have the form

$$\phi_n(x) = A_n\left\{(\sinh\gamma_n + \sin\gamma_n)\left[\cos\left(\frac{\gamma_n x}{L}\right) - \cosh\left(\frac{\gamma_n x}{L}\right)\right]\right.$$
$$\left. - (\cosh\gamma_n + \cos\gamma_n)\left[\sin\left(\frac{\gamma_n x}{L}\right) - \sinh\left(\frac{\gamma_n x}{L}\right)\right]\right\}$$

where the eigenvalues (eigenfrequencies) are given by

$$\omega_n = (\gamma_n/L)^2\sqrt{EI/\rho}$$

with **(10.1–5)**

$$\cosh\gamma_n\cos\gamma_n = -1.$$

(See Skudrzyk (1968), Wylie (1975), or Hildebrand (1976); also see Problem 2.)

The above reduced boundary-value problem can also be approached using Green's function. If $G(x,x')$ is the Green's function for the homogeneous problem

$$\begin{aligned} &(EI\,\phi'')'' = 0 && 0 \le x \le L \\ &\phi(0) = 0, && \phi''(L) = 0, \\ &\phi'(0) = 0, && (EI\phi'')'|_{x=L} = 0, \end{aligned} \qquad \textbf{(10.1–6)}$$

then the solution ϕ of (10.1–4), (10.1–6) must satisfy

$$\phi(x) = \int_0^L G(x,x')\,f(x')dx' - \omega^2\int_0^L G(x,x')\,\rho(x')\,\phi(x')dx'.$$

This is not an integral representation for ϕ since this unknown function also occurs on the right-hand side of the equation. Rather this expression is an integral equation, or more particularly, a *Fredholm integral equation of the second kind* for ϕ. When I is constant, the Green's function associated with (10.1–6) is given by

$$G(x,x') = \frac{1}{EI}\begin{cases} \dfrac{1}{6}x^3 - \dfrac{1}{2}x'x^2 & x < x' \\[2mm] \dfrac{1}{6}x'^3 - \dfrac{1}{2}xx'^2 & x > x'. \end{cases} \qquad \textbf{(10.1–7)}$$

We will see in Section 10.6 how this Green's function can be used in conjunction with the above integral equation in order to provide excellent approximations to the dominant eigenfrequencies mentioned earlier.

There exists yet another way in which the boundary-value problem consisting of (10.1–4), (10.1–6) can be analyzed using integral equations. If we define

$$\psi(x) \equiv \phi^{(4)}(x),$$

It follows readily that

$$\phi'''(x) = \phi'''(0) + \int_0^x \psi(x')dx',$$

$$\phi''(x) = \phi''(0) + x\phi'''(0) + \int_0^x (x - x')\,\psi(x')dx',$$

$$\phi'(x) = \phi'(0) + x\phi''(0) + \frac{1}{2}x^2\,\phi'''(0) + \frac{1}{2}\int_0^x (x - x')^2\,\psi(x')dx',$$

and

$$\phi(x) = \phi(0) + x\phi'(0) + \frac{1}{2}x^2\,\phi''(0) + \frac{1}{6}x^3\,\phi'''(0) + \frac{1}{6}\int_0^x (x - x')^3\,\psi(x')dx'.$$

Through this substitution, derivatives of $\phi(x)$ have been transformed into integrals of $\psi(x)$. Therefore, in view of these relationships, the differential equation (10.1–4) can be rewritten as the integral equation

$$\psi(x) = F(x) + \int_0^x K(x,x')\,\psi(x')dx', \tag{10.1–8}$$

where

$$F(x) \equiv (1/EI)\left\{A\left[\frac{1}{6}\rho\omega^2x^3 - x(EI)'' - 2(EI)'\right] + B\left[\frac{1}{2}\rho\omega^2x^2 - (EI)''\right] - f\right\}$$

and

$$K(x,x') \equiv (1/EI)\left\{\frac{1}{6}\rho\omega^2(x - x')^3 - (EI)''(x-x') - 2(EI)'\right\}.$$

Here we have made use of the homogeneous boundary conditions (10.1–6) at the origin $x = 0$ and have introduced the shorthand notation $A \equiv \phi'''(0)$, $B = \phi''(0)$.

Equation (10.1–8) is a *Volterra integral equation* of the second kind* for $\psi(x)$. It differs from the earlier integral equation in that the domain of integration is not fixed, but rather varies with the independent variable x. This feature provides certain intrinsic advantages, as we will see.

---

*Actually it is a *pseudo* Volterra integral equation, since the boundary conditions (10.1–6) at $x = L$ necessitate

$$A + \int_0^L \psi(x')dx' = 0 \qquad \text{and} \qquad B - \int_0^L x'\psi(x')dx' = 0.$$

10.2 Classification of equations and historical examples

As suggested in the previous section, linear integral equations fall naturally into a number of categories. For example, standardized types include

$$\psi(x) = \int_a^b K(x,y)\,\phi(y)dy,$$
$$\phi(x) = f(x) + \lambda \int_a^b K(x,y)\,\phi(y)dy, \tag{10.2–1}$$

$$\psi(x) = \int_a^x K(x,y)\,\phi(y)dy,$$
$$\phi(x) = f(x) + \lambda \int_a^x K(x,y)\,\phi(y)dy. \tag{10.2–2}$$

In each of these cases $K(x,y)$, $\psi(x)$, and $f(x)$ are known functions and λ is a complex-valued parameter. The equations (10.2–1) are generally known as *Fredholm integral equations of the first and second kinds,* respectively, for the unknown function $\phi(x)$. This terminology honors the Swedish geometer Ivar Fredholm (1866–1927) whose classic work at the turn of the century laid the foundations for much of the theoretical progress in integral equations that followed.

The two-variable function $K(x,y)$ is called the *kernel* of the integral equation. If the kernel vanishes identically for $y > x$, equations (10.2–1) reduce to (10.2–2). These latter equations are termed *Volterra integral equations* in recognition of the contributions of the Italian mathematician Vito Volterra (1860–1940) to our understanding of this special case.

For the most part in this chapter, the independent variable x, and hence y, will be restricted to a closed bounded interval $[a,b]$ of the real lin The kernel $K(x,y)$ and the known functions $\psi(x)$ and $f(x)$ will be assumed to be real-valued and continuous for such x,y.

Equations of the second kind are generally easier to handle than those of the first kind. If $f(x) \equiv 0$, the equations are called *homogeneous*. Regardless of the value of the parameter λ, homogeneous Volterra integral equations of the second kind have only the trivial solution $\phi(x) \equiv 0$. For certain critical values of λ, however, homogeneous Fredholm equations of the second kind possess nontrivial solutions. These special values of the eigenparameter are usually termed *characteristic values* (cv's) and the corresponding ϕ's *characteristic functions* (cf's). Later, both in the text and in the problems at the end of this chapter, we will make some clarifying remarks regarding the relationship between these cv's and cf's and the eigenvalues and eigenfunctions encountered in Chapters 2, 3 in connection with our study of differential boundary-value problems.

Fourier's and Abel's integral equations

In Chapter 7 we discussed the Fourier transform pair

$$F(\omega) = \frac{1}{\sqrt{2\pi}} \int_{-\infty}^{\infty} f(t)\, e^{i\omega t}\, dt$$

$$f(t) = \frac{1}{\sqrt{2\pi}} \int_{-\infty}^{\infty} F(\omega)\, e^{-i\omega t}\, d\omega .$$

Each of these equations can be viewed as an integral equation (of the first kind) with kernel $K(\omega,t) = \exp(\pm i\omega t)$. Construed in this manner, the other relation then provides the solution. If we adopt this interpretation, Fourier's and Cauchy's application of these transforms in the early 1800's represents the initial appearance of integral equations in a practical setting.

However, perhaps the first significant analysis of a problem by distinctly integral equation techniques was published by the young Norwegian mathematician, Niels Henrik Abel (1802–1829), in 1823. Abel was interested in the generalized *tautochrone* problem in which one seeks to determine the curve along which a point mass should be constrained to descend, under the sole influence of gravity, in order to "fall" a given height in a specified time. Mathematically, the problem is: Given $\psi(x)$, find $\phi(y)$ such that

$$\psi(x) = \int_0^x \frac{\phi(y)}{\sqrt{x - y}}\, dy. \qquad \textbf{(10.2–3)}$$

(See the discussion of the other "chrone" problem given earlier in Section 4.2; also see Cochran (1972) for more details). As Abel showed, the Volterra integral equation (10.2–3) can be solved by observing that

$$\int_0^y \frac{\psi(x)}{\sqrt{y - x}}\, dx = \int_0^y \frac{1}{\sqrt{y - x}} \left[\int_0^x \frac{\phi(y')}{\sqrt{x - y'}}\, dy' \right] dx$$

$$= \int_0^y \phi(y') \left[\int_{y'}^y \frac{dx}{\sqrt{(y - x)(x - y')}} \right] dy'$$

$$= \pi \int_0^y \phi(y') dy',$$

from which it follows that

$$\phi(y) = \frac{1}{\pi} \frac{d}{dy} \left[\int_0^y \frac{\psi(x)}{\sqrt{y - x}}\, dx \right]. \qquad \textbf{(10.2–4)}$$

(Alternatively, the transform techniques of Section 7.5 can be applied to obtain the representation (10.2–4) for the solution of (10.2–3); see Problem 3.) In the special case that $\psi(x) = \text{constant}$, $\psi(y)$ turns out to be proportional to $y^{-1/2}$ and the associated descent curve in the original problem is a portion of a cycloid.

10.3 Integral equations with degenerate kernels

Certain classes of integral equations can be solved, at least formally, in rather straightforward fashion. One such class consists of Fredholm integral equations with kernels that can be expressed as finite sums of products of functions of x alone by functions of y alone. These special kernels are said to be *degenerate* (*separable*, or *of finite rank*) and have the form

$$K(x,y) = \sum_{i=1}^{n} A_i(x)\, B_i(y) \qquad a \leq x,y \leq b. \tag{10.3–1}$$

When we employ such a representation, we shall assume that each of the sets $\{A_i\}$, $\{B_i\}$ is linearly independent over the interval $[a,b]$. the integer n is then unique; it is called the *rank* of K. As we shall see in the remainder of this section, the theory of Fredholm equations with degenerate kernels involves nothing more than the linear algebra developed in Chapter 1.

Equations of the first kind

If the kernel of the Fredholm integral equation

$$\psi(x) = \int_a^b K(x,y)\, \phi(y)\, dy \tag{10.3–2}$$

has the form (10.3–1), then

$$\psi(x) = \sum_{i=1}^{n} A_i(x)(\phi,B_i)$$

where we have used the inner-product notation first introduced in Chapter 2. Therefore, clearly a *necessary* condition for (10.3–2) to have a solution in this case is that the known function $\psi(x)$ is a linear combination of the A_i, $i=1,2,\ldots,n$. This condition also turns out to be *sufficient*. Indeed, if we assume

$$\psi(x) \equiv \sum_{i=1}^{n} \alpha_i A_i(x),$$

then one solution is

$$\phi(y) = \sum_{j=1}^{n} \beta_j B_j(y), \tag{10.3–3}$$

where the β_j are the unique solutions of the algebraic equations

$$\alpha_i = \sum_{j=1}^{n} (B_i,B_j)\beta_j \qquad i = 1,2, \ldots, n$$

(see Problem 5).

The solution (10.3–3) is not unique, however. There exist an infinite number of continuous functions $\omega(y)$ that are orthogonal to the B_i, $i = 1,2, \ldots, n$. If we take any one of these $\omega(y)$ and form $\Phi(y) \equiv \phi(y) + \omega(y)$, where ϕ is given by (10.3–3), then

$$\begin{aligned}\int_a^b K(x,y)\, \Phi(y)\, dy &= \int_a^b K(x,y)\, [\phi(y) + \omega(y)]\, dy \\ &= \int_a^b K(x,y)\, \phi(y) dy + \sum_{i=1}^{n} A_i(x)(B_i,\omega) \\ &= \psi(x).\end{aligned}$$

The function $\Phi(y)$ is therefore also a solution of the integral equation (10.3–2). (In fact, it is easy to see that every solution is encompassed by this recipe.)

Example In order to illustrate the above ideas, consider the equation

$$x - 1 = \int_{-1}^{1} (1 + y + 3xy)\, \phi(y) dy.$$

One choice for the components of the degenerate kernel is

$$\begin{aligned} A_1(x) &= 1 \qquad & B_i(y) &= 1 \\ A_2(x) &= 1+3x, \qquad & B_2(y) &= y, \end{aligned}$$

and obviously these sets of $\{A_i\}$, $\{B_i\}$ are linearly independent over $[-1,1]$ (see Problem 6 for an alternative choice). Now we note that

$$x - 1 = \frac{1}{3}A_2 - \frac{4}{3}A_1.$$

The underlying algebraic equations of interest thus become, in matrix notation,

$$\begin{bmatrix} -\dfrac{4}{3} \\ \dfrac{1}{3} \end{bmatrix} = \begin{bmatrix} 2 & 0 \\ 0 & \dfrac{2}{3} \end{bmatrix} \begin{bmatrix} \beta_1 \\ \beta_2 \end{bmatrix},$$

and the solution (10.3–3) has the form

$$\phi = -\frac{2}{3}B_1 + \frac{1}{2}B_2,$$

that is,

$$\phi(y) = -\frac{2}{3} + \frac{y}{2}.$$

It is worth observing that $\omega(y) = \frac{1}{2}(3y^2 - 1)$ is orthogonal to B_1, B_2. So also is $\omega(y) = \frac{1}{2}(5y^3 - 3y)$ and, in fact, so also are all the Legendre polynomials $P_n(y)$ for which $n \geq 2$ (recall the representation (3.4–1)). Therefore, additional (and all) solutions of the given integral equation can be obtained by adding to the above function any linear combination of Legendre polynomials with these higher indices. □

Equations of the second kind

For Fredholm integral equations of the second kind with degenerate kernels we find

$$\begin{aligned}\phi(x) &= f(x) + \lambda \int_a^b K(x,y)\, \phi(y)dy \\ &= f(x) + \lambda \sum_{i=1}^{n} A_i(x)(\phi,B_i). \qquad \textbf{(10.3–4)}\end{aligned}$$

Hence, if a solution exists, it must be of the form

$$\phi(x) = f(x) + \sum_{i=1}^{n} \alpha_i A_i(x) \qquad a \leq x \leq b,$$

where the coefficients α_i are yet to be determined. Substituting this expression into (10.3–4) yields

$$\begin{aligned}f(x) + \sum_{i=1}^{n} \alpha_i A_i(x) &= f(x) + \lambda \sum_{i=1}^{n} A_i(x) \left(f + \sum_{j=1}^{n} \alpha_j A_j, B_i\right) \\ &= f(x) + \lambda \sum_{i=1}^{n} A_i(x) \left[(f,B_i) + \sum_{j=1}^{n} \alpha_j (A_j,B_i) \right],\end{aligned}$$

which, in view of the linear independence of the A_i, is equivalent to the set of algebraic equations

$$\alpha_i = \lambda(f,B_i) + \lambda \sum_{j=1}^{n} (B_i,A_j)\, \alpha_j \qquad i=1,2, \cdots ,n. \qquad \textbf{(10.3–5)}$$

Therefore, just as in the case of equations of the first kind, obtaining the solution of a Fredholm integral equation of the second kind which has a kernel of finite rank reduces to a problem in matrix theory.

Some additional notation will simplify the ensuing analysis. We define two column vectors $\boldsymbol{\alpha}$, $\boldsymbol{\beta}$, and a square matrix C of order n by

$$\boldsymbol{\alpha} \equiv [\alpha_i],$$

$$\boldsymbol{\beta} \equiv [\beta_i] \equiv [(f,B_i)],$$

and

$$C \equiv [C_{ij}] \equiv [\delta_{ij} - \lambda(B_i,A_j)],$$

respectively. The set of equations (10.3–5) can now be written as the single matrix equation

$$C\boldsymbol{\alpha} = \lambda\boldsymbol{\beta}. \qquad \textbf{(10.3–6)}$$

If C is nonsingular, then this equation has a unique solution that may be expressed in terms of the inverse matrix C^{-1} by

$$\boldsymbol{\alpha} = \lambda C^{-1}\,\boldsymbol{\beta}$$

or, in terms of the cofactors C^{ij} of C_{ij} in C by

$$\alpha_i = [\lambda/D_n(\lambda)] \sum_{j=1}^{n} C^{ji}\,\beta_j \qquad \mathrm{i} = 1,2,\cdots,n$$

with

$$D_n(\lambda) \equiv \det C.$$

For invertible C, therefore, the original integral equation has the solution

$$\phi(x) = f(x) + \lambda \int_a^b R_K(x,y;\lambda)\, f(y)dy, \qquad \textbf{(10.3–7)}$$

where

$$R_K(x,y;\lambda) \equiv [1/D_n(\lambda)] \sum_{i,j=1}^{n} C^{ji}\, A_i(x)\, B_j(y). \qquad \textbf{(10.3–8)}$$

The existence of the solution (10.3–7) obviously requires that $D_n(\lambda) \neq 0$ for the particular value of the eigenparameter λ being considered. In this regard we note that $D_n\lambda)$ is in general a nontrivial polynomial in λ of at most nth degree* (hence, the notation) and thus there exist at most n distinct values of λ such that $D_n(\lambda) = 0$. For these special values of λ the original equation (10.3–4) may still have a solution. However, if it does, the solution is certainly not unique since these values of λ are actually characteristic values of the given equation and have associated with them nontrivial solutions of the homogeneous version of (10.3–6) and hence of (10.3–4) (see Problem 7). All other values of the eigenparameter λ are termed *regular* values. For these latter λ, the associated homogeneous integral equation has only the trivial solution, and the unique solution of (10.3–4) is thus given by (10.3–7).

The Fredholm alternative

The above results constitute the essence of a classical statement concerning the solvability of Fredholm integral equations of the second kind with degenerate kernels. We present the statement in a more general form, however, since it is equally valid for arbitrary (real-valued and continuous) kernels.

---

*Also note that $R_K(x,y;\lambda)$ given by (10.3–8) is a rational function of the eigenparameter λ.

The Fredholm Alternative:

Either λ is a regular value, in which case the inhomogeneous equations

$$\phi(x) = f(x) + \lambda \int_a^b K(x,y)\, \phi(y)\, dy$$

and **(10.3–9)***

$$\psi(x) = g(x) + \lambda \int_a^b K(y,x)\, \psi(y)\, dy$$

have unique (continuous solutions $\phi(x)$, $\psi(x)$ for any given (continuous) functions $f(x)$, $g(x)$;

Or λ is a characteristic value, in which case the homogeneous equations

$$\omega(x) = \lambda \int_a^b K(x,y)\, \omega(y) dy$$

and **(10.3–10)**

$$\Omega(x) = \lambda \int_a^b K(y,x)\, \Omega(y) dy$$

have nontrivial (continuous) solutions, with the number of linearly independent such solutions being the same finite number for both equations.

If the latter situation prevails, the inhomogeneous equations (10.3–9) have solutions (nonunique) if and only if f is orthogonal to every solution Ω and g is orthogonal to every solution ω of the non-respective homogeneous equations (10.3–10). □

The verification of this final portion of the alternative in the special case of degenerate kernels is left as an exercise for the reader (Problem 8).

An illustrative example

As the Fredholm Alternative emphasizes, solvability of a given integral equation depends primarily upon the nature of the kernel function $K(x,y)$. For example, in the case of

$$K(x,y) = \sin x \cos y \qquad 0 \leq x,y \leq \pi$$

we have $D_n(\lambda) \equiv 1$ and all values of λ are regular. Such is not the situation, however, in the following.

*The kernel $K(y,x)$ is the *adjoint* (actually transpose in this real-valued case) of the kernel $K(x,y)$. The equation involving the adjoint kernel is often referred to as the *adjoint equation*.

Example Consider the kernel

$$K(x,y) = x + y \qquad 0 \le x,y \le 1.$$

Here, say,

$$A_1(x) = x \qquad B_1(y) = 1$$
$$A_2(x) = 1, \qquad B_2(y) = y,$$

and

$$C = \begin{bmatrix} 1 - \frac{1}{2}\lambda & -\lambda \\ -\frac{1}{3}\lambda & 1 - \frac{1}{2}\lambda \end{bmatrix}.$$

It follows that $D_n(\lambda) = 1 - \lambda - \lambda^2/12$, so this kernel has the characteristic values $\lambda = -6 \pm 4\sqrt{3}$.

If λ is not a cv, the unique solution of

$$\phi(x) = x + \lambda \int_0^1 (x + y)\, \phi(y)\, dy$$

can be determined as

$$\phi(x) = \frac{6x(\lambda - 2) - 4\lambda}{\lambda^2 + 12\lambda - 12}.$$

For characteristic λ we have

$$\Omega(x) = \text{const.}\ (1 \pm \sqrt{3}\, x)$$

and

$$(f,\Omega) = \text{const.}\ (x, 1 \pm \sqrt{3}\, x)$$
$$= \text{const.} \left(\frac{1}{2} \pm \frac{1}{3}\sqrt{3}\right).$$

Therefore, when λ is a cv, this inhomogeneous equation is without solution. □

Volterra equations

We close this section with a remark about Volterra kernels. Unlike its Fredholm counterpart, the degeneracy of the kernel of a Volterra integral equation typically does not simplify its solution. The only exception occurs when the kernel is of rank 1. In this case it turns out that the solution of the inhomogeneous integral equation

$$\phi(x) = f(x) + \lambda \int_a^x A(x)\, B(y)\, \phi(y)\, dy$$

can be compactly expressed as

$$\phi(x) = f(x) + \lambda \int_a^x A(x)\, B(y) \exp\left\{\lambda \int_y^x A(z)\, B(z) dz\right\} f(y) dy. \qquad \textbf{(10.3–11)}$$

After the next section, it will be clear how this representation can be derived in a straightforward manner.

10.4 The resolvent kernel and Fredholm's identities

Even when the kernel $K(x,y)$ is not degenerate, there exists an important and useful expression for the solution of the inhomogeneous equation

$$\phi(x) = f(x) + \lambda \int_a^b K(x,y)\, \phi(y) dy \qquad \textbf{(10.4–1)}$$

whenever the product λK is "small" in an appropriate sense. The key to the approach is Picard's method of successive approximations. We define the sequence of "approximate solutions"

$$\phi_0(x) = f(x)$$

$$\phi_1(x) = f(x) + \lambda \int_a^b K(x,y)\, \phi_0(y) dy$$

$$\vdots$$

$$\phi_m(x) = f(x) + \lambda \int_a^b K(x,y)\, \phi_{m-1}(y) dy.$$

Substituting in succession and utilizing the notion of the *iterated kernel*

$$K^1(x,y) \equiv K(x,y)$$

$$K^m(x,y) \equiv \int_a^b K(x,z)\, K^{m-1}(z,y) dz \qquad m \geq 2,$$

we can express these approximate solutions as

$$\phi_m(x) = f(x) + \lambda \int_a^b \left[\sum_{n=1}^{m} \lambda^{n-1} K^n(x,y)\right] f(y) dy.$$

If we now let m tend to infinity, we obtain formally

$$\phi_\infty(x) = f(x) + \lambda \int_a^b R_K(x,y;\lambda)\, f(y) dy, \qquad \textbf{(10.4–2)}$$

where

$$R_K(x,y;\lambda) \equiv \sum_{n=1}^{\infty} \lambda^{n-1} K^n(x,y). \qquad \textbf{(10.4–3)}$$

The function R_K is known as the *resolvent kernel* and the series expansion (10.4–3) is usually termed the *Neumann series for the resolvent*. Whenever the series converges, the expression (10.4–2) provides a convenient representation for the solution of the original integral equation (10.4–1).

Introducing the notion of the *norm* of the kernel K given by

$$\|K\| \equiv \left\{ \int_a^b \int_a^b [K(x,y)]^2 \, dx \, dy \right\}^{1/2}$$

simplifies the investigation of the convergence of the Neumann series (10.4–3). The Cauchy-Schwarz inequality implies that

$$\|K^2\| \leq \|K\|^2$$

and, in general,

$$\|K^n\| \leq \|K\|^n \qquad n \geq 2.$$

It then follows that for $n \geq 2$

$$\begin{aligned} |\lambda^{n-1} K^n(x,y)| &= |\lambda|^{n-1} |KK^{n-2} K(x,y)| \\ &\leq |\lambda|^{n-1} k_1(x) \|K^{n-2}\| k_2(y) \\ &\leq |\lambda|^{n-1} \|K\|^{n-2} k_1(x) k_2(y), \end{aligned}$$

where

$$k_1(x) \equiv \left\{ \int_a^b [K(x,y)]^2 dy \right\}^{1/2}, \qquad k_2(y) \equiv \left\{ \int_a^b [K(x,y)]^2 \, dx \right\}^{1/2}.$$

Since $k_1(x)$ and $k_2(y)$ are well-defined continuous functions and $\|K\|$ is a constant for any continuous kernel $K(x,y)$, the Neumann series converges absolutely and uniformly, at least whenever

$$|\lambda| \, \|K\| < 1. \qquad \textbf{(10.4–4)}$$

The nature of the resolvent kernel

In Section 10.3 we represented the solution of a Fredholm integral equation with degenerate kernel in terms of a function $R_K(x,y;\lambda)$. Now we have just employed in (10.4–2) the same notation for the solution of an integral equation with a possibly more general kernel. Some explanation is in order.

It turns out that whenever the eigenparameter λ is not one of the characteristic values of the given kernel K, the unique solution of the inhomogeneous Fredholm equation (10.4–1) asserted by the Fredholm Alternative can always be expressed in terms of a resolvent kernel function R_K as in (10.3–7), (10.4–2), that is, in the form

$$\phi(x) = f(x) + \lambda \int_a^b R_K(x,y;\lambda) f(y) dy. \qquad \textbf{(10.4–5)}$$

If the kernel $K(x,y)$ is degenerate, the resolvent kernel can be "calculated" as in (10.3–8). However, no such closed-form expression is uniformly applicable in the general case.

When the kernel K of (10.4–1) is nondegenerate, the associated resolvent kernel typically is no longer a rational function of λ. On the other hand, as first shown by Fredholm for continuous kernels, R_K can be expressed as the quotient of two functions both of which are analytic in λ throughout the complex λ-plane (see Smithies (1962) or Cochran (1972), for details.* The Neumann series (10.4–3) then gives the power series expansion for the resolvent that is valid for small λ. Taking the opposite point of view, the resolvent kernel in general is the analytic continuation to the entire complex λ-plane of the small λ expression (10.4–3). (See Problem 11 for an example involving degenerate kernels.)

The singularities of $R_K(x,y;\lambda)$, as a function of λ, are the characteristic values of the given kernel K. They are isolated and have no finite limit point, and corresponding to each is a finite number of linearly independent characteristic functions (recall the Fredholm Alternative). The Neumann series for the resolvent, which we know converges for λ satisfying (10.4–4), actually is valid for $|\lambda| < |\lambda_1|$ where λ_1 is the characteristic value of smallest absolute value. Since the series representation (10.4–3) can be shown to be convergent for all finite λ in the case of Volterra kernels (see Problem 12), such kernels must perforce have no cv's. Solutions of Volterra integral equations of the second kind, therefore, can always be expressed as in (10.4–5) with R_K given by the classical Neumann series.

The Fredholm identities

As suggested above, the kernel of any integral equation and the resolvent kernel essential for the representation of solutions of that equation have an extraordinary interrelationship. Whenever the associated resolvent $R_K(x,y;\lambda)$ exists, the solution of (10.4–1) is given by (10.4–5). Owing to the arbitrariness of f, it follows that

$$R_K(x,y;\lambda) = K(x,y) + \lambda \int_a^b K(x,z)\, R_K(z,y;\lambda)dz. \qquad \textbf{(10.4–6a)}$$

At the same time, since the solution of (10.4–1) is known to be *unique* by the Fredholm Alternative whenever λ is noncharacteristic, the various expressions can be rewritten to reveal also that

$$R_K(x,y;\lambda) = K(x,y) + \lambda \int_a^b R_K(x,z;\lambda)\, K(z,y)dz. \qquad \textbf{(10.4–6b)}$$

These two relations are termed the *Fredholm Identities*. From a theoretical point of view they show that, given a kernel K and a value of the eigenparameter

*This was later extended to more general kernels by David Hilbert (1862–1943) and Torsten Carleman (1892–1949). Erhard Schmidt, a student of Hilbert at Göttingen, used an intriguing method of interplaying the degenerate and "small" kernel expressions in order to arrive at the general result in alternative fashion.

λ, the associated R_K (if it exists) is the unique solution of each of a pair of integral equations. More practically, the identities (10.4–6) provide a convenient starting point from which to develop alternative representations for and approximations to the solution of the general integral equation (10.4–1). Indeed, much of the early work of Hilbert and his student, Schmidt, found its starting point in these identities as have more recent efforts of Weinstein, Aronszajn, Rellich, Kato, and others. (See Cochran (1972) for a survey of some of these contributions.) We content ourselves with notice that since the Neumann series (10.4–3) for the resolvent is an exact small-λ solution of the Fredholm Identities, we are assured that this power series expansion is valid whenever λ is small enough.

10.5 Symmetric kernels and series representations

A real-valued kernel is said to be symmetric if

$$K(y,x) = K(x,y).$$

Analogous to the situation in matrix theory, integral equations with such kernels occur over and over again in applied problems (recall the kernel (10.1–7) associated with the earlier beam vibration analysis). In further agreement with the discrete case, symmetric kernels possess enough pleasant properties that a theoretically complete analysis of linear integral equations of the second kind with these functions as kernels is possible. Some of the principal results are as follows (see Smithies (1962) or Cochran (1972), for example; also compare with the results of Section 1.5)*:

Property 1
The characteristic values of symmetric kernels are real.

Property 2
If $K(x,y)$ is a nonnegative definite kernel, that is, if

$$\int_a^b \int_a^b K(x,y)\, \phi(x)\, \phi(y)\, dx\, dy \geq 0$$

for all continuous functions ϕ, then its characteristic values are all positive.

Remark: It should be noted that the iterates of symmetric kernels are all symmetric and the even iterates K^{2n} are nonnegative definite. The cv's are of course the $2n$th powers of the cv's of the original kernel.

Property 3
The characteristic functions of symmetric kernels may be chosen to be real and to form an orthonormal set.

---

*In the case of complex-valued kernels, comparable results can be established for Hermitian kernels that satisfy $\overline{K(y,x)} = K(x,y)$.

Customarily, the characteristic values associated with a given kernel are indexed according to the size of their absolute values, beginning with the smallest;

$$0 < |\lambda_1| \leq |\lambda_2| \leq \cdots \leq |\lambda_n| \leq \cdots .$$

In the case of cv's of multiplicity greater than one, a value of λ is included in this enumeration once for each of the (finitely many) linearly independent characteristic functions corresponding to the cv (recall the Fredholm Alternative). This prompts the following definition:

Definition 1 An orthonormal set of characteristic functions $\{\phi_n\}$, with associated characteristic values $\{\lambda_n\}$ enumerated as above and such that every continuous cf of the given kernel $K(x,y)$ is a finite linear combination of the functions ϕ_n, is termed a **full characteristic system** for K.

As examples we observe that the full characteristic system associated with the familiar kernel

$$K(x,Y) = \min(x,y) - xy \equiv \begin{cases} x(1-y) & 0 \leq x \leq y \leq 1 \\ y(1-x) & 0 \leq y \leq x \leq 1 \end{cases}$$

is $\{\sqrt{2}\sin n\pi x; n^2\pi^2\}$. For the related kernel

$$K(x,y) = \frac{1}{3} + \frac{1}{2}(x^2 + y^2) - \max(x,y) \qquad \textbf{(10.5–1)}$$

on the same domain, the full characteristic system turns out to be $\{\sqrt{2}\cos n\pi x; n^2\pi^2\}$ (see Problem 15).

In Section 5 of Chapter 2 we considered the problem of representing continuous real-valued functions by series expansions in terms of various sets of orthonormal functions. We can apply all of our earlier analysis in the case when the expansion functions are the cf's ϕ_n associated with some given symmetric kernel K.* In the remainder of this section we explore a number of the consequences of this fact.

Schur's inequality

If, in Bessel's inequality

$$\sum_n (f,y_n)^2 \leq \|f\|^2,$$

we let the set of orthonormal characteristic functions $\{\phi_n\}$ associated with a given symmetric kernel $K(x,y)$ play the role of the y_n and choose K itself, viewed as a function of y with x as a parameter, for the function f, we obtain

$$\sum_{n=1}^{\infty} \phi_n^2(x)/\lambda_n^2 = \sum_{n=1}^{\infty}\left[\int_a^b K(x,y)\,\phi_n(y)dy\right]^2 \leq \int_a^b [K(x,y)]^2 dy.$$

*The set $\{\phi_n\}$ need not be complete, however. (For example, it will not be when $K(x,y)$ is degenerate.)

Integrating then leads to the inequality

$$\sum_{n=1}^{\infty} 1/\lambda_n^2 \le \|K\|^2, \tag{10.5–2}$$

a result that allows us to get some feel for the growth of the characteristic values with increasing n, even if we cannot determine the cv's explicitly.

Completeness

Hilbert first showed (1904) that in the case of expansions in terms of the (not necessarily complete) sets of cf's of symmetric kernels, the earlier result (2.5–1) could be replaced by

$$f = h + \sum_{n=1}^{\infty} (f,\phi_n)\phi_n\,, \tag{10.5–3}$$

where h is such that

$$\int_a^b K(x,y)\, h(y)dy = 0.$$

Here, of course, the convergence of the series is meant in the same limiting average sense as in Section 2.5. As an immediate implication we have

> Corollary:
> If the symmetric kernel $K(x,y)$ is such that only trivial functions (with $\|h\| = 0$) satisfy
>
> $$\int_a^b K(x,y)\, h(y)dy = 0,$$
>
> then the (full) set of characteristic functions associated with K is complete.

This latter result allows us to give a quick independent proof of Property 2 of Section 2.5 concerning the completeness of the set of eigenfunctions of the Sturm-Liouville problem (2.3–1). We begin by reexpressing the eigenvalue problem in the form

$$u(x) = \lambda \int_a^b G(x,y)\, r(y)\, u(y)\, dy, \tag{10.5–4}$$

where G is the appropriate Green's function. We then recall from our discussion in Chapter 8 that G is symmetric in its arguments, and hence if $r(x) > 0$, (10.5–4) can be written as the integral equation

$$\sqrt{r(x)}\, u(x) = \lambda \int_a^b K(x,y)\, \sqrt{r(y)}\, u(y)\, dy$$

with the symmetric kernel

$$K(x,y) \equiv \sqrt{r(x)}\, G(x,y)\, \sqrt{r(y)}.$$

Accordingly, to apply the above Corollary we need to analyze the solutions of

$$\int_a^b G(x,y)\,\sqrt{r(y)}\,h(y)\,dy = 0.$$

But this is easy to do since, owing to Theorem 1 of Section 8.2 and Theorem 2 of Section 8.3, it readily follows that

$$\sqrt{r(x)}\,h(x) = 0$$

and hence $h = 0$, which completes the demonstration.

The problems at the end of this chapter contain oher applications of the above Corollary. Both here and there, however, we are making use of the fact that, barring possible notational differences, the eigenfunctions (and eigenvalues) associated with a given Sturm-Liouville problem are *identical* to the characteristic functions (and characteristic values) associated with the corresponding Fredholm integral equation.*

Kernel expansions

Given a symmetric kernel K, if in the Hilbert result (10.5–3) we take $f(x) = K(x,y)$ (treating y as a parameter), it can be shown that $\|h\| = 0$ and hence

$$K(x,y) = \sum_{n=1}^{\infty} \frac{\phi_n(x)\,\phi_n(y)}{\lambda_n}.$$

It follows then that for positive integer p the iterated kernels have the representations

$$K^p(x,y) = \sum_{n=1}^{\infty} \frac{\phi_n(x)\,\phi_n(y)}{\lambda_n^{\,p}},$$

and from this latter expression we formally deduce that

$$R_K(x,y;\lambda) = \sum_{n=1}^{\infty} \frac{\phi_n(x)\,\phi_n(y)}{\lambda_n - \lambda}. \qquad \textbf{(10.5–5)}$$

Although this expansion has an importance of its own for integral equations, the similarity of (10.5–5) to the earlier representation (9.6–2) of the Green's function associated with the Sturm-Liouville eigenvalue problem (9.6–1) should not go unnoticed. In that previous section we distinguished between the Green's function $G(x,x') \equiv G(x,x';0)$ associated with the differential operator L and the Green's function $G(x,x';\lambda)$, which was appropriate for the differential operator $(L + \lambda r)$ with fixed $\lambda \neq 0$. Viewed in the context of integral equations theory, the relationship between these two Green's functions is now clear. If $G(x,x')$ is interpreted as the kernel of a Fredholm integral equation, then $G(x,x';\lambda)$ is its resolvent kernel. This fact goes a long way towards explaining why the explicit determination of these latter Green's functions is generally a nontrivial undertaking in most practical problems.

---

*Indeed, for appropriate kernels the application of this fact (in the reverse direction) provides the simplest method for analytically determining the underlying cv's and cf's (see Problems 15, 16).

10.6 The approximation of characteristic values and functions

Many of our earlier considerations in this chapter have been theoretical in nature and may have led the casual reader to assume that finding solutions to integral equations, particularly those where the equation is of the second kind, is a straightforward task. It is true that in some providential situations a given integral equation may be transformed into an equivalent boundary-value problem in differential equations, and the latter can either be solved explicitly (see Problems 15, 16) or at least studied qualitatively (and perhaps quantitatively), as suggested in Chapter 2. More typically, however, the integral equation of interest does not lend itself to such an approach, and approximate methods developed (or adapted) for general integral equations must be employed. In this closing section we discuss three such approximation techniques, each of distinct practical importance. In order to indicate how these techniques are used in practice, we will apply them to determine the natural resonances associated with the beam vibration problem discussed in Section 10.1. We concentrate on the homogeneous problem since the related inhomogeneous equation with fixed eigenparameter λ can usually be treated using Picard's method or some similar iterative technique. (For a general survey of numerical methods applicable to Fredholm integral equations see Atkinson (1976).)

Kellogg's method

Perhaps the most popular iterative procedure for the "solution" of homogeneous Fredholm integral equations of the second kind is an extension of the method of power iteration (recall Section 5 of Chapter 1) first suggested by O. D. Kellogg (1878–1932). In this approach we begin with an arbitrary (continuous) function $\psi_0(x)$ that has been normalized so that $\|\psi_0\| = 1$. From ψ_0 we construct

$$\psi_1(x) \equiv \mu_1 \int_a^b K(x,y)\, \psi_0(y)\, dy,$$

taking care to select the parameter μ_1 so that ψ_1 has unit norm also. Continuing in like manner, we then successively form $\psi_2, \psi_3, \ldots$, where, in each instance,

$$\psi_n(x) \equiv \mu_n \int_a^b K(x,y)\, \psi_{n-1}(Y)\, dy$$

with $\|\psi_n\| = 1$. It can be shown that for symmetric kernels K, as n tends to infinity, the μ_n and ψ_n generally converge toward a cv (usually the smallest) and the corresponding cf associated with the given integral equation. Furthermore, the convergence oftentimes is sufficiently rapid that only a few iterations yield approximations of acceptable accuracy. (See Cochran (1972) for details regarding the procedure as well as remarks concerning the utility of this approach when the kernel is not symmetric.)

We illustrate Kellogg's method by considering the special case of the vibrating (uniform and homogeneous) beam of unit length as detailed in Section 10.1. The pertinent integral equation is

$$\phi(x) = \lambda \int_0^1 K(x,y)\, dy \qquad 0 \leq x \leq 1$$

in which the kernel K is the symmetric function

$$K(x,Y) = \begin{cases} \frac{1}{2}x^2y - \frac{1}{6}x^3 & x < y \\ \frac{1}{2}xy^2 - \frac{1}{6}y^3 & x > y \end{cases} \tag{10.6–1}$$

(recall (10.1–7)). The characteristic parameter λ is related to the vibration frequencies for this problem through the equation

$$\lambda = \frac{\rho\,\omega^2}{EI}.$$

The initial choice of $\psi_0(x) = 1$ leads to

$$\mu_1 = 18\sqrt{10/13} \doteq 15.78; \qquad \psi_1 = \frac{\mu_1}{24}[x^4 - 4x^3 + 6x^2]$$

and

$$\mu_2 \doteq 12.364; \qquad \psi_2 = \frac{\mu_1\,\mu_2}{40320}[x^8 - 8x^7 + 28x^6 - 336x^3 + 728x^2].$$

This second approximation for the least cv is remarkably close to the true value $\lambda_1 \doteq 12.362$. The function ψ_2 is likewise a rather good approximation to the corresponding normal mode.*

Successive substitutions

As mentioned earlier, a standard technique for the approximate solution of an inhomogeneous equation of the second kind is by means of the iterative method of Picard outlined in Section 10.4 above. Occasionally, homogeneous problems can be recast so as to appear inhomogeneous and, whenever this is so, Picard's method should be applicable to these problems as well. In view of (10.1–8) our beam oscillation problem is just such a case, so the technique of Picard offers an alternate iterative approach to the approximation of cv's and cf's for this practical example.

The normal modes of vibration and the corresponding eigenfrequencies for the special case of the uniform homogeneous beam can be derived from the solutions of the equation

---

*The quality of these approximations after so few iterations is due principally to the fact that $|\lambda_1/\lambda_2| << 1$. (Compare with the discussion in Section 1.5.)

$$\psi(x) = f(x) + (\lambda/3!) \int_0^x (x - y)^3 \, \psi(y) \, dy \tag{10.6–2a}$$

where f is given by

$$f(x) = \lambda(Ax^3/3! + Bx^2/2!), \tag{10.6–2b}$$

and the constants A and B are such that

$$A + \int_0^1 \psi(x) \, dx = 0, \qquad B - \int_0^1 x\psi(x) \, dx = 0. \tag{10.6–3}$$

(Recall (10.1–8) and (10.6–1)). Notice that the eigenparameter λ appears in *two* places in (10.6–2).

In the Picard procedure, the function $f(x)$ itself provides the initial approximation to the solution of the integral equation. In this case, substituting $\psi_0 = f$ into (10.6–3) leads to a set of algebraic equations in A and B, namely

$$A + \lambda A/4! + \lambda B/3! = 0$$

$$B - \lambda A/5 \cdot 3! - \lambda B/4 \cdot 2! = 0.$$

However, these algebraic equations are satisfied by nontrivial A,B only if

$$\lambda^2 - 240\lambda + 2880 = 0,$$

and thus we arrive at the approximation

$$\lambda_1 \doteq 12.67$$

for the smallest cv. (The larger root of this quadratic equation provides an estimate of λ_2.)

Improved approximations to the solution of (10.6–2) can be obtained from further iterations in Picard's method. For the next "approximate solution" we find

$$\begin{aligned}\psi_1(x) &= f(x) + (\lambda/3!) \int_0^x (x - y)^3 f(y) \, dy \\ &= \lambda(Ax^3/3! + Bx^2/2!) + \lambda^2(Ax^7/7! + Bx^6/6!),\end{aligned}$$

and hence, owing to (10.6–3),

$$A(1 + \lambda/4! + \lambda^2/8!) + B(\lambda/3! + \lambda^2/7!) = 0$$

$$-A(\lambda/5 \cdot 3! + \lambda^2/9 \cdot 7!) + B(1 - \lambda/4 \cdot 2! - \lambda^2/8 \cdot 6!) = 0.$$

It follows that λ must be a root of

$$\lambda^4 - 672\lambda^3 + 8 \cdot 9!\lambda^2 - 6 \cdot 7!8!\lambda + 8!9! = 0$$

from which we calculate

$$\lambda_1 \doteq 12.363,$$

which is another extremely accurate estimate of the smallest cv.

The Rayleigh-Ritz procedure

In contradistinction to the above two iterative techniques, our last procedure is a *direct method** that was first used in some theoretical investigations in acoustics by Lord Rayleigh and subsequently applied with considerable success by the young Swiss physicist, Walther Ritz (1878–1909). The method is designed principally for symmetric kernels that are nonnegative definite. It is based upon the fact that for any test function ϕ the so-called *Rayleigh quotient*

$$R[\phi] \equiv \frac{\int_a^b \int_a^b K(x,y)\, \phi(y)\, \phi(x)\, dy\, dx}{\int_a^b [\phi(x)]^2\, dx} \tag{10.6–4}$$

always provides a *lower* bound for the reciprocal of the smallest cv associated with such kernels.

The approach of Rayleigh and Ritz proceeds as follows. We consider *any* orthonormal set of functions $\{\psi_i\}$ and form the test function

$$\phi(x) = \sum_{i=1}^{n} \alpha_i\, \psi_i(x), \tag{10.6–5}$$

where the α_i are as yet undetermined real coefficients. If the function ϕ is substituted into (10.6–4), the Rayleigh quotient becomes

$$R[\phi] = \frac{\sum_{i,j=1}^{n} c_{ij}\, \alpha_i\, \alpha_j}{\sum_{i=1}^{n} \alpha_i^2} \tag{10.6–6}$$

with

$$c_{ij} \equiv \int_a^b \int_a^b K(x,y)\, \psi_i(y)\, \psi_j(x)\, dy\, dx.$$

In view of the lower bound character of $R[\phi]$, the "best" test function of the form (10.6–5) is the one that gives rise to the *largest* value in (10.6–6). It is a well-known result in matrix theory that the quadratic form (10.6–6) is largest when the coefficients α_i are the components $\alpha_i^{(n)}$ of an eigenvector corresponding to the largest eigenvalue of the symmetric matrix $C \equiv [c_{ij}]$. Moreover, the maximum of $R[\phi]$ is precisely this largest eigenvalue $\gamma_1^{(n)}$.

The Raleigh-Ritz procedure obviously owes a substantial portion of its popularity to its matrix-theoretic nature. It is a fool-proof method that provides

*A method is said to be *direct* if it reduces the task of solving the given integral equation to that of solving a *finite* approximating system of algebraic equations.

increasingly better approximations as the number of terms in the expansion (10.6–5) is increased. Indeed, if the $\{\psi_i\}$ form a complete set (recall Section 2.5), not only does $\gamma_1^{(n)}$ approach $1/\lambda_1$ monotonically from below, but

$$\phi^{(n)}(x) = \sum_{i=1}^{n} \alpha_i^{(n)} \psi_i(x)$$

converges to an associated cf of K as well. Furthermore, as a bonus, the other eigenvalues and eigenfunctions of the matrix C converge towards higher-order cv's and cf's of K.

A final look at beam vibration

We close this chapter with an application of the Rayleigh-Ritz procedure to our oscillating beam problem. Since the kernel (10.6–1) can be shown to be nonnegative definite, the technique is suitable.

The specifics of the orthonormal set $\{\psi_i\}$ are at our disposal; a simple choice is to take each ψ_i to be a polynomial of degree $i - 1$. This leads to

$$\begin{aligned} \psi_1(x) &= 1 \\ \psi_2(x) &= \sqrt{3}\,(1 - 2x) \\ \psi_3(x) &= \sqrt{5}\,(1 - 6x + 6x^2) \\ &\text{etc.} \end{aligned}$$

The $K\psi_i \equiv \int_a^b K(x,y)\,\psi_i(y)dy$ then turn out to be

$$\begin{aligned} K\psi_1(x) &= (1/24)\,(x^4 - 4x^3 + 6x^2) \\ K\psi_2(x) &= (\sqrt{3}/120)\,(-2x^5 + 5x^4 - 10x^2) \\ K\psi_3(x) &= (\sqrt{5}/120)\,(2x^6 - 6x^5 + 5x^4) \\ &\text{etc.} \end{aligned}$$

Hence, for $n = 3$,

$$C = \begin{bmatrix} 1/20 & -\sqrt{3}/45 & \sqrt{5}/420 \\ -\sqrt{3}/45 & 13/420 & -\sqrt{15}/720 \\ \sqrt{5}/420 & -\sqrt{15}/720 & 1/504 \end{bmatrix}.$$

Designating the reciprocals of the eigenvalues $\gamma_1^{(n)} \geq \gamma_2^{(n)} \geq \ldots$ by $\lambda_1^{(n)} \leq \lambda_2^{(n)} \leq \ldots$, respectively, we can determine from C that

$$\begin{aligned} &\lambda_1^{(1)} \doteq 20, \quad && \lambda_1^{(2)} \doteq 12.48, \quad && \lambda_1^{(3)} \doteq 12.37 \\ & && \lambda_2^{(2)} \doteq 1212, && \lambda_2^{(3)} \doteq 494.3 \\ & && && \lambda_3^{(3)} \doteq 13{,}960. \end{aligned}$$

The final three values are approximations for the cv's $\lambda_1 \doteq 12.362$, $\lambda_2 \doteq 485.5$, and $\lambda_3 \doteq 3807$.

As an alternative we could have taken the single function

$$\psi_1(x) \sqrt{45/104}\,(x^4 - 4x^3 + 6x^2).$$

This function is the simplest polynomial of unit norm that satisfies all of the boundary conditions (10.1–6). (Compare with the first approximation obtained in the earlier application of Kellogg's method to this problem.) With this choice we find that

$$K\psi_1(x) = \frac{\sqrt{45/104}}{1680}\,(x^8 - 8x^7 + 28x^6 - 336x^3 + 728x^2)$$

from which we deduce that $c_{11} \doteq 0.080875$, and hence

$$\lambda_1^{(1)} \doteq 12.365.$$

The exceptional accuracy of this easily-determined result is a consequence of the fact that better approximations are generally obtained at each stage of the Rayleigh-Ritz procedure when the orthonormal set $\{\psi_1\}$ shares as many properties as possible with the cf's under investigation.

References

Atkinson, K. E. (1976): *A Survey of Numerical Methods for the Solution of Fredholm Integral Equations of the Second Kind*, Soc. Indust. Appl. Math., Philadelphia.

Cochran, J. A. (1972): *The Analysis of Linear Integral Equations*, McGraw-Hill, New York

Courant, R. and D. Hilbert (1953): *Methods of Mathematical Physics, Vol. I*, Interscience, New York; Chapter III.

Hildebrand, F. B. (1976): *Advanced Calculus for Applications*, 2nd ed., Prentice-Hall, Englewood Cliffs, N. J.; Chapter 5.

Kanwal, R. P. (1971): *Linear Integral Equations*, Academic Press, New York.

Lovitt, W. V. (1950): *Linear Integral Equations*, Dover, New York; McGraw-Hill, New York (1924).

Mikhlin, S. G. (1964): *Integral Equations*, 2nd rev. ed., Pergamon Press, Macmillan, New York; Transl. of Russian edition, Fizmatgiz, Moscow (1959).

Skudrzyk, E. (1968): *Simple and Complex Vibratory Systems*, Pennsylvania State Univ. Press, Univ. Park; Chapter VI.

Smithies, F. (1962): *Integral Equations*, Cambridge Univ. Press, London.

Tricomi, F. G. (1957): *Integral Equations*, Interscience, New York.

Wylie, C. R. (1975): *Advanced Engineering Mathematics*, 4th ed., McGraw-Hill, New York; Chapters 2, 8.

Problems

Sections 10.1, 10.2

★1. A differential operator L such that $uLv - vLu$ can be written as a perfect derivative is termed self-adjoint (recall Problem 2, Chapter 2 and Section 8.2).

(a) Show that the fourth-order operator

$$L\phi \equiv \frac{d^2[p(x)\,\phi'']}{dx^2} - q(x)\phi$$

is self-adjoint.

(b) Extend to this operator L the notion of regular boundary conditions as originally given for second-order operators by Definition 2 of Section 2.2.

(c) Use the results of parts (a) and (b) to demonstrate that eigensolutions, corresponding to different eigenvalues, of the generalized eigenvalue problem

$$(p\phi'')'' + (\lambda r - q)\phi = 0 \qquad a \le x \le b$$

$$[A_1\,\phi - a_1(p'')']_{x=a} = 0$$

$$[A_2\,\phi' - a_2(p\,\phi'')]_{x=a} = 0$$

$$[B_1\,\phi - b_1(p\,\phi'')']_{x=b} = 0$$

$$[B_2\,\phi' - b_2(p\,\phi'')]_{x=b} = 0,$$

(where neither A_1 and a_i nor B_i and b_i both vanish) are orthogonal over $[a,b]$ with respect to the weight function $r(x)$.

2. Verify that when I and ρ are constant the eigenvalues and associated eigenfunctions of the boundary-value problem (10.1–3), (10.1–4), namely

$$\frac{d^4\phi}{dx^4} - \frac{\rho\omega^2}{EI}\phi = 0 \qquad 0 \le x \le L$$

$$\phi(0) = 0 \qquad \frac{d^2\phi(L)}{dx^2} = 0$$

$$\frac{d\phi(0)}{dx} = 0 \qquad \frac{d^3\phi(L)}{dx^3} = 0,$$

have the form given in Section 10.1.

3. The Abel integral equation (10.2–3) can be rewritten as a Volterra equation with a quotient kernel. Use the transform techniques of Section 7.5 to provide an alternative derivation of the solution formula (10.2–4).

4. (a) Imitate the analysis of Section 10.2 to establish that the solution of the *generalized Abel equation*

$$\psi(x) = \int_0^x \frac{\phi(y)}{(x-y)^\alpha}\, dy \qquad 0 < \alpha < 1$$

can be formally represented as

$$\phi(y) = \frac{\sin \pi\alpha}{\pi} \frac{d}{dy}\left[\int_0^y \frac{\psi(x)}{(y-x)^{1-\alpha}}\, dx\right].$$

(b) Alternatively, use the transform techniques of Section 7.5, applicable to integral equations with quotient kernels, to obtain this inversion formula.

Section 10.3

5. Consider the Fredholm integral equation of the first kind

$$\psi(x) = \int_a^b K(x,y)\, \phi(y)\, dy$$

where the kernel K is of finite rank and has the form

$$K(x,y) = \sum_{i=1}^{n} A_i(x)\, B_i(x) \qquad a \leq x,y \leq b.$$

If ψ is a linear combination of the A_i, $i = 1, 2, \ldots, n$, show that there exists an appropriate liner combination of the B_i that solves the integral equation.

6. Solve the Fredholm integral equation

$$x - 1 = \int_{-1}^{1} (1 + y + 3xy)\, \phi(y)\, dy$$

using the component decomposition

$$A_1(x) = 1 \qquad B_1(y) = 1 + y$$

$$A_2(x) = 3x \qquad B_2(y) = y.$$

★7. Consider the homogeneous Fredholm equations

$$\omega(x) = \lambda \int_a^b K(x,y)\, dy\, \omega(y)\, dy \qquad \text{and} \qquad \Omega(x) = \lambda \int_a^b K(y,x)\, \Omega(y)\, dy$$

where $K(x,y)$ is the degenerate kernel of rank n

$$K(x,y) \equiv \sum_{i=1}^{n} A_i(x)\, B_i(y).$$

Define the matrix C by

$$C \equiv [\delta_{ij} - \lambda(b_i, A_j)],$$

where δ_{ij} is the familiar *Kronecker* δ, and let $D_n(\lambda) \equiv \det C$. Satisfy yourself that if λ is such that $D_n(\lambda) = 0$, both of these integral equations have nontrivial solutions and the number of linearly independent solutions of each equation is equal to the nullity of the matrix C.

★8. For the case of a degenerate kernel, demonstrate the validity of the final portion of the Fredholm Alternative given in Section 10.3. [Courant and Hilbert (1953) has a concise statement of the underlying fundamental matrix-theoretic results.]

9. Explicitly determine the solutions of the Fredholm integral equation

$$\phi(x) = c + \lambda \int_{-1}^{1} (1 + y + 3xy)\, \phi(y)\, dy$$

where λ and c are parameters.

Section 10.4

10. Consider the homogeneous Fredholm integral equation

$$\phi(x) = \lambda \int_a^b K(x,y)\, \phi(y)\, dy$$

with a continuous kernel $K(x,y)$ satisfying

$$|K(x,y)| \leq m \qquad a \leq x,y \leq b.$$

(a) Show that this equation has only the trivial solution if

$$|\lambda| < [M(b - a)]^{-1}.$$

(b) Compare this result with what you expect to be the case on the basis of the discussion in Section 10.4 regarding the Neumann series for the resolvent kernel (10.4–3). In particular, is there any contradiction between this result and the convergence criterion (10.4–4)?

11. (a) For the degenerate Fredholm kernel $K(x,y) = a(x)\, b(y)$, calculate the Neumann series representation for the associated resolvent kernel.

(b) Relate the result obtained in part (a) with the expression which would be obtained from degenerate kernel theory using (10.3–8).

12. Consider a Volterra kernel $K(x,y)$ which is continuous for $a \leq x,y \leq b$ and vanishes whenever $y > x$.

(a) Show that

$$K^i(x,y) = \begin{cases} \displaystyle\int_y^x K(x,z)\, K^{i-1}(z,y)\, dz & y \leq x \\ 0 & y > x \end{cases}$$

for the iterated kernels K^i with $i \geq 2$.

(b) If K is degenerate and of rank 1, that is

$$K(x,y) = \begin{cases} A(x)\, B(y) & y \leq x \\ 0 & y > x, \end{cases}$$

use an inductive argument to verify that

$$K^i(x,y) = \frac{A(x)\,B(y)}{(i-1)!}\left[\int_y^x A(z)\,B(z)dz\right]^{i-1}$$

and hence

$$R_K(x,y;\lambda) = A(x)\,B(y)\exp\left[\lambda\int_y^x A(z)\,B(z)\,dz\right].$$

(c) For the general case, demonstrate that

$$|K^i(x,y)| \le M^i(x-y)^{i-1}/(i-1)!,$$

where $M = \max|K(x,y)|$ over $a \le y \le x \le b$, and thus that

$$|R_K(x,y;\lambda)| \le M\,e^{|\lambda|M(x-y)}.$$

13. For the Volterra kernel

$$K(x,y) = \begin{cases} y-x & 0 \le y < x \\ 0 & x < y \end{cases}$$

(a) determine a closed form expression for the resolvent $R_K(x,y;\lambda)$, and

(b) use this resolvent kernel in order to derive an explicit solution for the integral equation

$$\phi(x) = 1 + \int_0^x K(x,y)\,\phi(y)dy.$$

14. Let $R_2(\lambda)$ be the resolvent associated with the second iterate K^2 of a given Fredholm kernel K.

(a) Suppressing the dependence upon the independent variables x,y, show that

$$2\lambda\,R_2(\lambda^2) = R_K(\lambda) - R_K(-\lambda).$$

(b) Reciprocally, verify that

$$R_K(\lambda) = K + \lambda R_2(\lambda^2) + \lambda^2 K R_2(\lambda),$$

where

$$KR_2(\lambda^2) \equiv \int_a^b K(x,z)\,R_2(z,y;\lambda^2)\,dz.$$

Section 10.5

15. (2) Reexpress the homogeneous integral equation with the symmetric kernel (10.5–1) as a differential eigenvalue problem on the interval [0,1].

(b) Solve this BVP and establish that the full characteristic system associated with the original kernel is as given in the text.

16. Determine a full characteristic system for the symmetric kernel

$$K(x,y) = |x - y| \qquad -\tfrac{1}{2} \leq x,y \leq \tfrac{1}{2}\,.$$

[Hint: Follow the procedure suggested in Problem 15.]

17. (a) Apply the Corollary of Section 10.5 to the classical kernel

$$K(x,y) = \begin{cases} x(1-y) & x \leq y \\ y(1-x) & y \leq x \end{cases}$$

in order to show directly that the characteristic functions $\{\sin n\pi x\}$ form a complete set.

(b) Use a similar approach to demonstrate formally that the set of characteristic functions associated with the symmetric kernel

$$K(x,y) = \ln\,[1 - \min(x,y)]\,[1 + \max(x,y)]$$

is complete over $(-1,1)$. In this case the cf's are the Legendre polynomials $P_n(x)$, $n = 0,1,2,\ldots$ (see Appendix 2).

Section 10.6

18. Apply Kellogg's method to the symmetric kernel

$$K(x,y) = \min(x,y) \qquad 0 \leq x,y \leq 1$$

to determine approximations to the least cv and its associated cf.

19. Repeat Problem 18 for the kernel

$$K(x,y) = |x - y| \qquad -\tfrac{1}{2} \leq x,y \leq \tfrac{1}{2}.$$

20. Use the Rayleigh-Ritz procedure with simple polynomials in order to obtain estimates for the least cv of the kernel of Problem 18. Try this first without using any a priori information about the boundary behavior of the cf's.

21. Repeat Problem 20 for the symmetric kernel of Problem 19.

22. (a) Show that the homogeneous integral equation

$$\phi(x) = \lambda \int_0^1 K(x,y)\,\phi(y)\,dy$$

with kernel $K(x,y) = \min(x,y)$ is equivalent to the Sturm-Liouville eigenvalue problem

$$\begin{aligned} \phi'' + \lambda\phi &= 0 \qquad 0 \leq x \leq 1 \\ \phi(0) &= 0 \\ \phi'(1) &= 0. \end{aligned}$$

(b) Follow the procedure suggested in Section 10.1 and recast this BVP as the (pseudo) Volterra integral equation

$$\psi(x) = -\lambda A x + \lambda \int_0^x (y - x)\,\psi(y)\,dy$$

where A is a constant such that

$$A + \int_0^1 \psi(y)\, dy = 0.$$

(c) Now perform several iterations of the successive substitution method discussed in Section 10.6 for this Volterra integral equation and determine reasonable approximations to the smallest allowable value of the characteristic parameter λ.

11

Asymptotics

11.1 The dispersion of water waves

Twice before we have demonstrated the use of varied mathematical techniques for the resolution of problems in fluid mechanics. In this final chapter we turn one last time to a fluid application (adapted from Jeffreys and Jeffreys (1962). However, because of the nature of the particular problem, our interests are going to be somewhat different than heretofore; we will be more readily content with approximate, as opposed to exact, answers: a situation that typifies the entire area of asymptotic analysis.

Consider a collection of one-dimensional waves propagating along the x-axis with period $2\pi/\gamma$ and wavelength $2\pi/k$, where γ is the frequency and k is the so-called individual wavenumber (recall Section 3.1). Each individual wave has a waveform proportional to, say, $e^{i(\gamma t - kx)}$, so the composite contribution of a wave train with components varying in amplitude as $A(k)$ can be expressed as

$$I = \int_{-\infty}^{\infty} A(k)\, e^{i(\gamma t - kx)}\, dk. \qquad \textbf{(11.1–1)}$$

If $\gamma = ck$ with c a constant, this representation reduces to a Fourier integral with which we already have some familiarity. On the other hand, if γ is not a linear function of k, the analysis of the expression (11.1–1) is considerably more difficult. In physical terms what happens in this case is that the individual waves making up the wave train each travel with a *different* velocity $c(k) \equiv \gamma(k)/k$, and they thus *disperse* as time evolves. In this way, a once-concentrated hump may become so spread out as to make extremely complicated the exact evaluation of the unified representation (11.1–1), and some approximations may be necessary.

When confronted with just this problem, Lord Kelvin (William Thomson, 1824–1907) contended that the dominant contribution to (11.1–1) as $t \to \infty$ comes from the vicinity of points where

$$\frac{d}{dk}(\gamma t - kx) = 0, \, *$$

*Since these are the points where the phase of the integrand of (11.1–1) is stationary, Kelvin's assertion is called the *stationary phase principle*, and application of this approach is termed the *method of stationary phase*. We analyze this technique in Section 11.4.

that is, where

$$\frac{d\gamma}{dk} = \frac{x}{t}.$$

The quantity $C(k) \equiv d\gamma/dk$ is called the *group velocity*; it turns out to be the central (and natural) construct for a "group" of waves with varying wavenumbers. An observer moving with velocity $C(k_0)$ always sees waves with wavenumber k_0 and frequency $\gamma(k_0)$. In other words, waves with different wavenumber all propagate with the group velocity. Moreover, as $t \to \infty$ the energy present in the wave train described by (11.1–1) propagates with this velocity (see Whitham (1974), for example).

Water waves provide a nice illustration of just such a dispersive phenomenon and one that is mathematically tractable. When both gravity and capillary attraction are taken into account, the empirical relation between γ and k in water of at least moderate depth is given by

$$\gamma^2 = (gk + Tk^3) \tanh (kh). \quad \textbf{(11.1–2)}$$

In this expression g is the gravitational acceleration constant, $T\rho$ is the surface tension (with ρ the fluid density), and h is the depth of the water. We note that for small k

$$\gamma \approx k\sqrt{gh}\,[1 + \tfrac{1}{2}k^2 (T/g - h^2/3)] \quad \textbf{(11.1–3)}$$

(the first term here was seen earlier in Chapter 2), while for $k >> 1$

$$\gamma \approx \sqrt{gk + Tk^3}.$$

However, the most significant feature of the general relationship (11.1–2) is that for deep enough water, as k increases from 0, the group velocity $C = d\gamma/dk$ initially *decreases* to a minimum value before finally increasing like $\sqrt{k}$ (see Problem 1).

This group velocity minimum has substantial influence on the dispersion of water waves. For example, if we consider at some particular time a disturbance (like the splash of a thrown rock or a rising fish) initiated at some earlier time, we find

(i) smooth water near the point of impact;
(ii) a mixing of both shorter (essentially capillarity) and longer (gravity) waves further out; and
(iii) only the shorter waves at the larger distances. □

A complete analysis of (11.1–1) with γ given by (11.1–2) is a substantial undertaking. Therefore, we simplify the expression by using the long wavelength approximation (11.1–3) and assuming $T = 0$ and $A(k) \equiv 1$. It follows then that

$$I = 2\int_0^\infty \cos k\,[\sqrt{gh}\,(1 - k^2h^2/6)t - x]dk$$

$$= \frac{2z}{x - \alpha t}\int_0^\infty \cos (sz + \tfrac{1}{3}s^3)ds, \quad \textbf{(11.1–4)}$$

where

$$\alpha \equiv \sqrt{gh}, \qquad z \equiv (x - \alpha t)/(\tfrac{1}{2}\alpha h^2 t)^{1/3}, \qquad s \equiv k(\tfrac{1}{2}\alpha h^2 t)^{1/3}.$$

Observe that α is the maximum (small k) group velocity while $z = 0$ separates the waves moving "faster" than the wave train from those moving "slower."

Apart from the terms in front of the integral, the representation (11.1–4) first appeared in the investigations of Sir George Airy (1801–1892) concerning light propagating in the vicinity of a caustic. Stokes later observed that Airy's integral formally satisfies the differential equation

$$\frac{d^2W}{dz^2} = zW,$$

an equation that we briefly considered in Chapter 3. This particular equation is the simplest example of a differential equation exhibiting turning-point behavior and is fundamental to a thorough understanding of the asymptotics of special functions. (See Erdélyi (1956), Sirovich (1971), Olver (1974).)

The analysis of Airy's integral for small values of z is a relatively straightforward task (see Problem 2). However, aside from saying that the integral tends to zero as z tends to $+\infty$, a precise investigation for large z is more difficult. The point at infinity is a singularity of the analytic function represented by the integral, and in order to handle this case, we will need some background in the techniques of asymptotic analysis. Therefore, we devote the next several sections to this important subject.

11.2 Asymptotics: an introduction

In analytic function theory the emphasis is on well-behaved functions and their points (domains) of analyticity. In asymptotics, on the other hand, the singularities of functions play the key role. Indeed, asymptotics can be characterized as the study of a function in the vicinity of its singularities, a study usually carried out by comparing the given function with other, presumably more elementary, functions whose behavior near the singularities is well-known.

The subject has its origins in the work of Thomas Stieltjes (1856–1894) and Henri Poincaré (1854–1912) on divergent series. Significant contributions were subsequently made by Lord Kelvin, G. H. Hardy (1877–1947), G. N. Watson of Bessel function fame (1886–1965), and Jacques Hadamard. A number of more modern efforts are detailed in the general references listed at the end of this chapter.

Order relations

In the following, z will represent a complex variable restricted to some region R of the complex plane and z_0 a limit point of R, while $f(z)$, $g(z)$, $\phi(z)$, etc. will denote complex functions of z.

We begin by defining the classical order relations O and o .

Definition 1 $f = O(g)$ for z in R if and only if (iff)

$$|f(z)| \leq C|g(z)|$$

for some constant C and all z in R.

Definition 2 $f = O(g)$ as z (in R) $\rightarrow z_0$ iff there exists a neighborhood N of z_0 and a constant C such that

$$|f(z)| \leq C|g(z)|$$

for all z in both N and R.

Definition 3 $f = o(g)$ as z (in R) $\rightarrow z_0$ iff for each $\epsilon > 0$ there exists a neighborhood $N_\epsilon(z_0)$ such that

$$|f(z)| \leq \epsilon|g(z)|$$

for all z in both N_ϵ and R.*

It should be noted that the point z_0 need not be in R. The case where z_0 is on the boundary of R is important for numerous applications. Whenever the constants and neighborhoods appearing in these definitions are independent of any auxiliary parameters that may be present in the problem under consideration, then we will say that the order relations are valid *uniformly*.

As an elementary example of the application of the order relations O and o, consider a function $f(z)$ which is analytic at a point $z = z_0$. In view of its analyticity, the function f has a series representation

$$f(z) = \sum_{n=0}^{\infty} a_n (z - z_0)^n$$

which is convergent for all z in some neighborhood R of z_0. As a consequence, if $f(z)$ has a zero of order $p \geq 0$ at z_0, then $a_0 = a_1 = \cdots = a_{p-1} = 0$, but $a_p \neq 0$ and thus

(i) $f(z) = O\left(\sum_{n=0}^{N} a_n (z - z_0)^n\right)$ as $z \rightarrow z_0$ for each $N \geq p$; on the other hand,

(ii) $f(z) - \sum_{n=0}^{N} a_n (z - z_0)^n = o((z - z_0)^N)$ as $z \rightarrow z_0$ for arbitrary $N \geq 0$.

*The order relation O is reflexive and transitive and thus can be used to generate partial orderings; the relation o obviously is only transitive.

Asymptotic representations and expansions

In order to characterize the behavior of functions near singular points, we need to generalize the notions illustrated in the above special case. The definitions below suffice in this regard:

Definition 4 Let the complex functions $\phi_n(z)$, $n = 0,1,2, \ldots$ be defined in some region R of the complex plane. If for all $n \geq 0$

$$\phi_{n+1} = o(\phi_n) \text{ as } z \to z_0,$$

the sequence $\{\phi_n(z)\}$ is said to form an **asymptotic sequence**. Whenever the order relation is uniform in n, $\{\phi_n(z)\}$ is designated a **uniform** asymptotic sequence.

In addition to the sequence $\{(z - z_0)^n\}$, the following are representative examples of asymptotic sequences:

(i) $\{z^{-n}\}$ as $z \to \infty$;
(ii) $\{z^{-\lambda n}\}$ as $z \to \infty$, $|\arg z| < \pi/2 - \delta$, with arbitrary $\delta > 0$, and $\text{Re } \lambda_{n+1} > \text{Re } \lambda_n$ for each n; and
(iii) $\{\Gamma(z)/\Gamma(z + n)\}$ as $z \to \infty$, $|\arg z| < \pi - \delta$, with arbitrary $\delta > 0$.

(Observe that in these examples, the limit point of interest is the point at infinity.) We also note that if $\{\phi_n (z)\}$ is an asymptotic sequence and α is a positive constant, then $\{|\phi_n (z)|^\alpha\}$ is likewise an asymptotic sequence. So also is the sequence $\{\phi_n \psi_n\}$ formed by taking the term-by-term product of $\{\phi_n(z)\}$ with any other asymptotic sequence $\{\psi_n (z)\}$ (see Problems 3,4).

Definition 5 Given a complex function $f(z)$ and an asymptotic sequence $\{\phi_n(z)\}$ defined in some region R of the complex plane, the expression

$$\sum_{n=0}^{N} a_n \phi_n(z)$$

constitutes an **asymptotic representation** (to $N + 1$ terms) of f near z_0 if

$$f(z) - \sum_{n=0}^{j} a_n \phi_n (z) = o(\phi_j)$$

as z (in R) $\to z_0$ for each $j = 0,1,2, \ldots, N$. If $N = \infty$, we call the infinite series an **asymptotic series** or an **asymptotic expansion** for f.

The following should serve to fix some of the above ideas.

Example Consider the function

$$f(z) = \int_0^\infty \frac{e^{-t}}{1 + zt} dt, \qquad |\arg z| < \pi/2. \tag{11.2–1}$$

Now

$$f(z) = \int_0^\infty e^{-t} \left[\sum_{n=0}^{N} (-zt)^n + \frac{(-zt)^{N+1}}{1+zt} \right] dt$$

$$= \sum_{n=0}^{N} (-1)^n n!\, z^n + z^{N+1} \int_0^\infty \frac{(-t)^{N+1} e^{-t}}{1+zt} dt.$$

The integral in the second term of this last expression is bounded for z in the right half-plane. If we select $\{z^n\}$ as our underlying asymptotic sequence, therefore, with $z_0 = 0$, we see that the second term is o of the last term retained (indeed, it is O of the first term neglected). It follows that

$$\sum_{n=0}^{\infty} (-1)^n n!\, z^n \qquad \textbf{(11.2–2)}$$

is an asymptotic expansion for the function $f(z)$ given by (11.2–1) as $z \to 0$. □

The analysis we have just gone through with this example typifies, in a sense, that encountered in much of asymptotics. An appropriate sequence for expressing the desired result is usually unspecified a priori; it appears as a byproduct of the analytical procedure.* Moreover, the expansion that we obtain in terms of this asymptotic sequence generally turns out to be *divergent*. Some care, therefore, needs to be exercised in using this asymptotic representation to determine approximate numerical values of the function under consideration. Since such numerical calculations constitute an important area of applications in asymptotic analysis, we shall return to these matters later.

Notation

A word or two about notation is in order at this point. The relationship expressed in Definition 5 can be written

$$f(z) = \sum_{n=0}^{N} a_n \phi_n(z) + o(\phi_N)$$

as $z \to z_0$. If $N = \infty$, it is convenient to express this alternatively as

$$f(z) \sim \sum_{n=0}^{\infty} a_n \phi_n(z)$$

and say that f is *asymptotically equivalent* to $\Sigma\, a_n \phi_n$ as $z \to z_0$.

*If the asymptotic sequence is specified beforehand, the task of asymptotic representation becomes one of determining the (unique) coefficients a_n appropriate for the given function.

In actuality, the notion of asymptotic equivalence has a more general rendition:

Definition 6 Given two complex functions $f(z)$, $g(z)$ and an asymptotic sequence $\{\phi_n(z)\}$ well-defined in some region R of the complex plane, f is said to be **asymptotically equivalent** to g as $z \to z_0$ (written $f \sim g$ with respect to $\{\phi_n\}$), if

$$f(z) = g(z) + o(\phi_n) \text{ as } z \to z_0,$$

for all n.

It follows then that $f \sim g$ with respect to $\{\phi_n\}$ if and only if the asymptotic expansions of both f and g in terms of the $\phi_n(z)$ have precisely the *same* coefficients (see Problem 6). Thus, given an asymptotic sequence $\{\phi_n(z)\}$ and a set of coefficients a_n, $\Sigma a_n \phi_n$ defines an *equivalence* class. As examples we note that

(i) $e^{-1/z^2} \sim 0$ with respect to $\{z^n\}$ as $z \to 0$, $|\arg z| < \pi/4 - \delta$ with arbitrary $\delta > 0$; and

(ii) $\dfrac{1}{1+z} + e^{-z} \sim \dfrac{1}{1+z} \sim \displaystyle\sum_{n=1}^{\infty} \frac{(-1)^{n+1}}{z^n}$ with respect to $\{z^{-n}\}$ as $z \to \infty$, $|\arg z| < \pi/2 - \delta$ with arbitrary $\delta > 0$.

11.3 Asymptotic expansion of integrals

Motivated by the expression (11.1–4), we shall be concerned in the remainder of this chapter with the derivation of asymptotic expansions for functions defined by integrals. In many cases where the expansion parameter occurs only in the limits of integration, an asymptotic representation can be easily derived. For example, it is relatively simple to show that

$$\int_x^{\infty} \frac{dt}{e^t + t^2} = e^{-x} - \frac{1}{4}(2x^2 + 2x + 1)e^{-2x} + O(x^4 e^{-3x})$$

as $x \to \infty$. In other situations the expansion parameter is embedded in the integrand and more specialized techniques are needed.

Although certain procedures are applicable to rather large classes of integrals, there does *not* exist a single, universally valid, algorithm. In this section and the next we discuss three well-known approaches: integration by parts, the method of stationary phase, and Laplace's method. We also indicate those types of integrals for which these techniques are particularly useful. For simplicity, we will restrict the expansion parameter z to real values and assume that $z_0 = \infty$ is the limit point of interest. All functions are presumed to be real-valued throughout Sections 11.3 and 11.4.

Integration by parts

Consider the integral

$$I(x) = \int_a^b g(t)\, h(xt) dt, \qquad \textbf{(11.3–1)}$$

where g is differentiable and h is integrable in $[a,b]$. Integration by parts gives rise to

$$I(x) = \left[g(t) \int_{t_o}^t h(x\tau) d\tau \right]_a^b - \int_a^b g'(t) \int_{t_o}^t h(x\tau) d\tau\, dt$$

$$= \frac{g(b)}{x} \int_{xt_o}^{xb} h(\xi) d\xi - \frac{g(a)}{x} \int_{xt_o}^{xa} h(\xi)\, d\xi - \frac{1}{x} \int_a^b g'(t) \int_{xt_o}^{xt} h(\xi) d\xi\, dt.$$

Since we want to integrate by parts several times, the unifying notation

$$f^{(0)}(t) \equiv f(t)$$

$$f^{(-n)}(t) \equiv \int_{t_o}^t f^{(1-n)}(\tau) d\tau \qquad n = 1,2, \ldots$$

will be helpful. With this convention, if g is N times continuously differentiable,

$$I(x) = \sum_{n=0}^{N-1} (-1)^n x^{-n-1} [g^{(n)}(b) h^{(-n-1)}(xb) - g^{(n)}(a) h^{(-n-1)}(xa)]$$

$$+ (-1)^N x^{-N} \int_a^b g^{(N)}(t) h^{(-N)}(xt) dt\,. \qquad \textbf{(11.3–2)}$$

We recognize that $\{x^{-n}\}$ forms an asymptotic sequence as $x \to \infty$. However, whether (11.3–2) constitutes an asymptotic representation for $I(x)$ depends upon whether the following two conditions hold as $x \to \infty$:

(i) $x^{-n-1} h^{(-n-1)}(xt)$ is an asymptotic sequence for $t = a,b$, and

(ii) $\int_a^b g^{(N)}(t)\, h^{(-N)}(xt) dt = o(1)$.

In many applications these conditions turn out to be satisfied. Such is the situation, for example, when $h(xt)$ is the Fourier kernel

$$h(xt) = e^{ixt}$$

and g is sufficiently smooth. The validity of (i) may be trivially verified, while (ii) is a consequence of the Riemann-Lebesgue theorem (see Section 7.2).

Asymptotic representations may still be obtained through integration by parts for integrals of the form (11.3–1) even when the integrands are more complicated. An important example is

$$I(x) = \int_a^b g(t)\,(t-a)^{\lambda-1}\,e^{ixt}\,dt \tag{11.3–3}$$

where $0 < \lambda < 1$. If we define

$$h(t) \equiv (t-a)^{\lambda-1}\,e^{ixt},$$

then

$$h^{(-n-1)}(t) = \frac{(-1)^{n+1}}{n!}\int_t^{t_o} (\tau - t)^n(\tau - a)^{\lambda-1}\,e^{ix\tau}\,d\tau\,.$$

Choosing $t_o = i\infty$, we can easily show that

$$h^{(-n-1)}(t) = (t-a)^{\lambda-1}\,x^{-n-1}\,O(1)$$

(see Problem 8) and hence, for a bounded interval,

$$\int_a^b g^{(N)}(t)\,h^{(-N)}(t)dt = x^{-N}\,O(1).$$

With this choice of t_o, it also follows that

$$h^{(-n-1)}(a) = \frac{(-1)^{n+1}}{n!}\left[\frac{i}{x}\right]^{n+\lambda} e^{ixa}\,\Gamma(n+\lambda)$$

(see Appendix 5). As a consequence, we find that if $g(t)$ is N-times continuously differentiable in the bounded interval $[a,b]$ and is such that $g(b) = g'(b) = \ldots = g^{(N-1)}(b) = 0$, then the integral given by (11.3–3) has the asymptotic representation as $x \to \infty$

$$I(x) = \sum_{n=0}^{N-1} \frac{\Gamma(n+\lambda)}{n!}\left[\frac{i}{x}\right]^{n+\lambda} e^{ixa}\,g^{(n)}(a) + O(x^{-N})\,. \tag{11.3–4}$$

This result has a natural analogue for integrals in which the integrand displays a comparable singularity at the upper endpoint $t = b$. Indeed, the two expressions can be combined for the case when both types of singularities are present, and the awkward endpoint conditions removed. Since Copson (1965) has the details we merely state the result:

Theorem 1:

Let $g(t)$ be N times continuously differentiable in the bounded interval $[a,b]$. If $0 < \lambda,\mu < 1$, then

$$\int_a^b g(t)\,(t-a)^{\lambda-1}\,(b-t)^{\mu-1}\,e^{ixt}\,dt$$

$$= \sum_{n=0}^{N-1} \frac{\Gamma(n+\lambda)\,i^{n+\lambda}}{n!\,x^{n+\lambda}}\,e^{ixa}\,[(b-t)^{\mu-1}\,g(t)]^{(n)}\Big|_{t=a}$$

$$+ \sum_{n=0}^{N-1} \frac{\Gamma(n+\mu)\,i^{n-\mu}}{n!\,x^{n+\mu}}\,e^{ixb}\,[(t-a)^{\lambda-1}\,g(t)]^{(n)}\Big|_{t=b} + O(x^{-N}) \text{ as } x \to \infty\,. \tag{11.3–5}$$

□

An application

The example of Section 11.2 can be treated using integration by parts (Problem 9). However, a more interesting case in which this procedure can be applied is the following.

Example Consider

$$I(x) = \int_{-1}^{1} (t + 1)^{\nu - 1/2} (1 - t)^{\nu - 1/2} e^{ixt} \, dt , \qquad \textbf{(11.3–6)}$$

where ν is assumed real. Recall that this integral appears in the important representation (3.6–6) for the Bessel function $J_\nu(x)$ with $\nu > -1/2$, so an asymptotic representation for (11.3–6) will provide valuable information concerning the behavior of this function for large values of its argument x.

In order to apply Theorem 1 we let $\alpha \equiv [\nu + 1/2]$, the integer part of $\nu + 1/2$, and set $g(t) = (1 - t^2)^\alpha$, thereby specifying $\lambda = \mu = \nu + 1/2 - \alpha$.

It follows by virtue of Leibniz's formula for the derivative of a product that

$$\begin{aligned}\left[(b - t)^{\mu-1} g(t)\right]^{(n)}\bigg|_{t=a} &= \left[(1 - t)^{\nu - 1/2} (1 + t)^\alpha\right]^{(n)}\bigg|_{t=-1} \\ &= \sum_{k=0}^{n} \binom{n}{k} \left[(1 - t)^{\nu - 1/2}\right]^{(k)} \left[(1 + t)^\alpha\right]^{(n-k)}\bigg|_{t=-1} \\ &= (-1)^{n-\alpha} 2^{\nu - 1/2 + \alpha - n} \frac{\Gamma(n + 1)\, \Gamma(\nu + 1/2)}{\Gamma(n - \alpha + 1)\, \Gamma(\nu + 1/2 + \alpha - n)} .\end{aligned}$$

Similarly,

$$\begin{aligned}\left[(t - a)^{\lambda-1} g(t)\right]^{(n)}\bigg|_{t=b} &= \left[(1 + t)^{\nu - 1/2} (1 - t)^\alpha\right]^{(n)}\bigg|_{t=1} \\ &= \sum_{k=0}^{n} \binom{n}{k} \left[(1 + t)^{\nu - 1/2}\right]^{(k)} \left[(1 - t)^\alpha\right]^{(n-k)}\bigg|_{t=1} \\ &= (-1)^{\alpha} 2^{\nu - 1/2 + \alpha - n} \frac{\Gamma(n + 1)\, \Gamma(\nu + 1/2)}{\Gamma(n - \alpha + 1)\, \Gamma(\nu + 1/2 + \alpha - n)} .\end{aligned}$$

Note that both of these expressions are zero if $n < \alpha$. If we substitute these two results into (11.3–5) and simplify, we then find that

$$I(x) \sim \Gamma(\nu + 1/2) \frac{(2/x)^\nu}{\sqrt{2x}} \sum_{n=0}^{\infty} \left(\frac{i}{2x}\right)^n \frac{\Gamma(\nu + 1/2 + n)}{\Gamma(\nu + 1/2 - n)} \left[\frac{(-1)^n i^{\nu + 1/2} e^{-ix} + i^{-\nu - 1/2} e^{ix}}{n!}\right].$$

This representation can be incorporated into (3.6–6). Splitting the summation into even and odd terms, we deduce that

$$J_\nu(x) \sim \sqrt{\frac{2}{\pi x}}\left\{\cos[x - \tfrac{1}{2}\pi(\nu + \tfrac{1}{2})] \sum_{m=0}^{\infty} (-1)^m \frac{\Gamma(\nu + \frac{1}{2} + 2m)}{\Gamma(\nu + \frac{1}{2} - 2m)} \frac{(2x)^{-2m}}{(2m)!}\right.$$

$$\left. + \sin[x - \tfrac{1}{2}\pi(\nu + \tfrac{1}{2})] \sum_{m=0}^{\infty} (-1)^{m+1} \frac{\Gamma(\nu + \frac{1}{2} + 2m + 1)}{\Gamma(\nu + \frac{1}{2} - 2m - 1)} \frac{(2x)^{-2m-1}}{(2m+1)!}\right\}.$$

This expression provides the complete large-argument asymptotic expansion for the Bessel function of the first kind, the first terms of which we discussed earlier in Chapter 3. □

Laplace transforms and Watson's lemma

Another important class of integrals to which integration by parts is directly applicable is the class of so-called Laplace integrals

$$I(x) = \int_a^b g(t)\, e^{-xt}\, dt \quad \text{(a finite)}, \tag{11.3–7}$$

of which the Laplace transform (recall Section 7.3) is a special case. For these integrals the dominant contributions come from evaluations at the lower limit a, and the representation (11.3–2) takes the form

$$I(x) = e^{-ax}\left[\sum_{n=0}^{N-1} g^{(n)}(a)\, x^{-n-1} + o(x^{-N})\right] \tag{11.3–8}$$

as $x \to \infty$.

In analogy with the Fourier integral situation, there are results akin to this one that remain valid when the integrand of (11.3–7) is more complicated and/or less well-behaved. The best known of these is due to Watson:

Watson's Lemma:

Consider the integral

$$I(x) = \int_a^b g(t)\, e^{-xt}\, dt\,,$$

where a is finite and g is the restriction to $(a,b]$ of an analytic function. If $b = \infty$, we also assume that $g(t) = O(e^{\alpha t})$ as $t \to \infty$ for some positive constant α. Let g have the asymptotic development

$$g(t) = \sum_{n=1}^{N} a_n (t - a)^{\lambda_n - 1} + o\left((t - a)^{\lambda_N - 1}\right)$$

as $t \to a^+$ (here $0 < \mathrm{Re}\,\lambda_1 \le \mathrm{Re}\,\lambda_2 \le \cdots \le \mathrm{Re}\,\lambda_N$). Then as $x \to \infty$,

$$I(x) = e^{-ax}\left[\sum_{n=1}^{N} a_n\, \Gamma(\lambda_n)\, x^{-\lambda_n} + o(x^{-\lambda_N})\right]. \tag{11.3–9}$$

(See Sirovich (1971), pp. 66ff, for example.)

Since so many Laplace-type integrals occur in practical applications, Watson's lemma is probably the most frequently used result for deriving asymptotic representations.*

Example As an illustration consider the integral expression

$$J_0(x) = \frac{\sqrt{2}}{\pi} \operatorname{Re} \left\{ e^{i(x - \frac{1}{4}\pi)} \int_0^\infty \frac{e^{-xt}}{\sqrt{t}\sqrt{1 + \frac{1}{2}it}} \, dt \right\}$$

(adapted from Watson (1958), p. 168). In the notation of Watson's lemma we have $a = 0$, $\lambda_n = n - \frac{1}{2}$, and

$$a_n = (\tfrac{1}{2}i)^{n-1} \frac{\Gamma(\frac{1}{2})}{\Gamma(\frac{3}{2} - n)\,\Gamma(n)} .$$

Then it follows that

$$J_0(x) = \sqrt{\frac{2}{\pi x}} \operatorname{Re} \left\{ e^{i(x - \frac{1}{4}\pi)} \sum_{n=1}^{N} (\tfrac{1}{2}i)^{n-1} \frac{\Gamma(n - \frac{1}{2})}{\Gamma(\frac{3}{2} - n)\,\Gamma(n)} x^{1-n} + o(x^{1-N}) \right\}$$

as $x \to \infty$, in agreement with the result already derived in the immediately previous subsection. □

11.4 Critical points and rapidly varying integrands

The technique of integration by parts can be readily applied to integrals of the form

$$I(x) = \int_a^b g(t)\, e^{ixh(t)}\, dt \tag{11.4–1}$$

and

$$I(x) = \int_a^b g(t)\, e^{-xh(t)}\, dt \tag{11.4–2}$$

as long as $h'(t)$ is positive for t in $[a,b]$. We need only introduce the new independent variable $u = h(t) - h(a)$ and rewrite the integrals accordingly. The resulting asymptotic representations show, as expected, that the dominant contributions to the asymptotic developments of the generalized Fourier and Laplace integrals (11.4–1) and (11.4–2) continue to come from the endpoint(s) of the interval $[a,b]$.

However, the situation is radically altered if $h'(t)$ changes sign in $[a,b]$. In such cases the behavior of the integrand in the vicinity of the so-called *critical points*, where $h'(t) = 0$, becomes increasingly important, and analysis beyond that carried out heretofore is needed. The considerations appropriate to the generalized

*One noteworthy class of applications concerns the relation of system transfer functions $H(s)$ to system impulse-responses (recall the discussion in Section 5.1).

Fourier integral (11.4–1) can be traced principally to Lord Kelvin, and the approach is known today as the *stationary phase procedure* (recall the earlier discussion and the footnote on p. 332). Laplace himself suggested how to deal with the critical point behavior of integrals of the form (11.4–2).

The method of stationary phase

The underlying "stationary phase" principle that guides the analysis of (11.4–1) as $x \to \infty$ is the assertion that the dominant effect upon the asymptotics of $I(x)$ is the behavior of g in the immediate neighborhood of the points where the phase of the integrand is stationary. Since h and g are both real-valued, these points of stationary phase are the critical points of h.

In order to assess their contribution more carefully, let us look at the situation where $h(t)$ has a single point t_o in the interval at which $h'(t)$ vanishes.* If t_o is not at either endpoint and $h''(t_o)$ is positive, then, heuristically,

$$\int_{t_o-\epsilon}^{t_o+\epsilon} g(t)\, e^{ixh(t)}\, dt \approx g(t_o) \int_{t_o-\epsilon}^{t_o+\epsilon} e^{ix[h(t_o) + \frac{1}{2}h''(t_o)(t-t_o)^2]}\, dt$$

$$\approx g(t_o)\, e^{ixh(t_o)} \int_{-\infty}^{\infty} e^{\frac{1}{2}ixh''(t_o)(t-t_o)^2}\, dt$$

$$= g(t_o) \sqrt{\frac{2\pi}{xh''(t_o)}}\, e^{i[xh(t_o) + \pi/4]} .$$

We designate this last expression $I_o(x)$.

Whenever g and h are sufficiently well-behaved, a rigorous asymptotic analysis of (11.4–1) can be performed (see Sirovich (1971)). The result specific for the generalized Fourier integral that we are considering is

$$\int_a^b g(t)\, e^{ixh(t)}\, dt = I_o(x) + O(x^{-1}) \qquad \textbf{(11.4–3)}$$

as $x \to \infty$. The stationary phase principle for this situation is thereby substantiated. (In Problem 15 the method of stationary phase is applied to the original integral (11.1–1) for which Kelvin essentially developed the procedure.)

The asymptotic expression given in (11.4–3) can be extended to higher-order critical points and the sign restriction on the first nonvanishing derivative at the critical point removed. As a general result we find that if $h'(t) \neq 0$ for all $t \neq t_o$ with $a < t_o < b$, but $h'(t_o) = h''(t_o) = \cdots = h^{(q-1)}(t_o) = 0$, $h^{(q)}(t_o) \neq 0$, then for sufficiently smooth $g(t)$, $h(t)$

---

*If there is more than one critical point in the interval of interest, the contributions from the separate points are added together to produce the composite contribution.

$$I(x) = \int_a^b g(t)\, e^{ixh(t)}\, dt = 2A_q(x) \begin{cases} \cos \dfrac{\pi}{2q} & , q \text{ odd} \\ e^{\frac{i\pi}{2q} \operatorname{sgn} h^{(q)}(t_o)} & , q \text{ even} \end{cases} + O(x^{-2/q}) \tag{11.4–4}$$

as $x \to \infty$, where

$$A_q(x) \equiv g(t_o)\, e^{ixh(t_o)}\, \Gamma\left(\frac{1+q}{q}\right) \left[\frac{q!}{x|h^{(q)}(t_o)|}\right]^{1/q}.$$

If the stationary point is at the left-hand endpoint of the interval $[a,b]$, the analogue of (11.4–4) is

$$I(x) = A_q(x)\, e^{\frac{i\pi}{2q} \operatorname{sgn} h^{(q)}(a)} + O(x^{-2/q}) \tag{11.4–5}$$

as $x \to \infty$, independent of the parity of q. If $t_o = b$, we find in comparable fashion that

$$I(x) = A_q(x)\, e^{(-1)^q \frac{i\pi}{2q} \operatorname{sgn} h^{(q)}(b)} + O(x^{-2/q}) \tag{11.4–6}$$

as $x \to \infty$. (See Copson (1965) or Sirovich (1971) for details of the above deriviations.)

Large order Bessel functions

Example One of the classic applications of the stationary phase method is to Bessel functions whose argument and order are both large. The specific integral representation that we want to analyze is

$$\begin{aligned} J_\nu(\nu \sec \beta) = {} & (1/\pi) \int_0^\pi \cos[\nu(t - \sec \beta \sin t)]dt \\ & - (\sin \nu\pi)/\pi \int_0^\infty e^{-\nu(t + \sec \beta \sinh t)}\, dt \end{aligned} \tag{11.4–7}$$

with ν large (Watson (1958); recall that we derived the integer-order case of this expression in Section 3.6). For real ν and $0 \leq \beta < \frac{1}{2}\pi$, it is easy to establish that the second integral in (11.4–7) is $O(\nu^{-1})$ as $\nu \to \infty$. The endpoint contributions from the first integral are of this same order, so we need concentrate only on the critical point behavior of this representation.

Case (i) $0 < \beta < \pi/2$

In this case $h(t)$ has a single simple ($q = 2$) critical point occurring at $t = \beta$ and is such that $h(\beta) = \beta - \tan \beta$, $h''(\beta) = \tan \beta$. Thence, using (11.4–4),

$$J_\nu(\nu \sec \beta) = \frac{2}{\pi} \Gamma\left(\frac{3}{2}\right) \left[\frac{2}{\nu \tan \beta}\right]^{1/2} \operatorname{Re}\left\{ e^{i[\nu(\beta - \tan \beta) + \pi/4]} \right\} + O(\nu^{-1})$$

$$= \sqrt{\frac{2}{\nu\pi \tan\beta}} \cos[\nu(\beta - \tan\beta) + \pi/4] + O(\nu^{-1}) \qquad \textbf{(11.4–8)}$$

as $\nu \to \infty$. Readers should note the nonuniformity with respect to β of this asymptotic representation when β is near 0. (For another example, see Problem 16.)

Case (ii) $\beta = 0$

In this case the critical point of $h(t)$ has coalesced with the endpoint $t = 0$, and consequently $h'(0) = h''(0) = 0$ and $h'''(0) = 1$. It follows then from (11.4–5) that

$$J_\nu(\nu) = \frac{1}{\pi}\Gamma\left(\frac{4}{3}\right)\left[\frac{6}{\nu}\right]^{1/3} \mathrm{Re}\,(e^{i\pi/6}) + O(\nu^{-2/3})$$

$$= \frac{\sqrt{3}}{2\pi}\Gamma\left(\frac{4}{3}\right)\left[\frac{6}{\nu}\right]^{1/3} + O(\nu^{-2/3}) \qquad \textbf{(11.4–9)}$$

as $\nu \to \infty$. This expression should be compared with that obtained above for $\beta > 0$. □

Laplace's method

Unlike the situation with Fourier integrals, two critical points, even when they are of the same order, need not contribute comparably to the asymptotic representation of the generalized Laplace integral

$$I(x) = \int_a^b g(t)\, e^{-xh(t)}\, dt \qquad \textbf{(11.4–2')}$$

as $x \to \infty$. The dominant terms in the asymptotic development of this integral come from the neighborhoods of only those points at which h attains its absolute minimum on $[a,b]$, and not every critical point of h need be associated with such a minimum of h. For a function h with a single interior minimum t_o at which $h'(t_o) = h''(t_o) = \cdots = h^{(2q-1)}(t_o) = 0$, while $h^{(2q)}(t_o) > 0$, we find an analogue of (11.4–4):

$$I(x) = e^{-xh(t_o)}\left[2B_{2q}(x) + O\left(x^{-3/(2q)}\right)\right] \qquad \textbf{(11.4–10)}$$

as $x \to \infty$, where

$$B_{2q}(x) \equiv g(t_o)\,\Gamma\left(\frac{1 + 2q}{2q}\right)\left[\frac{(2q)!}{{}_{xh}(2q)_{(t_o)}}\right]^{1/(2q)}$$

If the absolute minimum of h in (11.4–2) is at either endpoint of the interval, then (11.4–10) must be modified by dropping the 2 factor from in front of $B_{2q}(x)$ and changing the error term to $O(x^{-1/q})$.* If h has several relative minima t_i in $[a,b]$, then only that one (those) for which $h(t_i)$ is smallest need be considered.

*This assumes $h'(t)$ vanishes at the endpoint in question. If it does not, then integration by parts must be used.

The Gamma function of large argument

Example The well-known integral representation

$$\Gamma(1 + x) = \int_0^\infty e^{-\tau}\, \tau^x \, d\tau \qquad \textbf{(11.4–11)}$$

for the gamma function provides a nice illustration for Laplace's method. The substitution $\tau = xt$ transforms (11.4–11) into

$$\Gamma(1 + x) = e^{(x+1)\ln x} \int_0^\infty e^{-x(t - \ln t)} \, dt \, ,$$

which we recognize as having the form (11.4–2) with $g(t) = 1$ and $h(t) = t - \ln t$. Since $h(t)$ has a single minimum in $[0,\infty)$ at the interior point $t = 1$ with $h(1) = 1$ and $h''(1) = 1$, the expression (11.4–10) may be used in the form given. The result is the asymptotic representation

$$\Gamma(1 + x) = \sqrt{2\pi}\, e^{(x+\frac{1}{2})\ln x - x} \, [1 + O(x^{-1})] \qquad \textbf{(11.4–12)}$$

as $x \to \infty$. The essential term in this expression is equivalent to the dominant term in an infinite series for $\ln \Gamma(1 + x)$ first stated in 1730 by the British scientist James Stirling (1696–1770). (See Appendix 5.) □

11.5 The saddle-point method and paths of steepest descent

Heretofore we have confined our discussion to integrals of real-valued functions. Since this makes the analysis of a number of important examples unduly cumbersome, now we want to remove this restriction. In this section then, we will concern ourselves with the integral

$$I(x) = \int_C g(t)\, e^{-xh(t)} \, dt \qquad \textbf{(11.5–1)}$$

where $g(t)$ and $h(t)$ are complex functions of a complex variable t; indeed we assume that they are analytic functions of t in some region of the complex t-plane. We presume C to be a contour within this region.

The technique we shall consider is generally attributed to the Dutch physical chemist and Nobel laureate Peter Debye (1884–1966), although the approach was implicit in some earlier work of Riemann. Since $e^{-ix \,\mathrm{Im}\, h}$ oscillates and $e^{-x \,\mathrm{Re}\, h}$ grows/decays exponentially, our present analysis naturally intersects with our previous remarks regarding the stationary phase procedure and Laplace's method. The critical points of $h(t)$ and the behavior of both g and h in the vicinity of these special points are as essential to the asymptotic analysis of (11.5–1) as they were to our evaluation of the dominant contributions to the generalized Fourier and Laplace integrals (11.4–1), (11.4–2).

Steepest descent contours

Let us assume that at a point t_o, $h(t)$ has a critical point of order $q \geq 2$.* In other words, we presume $h(t)$ is such that

$$h'(t_o) = h''(t_o) = \cdots = h^{(q-1)}(t_o) = 0$$

while

$$h^{(q)}(t_o) = \alpha\, e^{i\beta}$$

with $\alpha > 0$, β real and q integral. If t is given by $t = t_o + \rho e^{i\phi}$ with $\rho > 0$, it follows that

$$h(t) = h(t_o) + (\rho^q/q!)\,\alpha\, e^{i(\beta + q\phi)} + O(\rho^{q+1}) \qquad \textbf{(11.5–2)}$$

as $\rho \to 0$. As a consequence, we can easily determine the directions of the curves Re h = constant and Im h = constant as they emanate from the point t_o. The directions of the former are given by the solutions of the equation cos $(\beta + q\phi) = 0$, namely,

$$\phi = -\beta/q + (m + \tfrac{1}{2})\pi/q \qquad m = 0, 1, \ldots, 2q - 1$$

(see Figure 11.1). In analogous fashion, the directions of the latter satisfy sin $(\beta + q\phi) = 0$, or

$$\phi = -\beta/q + m\pi/q \qquad m = 0, 1, \ldots, 2q - 1.$$

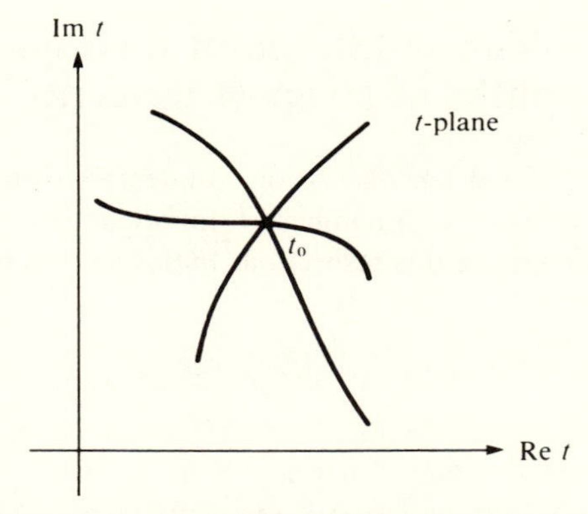

Figure 11.1 **Illustration of the projection in the t-plane of the curves of constant Re h near a critical point with $q = 3$**

However, in view of the form of the relationship (11.5–2), these are also the directions of the curves along which Im h and Re h, respectively, change most rapidly. For odd m these are the so-called contours of "steepest descent"; for even m they are the "steepest ascent" contours.

---

*If $q = 1$, there is no critical point.

The situation we have just described is the generalization of what normally occurs at a noncritical point. If $h'(t_o) \neq 0$, the curves Re h = constant and Im h = constant intersect orthogonally there. Since the gradient vector for each of Re h, Im h is then tangent to the level curves of the other function, it follows that, amongst all curves through the point of intersection, Re h (Im h) changes most rapidly along the curve of constant Im h (Re h).

If $h'(t_o) \neq 0$, the immediate neighborhood of t_o can be viewed as the union of two sectors, each with angular opening π at t_o; in one of the sectors Re h (Im h) is increasing, in the other it is decreasing. In the more general case of a critical point of order q, the immediate neighborhood of the point consists of $2q$ sectors, each with angular opening π/q at t_o. Re h (Im h) is alternately decreasing then increasing in adjoining sectors.

The case $q = 2$ can be easily displayed pictorially. If either Re h or Im h is graphed as a function of t in the neighborhood of t_o, the resultant three-dimensional figure resembles a saddle. In view of this, (complex) critical points are often termed *saddle points*. If $q = 3$, there are 3 ascent regions and 3 descent regions. The surface formed by Re h (Im h) as a function of t near t_o in this case is called a "monkey saddle" for obvious reasons (see Figure 11.2).

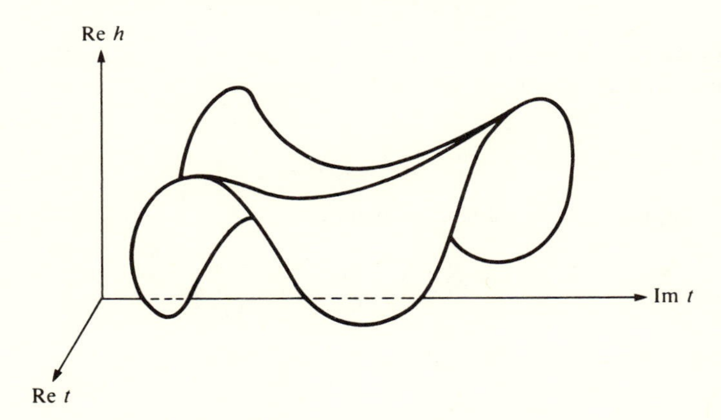

Figure 11.2 **Sketch of the real part of a typical h function in the vicinity of a critical point with $q = 3$**

The method of steepest descent

Now we are ready to describe the procedure proposed by Debye for the asymptotic representation, as $x \to \infty$, of the integral $I(x)$ given by (11.5–1). This technique is known variously as the *saddle-point method* or the *method of steepest descent*.* For

*Copson (1965) uses the term saddle-point method to refer to a slightly more general approach; see also Olver (1974).

the present we assume that the endpoints of the integration contour C are at infinity and $|h| \to \infty$ as $|t| \to \infty$ on C. Then the algorithm is as follows:

(i) find the saddle-points (critical points) of $h(t)$;

(ii) deform the contour of integration C into an (the) appropriate level curve(s) of Im h (steepest descent path for Re h) that pass(es) through one (or more) of these saddle points;

(iii) select a new (real) variable of integration along the new contour (Re h is often a convenient choice) and use Laplace's method.

An example should clarify the technique.

Example Consider the complex integral

$$I(x) = \int_{-\infty}^{\infty} e^{ix(T+t^3/3)}\,dt. \qquad \textbf{(11.5–3)*}$$

Here $h(t) = -i(t + t^3/3)$, and its saddle points are easily seen to be at $t = \pm i$. The level curve Im h = constant through the saddle point at $t = i$ is the hyperbolic arc

$$\operatorname{Im} t = \sqrt{1 + (\operatorname{Re} t)^2/3}\,.$$

If we shift the path of integration in (11.5–3) to this new contour and take τ as our integration variable, where

$$\begin{aligned}\tau^2 &= -\,i(t + t^3/3) - 2/3\\ &= -\,i(t - i)^2(t + 2i)/3,\end{aligned}$$

then the integral (11.5–3) can be rewritten as

$$I(x) = 2i\,e^{-2x/3}\int_{-\infty}^{\infty}\frac{\tau\,e^{-x\tau^2}}{1 + t^2\,(\tau)}\,d\tau \qquad \textbf{(11.5–4)}$$

(see Problem 19). Then it follows from (11.4–10) that as $x \to \infty$

$$\begin{aligned}I(x) &= e^{-2x/3}\left[4i\Gamma\left(\frac{3}{2}\right)\left(\frac{1}{x}\right)^{1/2}\lim_{\tau\to 0}\left(\frac{\tau}{1 + t^2}\right) + O(x^{-3/2})\right]\\ &= \sqrt{\pi/x}\,e^{-2x/3}\,[1 + O(x^{-1})]. \qquad \textbf{(11.5–5)}\end{aligned}$$

□

It turns out that $I(x)$ as given in the above example can easily be related to the Airy integral of (11.1–4) (which we encountered in our earlier discussion of water wave dispersion) and hence to the Airy function of the first kind. Indeed, making an obvious change of variables, we find

---

*Owing to the unboundedness of the domain of integration, an asymptotic representation for this integral cannot be obtained by merely changing the variable of integration and using integration by parts.

$$\pi Ai(z) = \int_0^\infty \cos(sz + \tfrac{1}{3} s^3)ds = \tfrac{1}{2} z^{1/2} I(z^{3/2}).$$

Then it follows that

$$Ai(z) = \frac{z^{-1/4}}{2\sqrt{\pi}} e^{-(2/3)z^{3/2}} [1 + O(z^{-3/2})]$$

as $z \to \infty$. Since large z is equivalent to large distance from the disturbance point, this last result has immediate and significant implications for the "far out" water waves arising from isolated disturbances, as modelled in Section 11.1. Not only do the shorter (capillarity) waves provide the dominant contribution to these dispersive wave trains, but the effect of any longer (gravity) waves is exponentially small.

General remarks

Two observations about the method we have just described and applied are in order. First, the technique may have to be modified if either of the endpoints of the integration contour C in (11.5–1) is finite. In this case, the finite endpoint(s) need not lie on steepest descent paths passing through a critical point of $h(t)$—in fact, $h(t)$ need not have any critical points. Nevertheless, the level curves of Re h and Im h can be ascertained and an alternative contour joining the endpoints can be determined. If this new contour is composed of segments of curves Re h = constant and Im h = constant, the asymptotic evaluation of the integral in question becomes nothing more than a series of stationary phase and Laplace calculations and (occasionally) a steepest descent evaluation (for further details, see Sirovich (1971), Section 2.6 or Bleistein and Handelsman (1975)).*

Second, regardless of the position of the endpoints of integration, care must be taken to account for any singularities of the integrand that are encountered in the process of shifting from one integration contour to another. The contributions to the integral arising from poles, branch points and branch cuts that the contour either "encloses" or intersects must also be taken into consideration.

11.6 Numerical evaluation using asymptotic expansions

Approximations obtained by the method of steepest descent or by other asymptotic procedures oftentimes turn out to be exceedingly accurate for the purpose of numerical calculation. In the final section of this chapter we explore briefly the utility of asymptotic representations of integrals as an alternative to numerical quadrature.

*Sometimes when the endpoints of the integration contour lie in the same "ascent" or "descent" region, a finite path of integration can be conveniently replaced by two infinite contours joined at infinity. Problem 22 provides a case in point.

In Section 11.2 we showed that the integral

$$f(x) = \int_0^\infty \frac{e^{-t}}{1 + xt}\,dt \tag{11.6–1}$$

has the asymptotic expansion

$$\sum_{n=0}^{\infty} (-1)^n\, n!\, x^n \tag{11.6–2}$$

as $x \to 0^+$. Now we note that, for small x, the magnitude of the terms in this divergent series initially decrease. It is reasonable to inquire, therefore, whether a partial sum of (11.6–2) through only the first few terms might yield a reasonably accurate numerical approximation of the value of (11.6–1), at least for some x. As the results in Table 1 suggest, such is indeed the case. If we terminate our summation at the smallest term, we are stopping with a value of S_N that is progressively better as x gets smaller, but is already acceptable for many purposes when x is as large as 0.1.

Table 1: Comparison of actual and approximate values of the integral (11.6–1)

x	$f(x)$	S_0	S_1	S_2	S_3	S_4	S_5
1	0.596	1	0	2	−4	20	−100
0.5	0.722	1	0.5	1	0.25	1.75	−2
0.25	0.825	1	0.75	0.875	0.781	0.875	0.758
0.1	0.916	1	0.90	0.92	0.914	0.916	0.915

General remarks for alternating series

The above behavior typifies what occurs with real-valued functions f whose asymptotic expansions have terms that alternate in sign. If we designate by E_N the error incurred by using as a numerical approximation for f the partial sum S_N of the function's asymptotic expansion through $(N + 1)$ terms, then we can write

$$f = \sum_{n=0}^{N} (-1)^n C_n + E_N \equiv S_N + E_N .$$

Here it is assumed that both f and the various C_ns are positive. If the expansion for f is at all meaningful, $S_0 = C_0$ is larger than f, $S_1 = C_0 - C_1$ is smaller, $S_2 = C_0 - C_1 + C_2$ is again larger, and so forth. It follows that the E_N alternate in sign and

$$|E_N| + |E_{N+1}| = C_{N+1}. \tag{11.6–3}$$

This last relation may be enunciated by saying that "the error incurred in using a partial summation is less than the first term neglected." (It is also obviously less than the last term retained in the summation.) If these terms are small, the numerical value S_N can be excellent approximation to the actual value of the function under consideration.

In many situations, near the smallest term in an asymptotic expansion the errors E_N and E_{N+1}, although of opposite sign, are of comparable magnitude. As a consequence, $|E_N| \approx \frac{1}{2} C_{N+1}$, so a better approximation to f should result from adding to S_N one-half of the next term. From an implementation point of view, this is equivalent to averaging S_N and S_{N+1}. If we were to use this approach with the function given by (11.6–1) when $x = 0.25$, we would have calculated $S_2 \doteq 0.875$, $S_3 \doteq 0.781$ and thence $\frac{1}{2}(S_2 + S_3) \doteq 0.828$. The latter serves as a substantially more accurate approximation to the actual value of 0.825.

Example An even more remarkable illustration of the efficacy of this approximate formula is provided by Stirling's representation for the logarithm of the gamma function of large argument:

$$\ln \Gamma(x) \sim (x - ½) \ln x - x + ½ \ln (2\pi) + \left[\frac{1}{12x} - \frac{1}{360x^3} + \frac{1}{1260x^5} - \frac{1}{1680x^7} + \cdots\right],$$

as $x \to \infty$ (recall (11.4–12); also see Appendix 5). For $x = 1$ we find as approximations for $\ln \Gamma(1) = 0$,

$$S_0 = -0.08106, \quad S_1 = 0.00227, \quad S_2 = -0.00051,$$
$$S_3 = 0.00029, \quad S_4 = -0.00031,$$

and

$$\frac{1}{2}(S_3 + S_4) = 0.00001.$$

The accuracy of this last result is rather surprising in view of the fact that few would consider 1 to be a large number in this context. □

General remarks for single-sign series

When a real-valued function f has an asymptotic expansion whose terms are of the same sign (assumed positive), early partial sums generally underestimate the function while later ones overestimate it. Thus, if

$$f = \sum_{n=0}^{N} C_n + E_N = S_N + E_N,$$

then the "error" E_N must undergo a change of sign for some value N_o of N. (This sign reversal often occurs at or near the term C_n of smallest size.) It follows that

(11.6–3) is still valid in this case for $N = N_o$, and we can say once again that stopping here incurs an error less than the first term neglected. Indeed, we have the precise expression

$$f = \sum_{n=0}^{N_o} C_n + \sigma C_{N_o} + 1 ,$$

where $0 \leq \sigma \leq 1$.

However, unlike the alternating series case, there is no convenient approximate expression (such as $\sigma \approx 1/2$) for the so-called "converging factor" σ, and significant effort has been expended since the time of Stieltjes in attempting to develop various formulas and/or calculation procedures that allow σ to be estimated in various practical situations. Unfortunately, the nature of these considerations is beyond the scope of this book, and we will have to refer the interested reader to discussions such as those found in Olver (1974) or Dingle (1973).

References

Abramowitz, M. and I. A. Stegun (1965): *Handbook of Mathematical Functions*, Dover, New York; NBS Applied Math Series 55, U.S. Dept. of Commerce, Washington, D.C. (1964); Chapter 10.

Bleistein, N. and R. A. Handelsman (1975): *Asymptotic Expansions of Integrals*, Holt, Rinehart and Winston, New York.

Copson, E. T. (1965): *Asymptotic Expansions*, University Press, Cambridge.

Dingle, R. B. (1973): *Asymptotic Expansions: Their Derivation and Interpretation*, Academic Press, London.

Erdélyi, A. (1956): *Asymptotic Expansions*, Dover, New York.

Erdélyi, A. et al (1953): *Higher Transcendental Functions, Vol. 2*, McGraw-Hill, New York; Chapter VII.

Jeffreys, H. (1962): *Asymptotic Approximations*, Clarendon Press, Oxford.

Jeffreys, H. and B. Jeffreys (1962): *Methods of Mathematical Physics*, 3rd ed., University Press, Cambridge; Chapter 17.

Olver, F. W. J. (1974): *Asymptotics and Special Functions*, Academic Press, London; Chapters 1, 3, 4, 14.

Sirovich, L. (1971): *Techniques of Asymptotic Analysis*, Applied Math Sciences Series Vol. 2, Springer-Verlag, New York.

Watson, G. N. (1958): *A Treatise on the Theory of Bessel Functions*, 2nd ed., University Press, Cambridge; Chapter VI.

Whitham, G. B. (1974): *Linear and Nonlinear Waves,* Wiley-Interscience, New York; Chapters 1, 11–15.

Problems

Sections 11.1, 11.2

1. Analyze the group velocity $C \equiv d\gamma/dk$ which is engendered by the empirical formula (11.1–2). In particular, show that for $k > 0$ there is a unique (positive) minimum at some point k_0 with C monotonically decreasing (increasing) for k to the left (right) of k_0.

★2. The definite integral appearing in the representation (11.1–4) can be expressed in terms of the Airy function of the first kind as

$$I(z) = \int_0^\infty \cos(sz + \tfrac{1}{3}s^3)ds = \pi Ai(z).$$

Use the fact that

$$I(z) = \text{Re} \lim_{\alpha \to 0} \int_0^\infty e^{-\alpha s^3} e^{i(sz + \frac{1}{3}s^3)} ds$$

to derive the first several terms in the power series expansion, about $z = 0$, of $Ai(z)$. [Compare with the ascending series given, say, in Abramowitz and Stegun (1965).]

3. (a) Show that the following are asymptotic sequences:
   (i) $\{z^{-n}\}$ as $z \to \infty$;
   (ii) $\{z^{-\lambda_n}\}$ as $z \to \infty$, $|\arg z| < \pi/2 - \delta$, with arbitrary $\delta > 0$, $\text{Re}\, \lambda_{n+1} > \text{Re}\, \lambda_n$ for each n;
   (iii) $\{\Gamma(z)/\Gamma(z + n)\}$ as $z \to \infty$, $|\arg z| < \pi - \delta$, with arbitrary $\delta > 0$.

   (b) Verify that if α is a positive constant, $\{\phi_n(z)\}$ is an asymptotic sequence if and only if $\{|\phi_n(z)|^\alpha\}$ is an asymptotic sequence.

4. (a) Establish that the term-by-term product $\{\phi_n \psi_n\}$ of two asymptotic sequences $\{\phi_n(z)\}$, $\{\psi_n(z)\}$ is an asymptotic sequence.

   (b) Two infinite sequences $\{\phi_n(z)\}$, $\{\psi_n(z)\}$ are said to be *equivalent* if $\phi_n = O(\psi_n)$ and $\psi_n = O(\phi_n)$ for each n. Prove that if $\{\phi_n\}$ and $\{\psi_n\}$ are equivalent, $\{\phi_n(z)\}$ is an asymptotic sequence if and only if $\{\psi_n(z)\}$ is an asymptotic sequence.

5. Consider the function

$$f(z) = (1 + z)^{1/z}.$$

   (a) If the underlying asymptotic sequence is chosen to be $\{z^n\}$, determine the asymptotic representation of $f(z)$ to 3 terms.

   (b) Develop the first several terms of an asymptotic representation for $f(z)$ which is valid for $z \to \infty$, $|\arg z| < \pi/2$.

6. (a) Show that if $\{\phi_n\}$ is an asymptotic sequence as $z \to z_0$ and

$$f(z) \sim \sum_{n=0}^{\infty} a_n \phi_n(z)$$

as $z \to z_0$, then for each N,

$$f(z) = \sum_{n=0}^{N} a_n \phi_n(z) + O(\phi_{N+1})$$

as $z \to z_0$.

(b) Prove that $f \sim g$ as $z \to z_0$, with respect to a given asymptotic sequence $\{\phi_n(z)\}$, if and only if the asymptotic expansions of both f and g in terms of the $\phi_n(z)$ have precisely the same coefficients.

Section 11.3

7. Find asymptotic expansions for

$$f(z) \int_0^z \frac{\sin t}{t} dt$$

that are valid when

(a) $z \to 0$; (b) $z = x \to \infty$.

★8. Carefully verify the intermediate steps that lead to the expression (11.3–4) as the asymptotic representation for the integral (11.3–3).

9. Apply integration by parts to

$$I(x) = x \int_0^\infty \frac{e^{-t}}{x + t} dt$$

to deduce an asymptotic expansion, valid as $x \to \infty$, which is equivalent to the expression (11.2–2).

10. The integral

$$I(x) = \int_0^x e^{i\xi^2} d\xi = \tfrac{1}{2}x \int_0^1 t^{-1/2} e^{ix^2 t} dt$$

is related to the so-called Fresnel integrals.

(a) Rewrite $I(x)$ as

$$I(x) = \tfrac{1}{2}x \int_0^\infty t^{-1/2} e^{ix^2 t} dt - \tfrac{1}{2}x \int_1^\infty t^{-1/2} e^{ix^2 t} dt$$

and then use the integration by parts formula (11.3–4) to derive an asymptotic representation for $I(x)$ valid when $|x| \to \infty$, x real.

(b) Compare the result obtained in part (a) with the one that follows from applying Theorem 1 of Section 11.3 directly to $I(x)$.

★11. Provide a formal proof of the validity of the Laplace integral representation (11.3–9).

12. Utilize (11.3–8)/(11.3–9) to derive asymptotic expansions for

(a) $\displaystyle\int_1^\infty t^\alpha e^{-xt} dt$ (b) $\displaystyle\int_0^1 e^{x(t - 1/t)} dt$

that are valid as $x \to \infty$.

Section 11.4

13. Use a heuristic argument to verify the form of the dominant terms in the asymptotic representations (11.4–4), (11.4–5), and (11.4–6).

14. Apply the method of stationary phase to the integral

$$I(x) = \int_0^1 e^{ixt^3}\,dt.$$

15. Analyze in general terms the integral (11.1–1) as $x,t \to \infty$ with x/t = constant, using the method of stationary phase. Note the role of the group velocity $C \equiv d\gamma/dk$.

16. (a) Obtain the leading term in the asymptotic representation of

$$I(x) = \int_{-\infty}^{\infty} e^{ix(at - \frac{1}{3}t^3)}\,dt \qquad (a \text{ real})$$

as $x \to \infty$. (Consider both $a = 0$ and $a \neq 0$.)

(b) Discuss the behavior of this lead term as the parameter $a \to 0$.

17. Apply Laplace's method to the integral

$$\int_0^1 e^t \left[\frac{t}{1 + t^2}\right]^x dt\,.$$

18. (a) Show formally that if

$$g(t) = \alpha(t - a)^\mu + o((t - a)^\mu)$$
$$h(t) = h(a) + \beta(t - a)^\nu + o((t - a)^\nu)$$

as $t \to a$, with $\nu,\beta > 0$ and $\mu > -1$, then whenever $h(a) < h(t)$ for all t in $(a,b]$,

$$\int_a^b g(t)\, e^{-xh(t)} \sim (\alpha/\nu)\Gamma\left[\frac{1 + \mu}{\nu}\right](\beta x)^{-(1+\mu)/\nu} e^{-xh(a)}$$

as $x \to \infty$.

(b) Use the result of part (a) to determine an asymptotic representation for the integral

$$\int_0^1 e^{(x/t)\sin t}\, t^\lambda\, dt\,,$$

where $\lambda > -1$.

Sections 11.5, 11.6

19. Carefully analyze the complex integral (11.5–3)

$$I(x) = \int_{-\infty}^{\infty} e^{ix(t + t^3/3)}\,dt$$

with an eye towards application of the method of steepest descent. In particular

(a) Sketch the level curves of Re h and Im h that pass through the saddle points of

$$h(t) = -i(t + t^3/3).$$

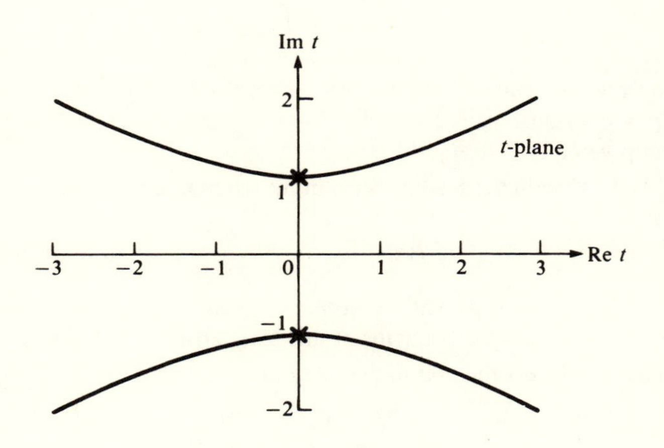

(b) Show that the path of integration can be deformed into the contour Im $h = 0$ passing through $t = i$, but that it cannot be shifted to the level curve of Im h passing through $t = -i$.

(c) Verify that the change of integration variable that leads to (11.5–4) is allowable and is, in essence, equivalent to selecting Re h as the variable of integration along the new contour.

20. Utilize the method of steepest descent in order to obtain the lead term in the asymptotic development of

$$I(x) = \int_0^{\pi i} \frac{e^{xt^2}}{\sqrt{t}}\, dt$$

as $x \to \infty$.

21. For positive order and argument the Hankel function of the first kind has the integral representation

$$H_{ka}^{(1)}(ka) = \frac{1}{\pi} \int_{-\pi+\delta+i\infty}^{\pi-\delta-i\infty} e^{ik[x \cos z + a(z-\pi/2)} \, dz \, ,$$

where $k > 0$, $x > 0$, $0 \le (a/x) < 1$, and $0 < \delta < \pi$. (See Erdélyi et al. (1953).) Apply the method of steepest descent to this integral to derive an asymptotic approximation for the Hankel function that is valid as $k \to \infty$.

22. Consider the integral

$$I(x) = \int_0^2 e^{ixt^2}\, dt \, .$$

(a) By means of the method of stationary phase, determine an asymptotic representation for $I(x)$, valid as $x \to \infty$.

(b) Alternatively, replace the finite contour of integration in $I(x)$ by two steepest descent paths joined at infinity, and then use the method of steepest descent to derive the same asymptotic approximation.

23. The first term of an asymptotic expansion for $Ai(x)$, the Airy function of the first kind, was derived in Section 11.5. More generally,

$$Ai(x) \sim \tfrac{1}{2}\pi^{-1/2} x^{-1/4} e^{-\xi} L(-\xi)$$

as $x \to \infty$, where $\xi = (2/3)x^{3/2}$ and

$$L(t) \equiv \sum_{n=0}^{\infty} a_n t^{-n} ,$$

wherein

$$a_n \equiv \frac{(2n + 1)(2n + 3)(2n + 5) \cdots (6n - 1)}{n!\ 6^{3n}} .$$

(a) Apply the ideas developed in Section 11.6 in order to assess numerically the quality of approximation provided by partial sums of this alternating sign expansion. Reasonable values to investigate include

$$Ai(1) \doteq 0.1353, \quad Ai(1.5) \doteq 0.0717, \quad \text{and} \quad Ai(2) \doteq 0.0349.$$

(b) $Bi(x)$, the Airy function of the second kind, has the related expansion

$$Bi(x) \sim \pi^{-1/2} x^{-1/4} e^{\xi} L(\xi)$$

as $x \to \infty$, where ξ and L are given above. Perform a comparable analysis with this single-sign series, and contrast the results so obtained with those determined in part (a). For $x = 1, 1.5$, and 2 the corresponding values of $Bi(x)$ are 1.207, 1.879, and 3.298, respectively.

Appendix 1

Matrix fundamentals

Basic definitions

Definition 1 A **matrix** $A \equiv [a_{ij}]$ is an $m \times n$ rectangular array of (real or complex) numbers consisting of m rows and n columns. The element a_{ij} sits in the ith row, $1 \leq i \leq m$, and the jth column, $1 \leq j \leq n$. If $m = n$, the matrix is said to be **square** and of **order** n. A matrix for which $n = 1$ is occasionally termed a **column vector**.

Definition 2 Two $m \times n$ matrices A and B are equal if $a_{ij} = b_{ij}$ for all i, j. We write $A = B$ in this case. Similarly, $A \geq B$ ($A \leq B$) if $a_{ij} \geq b_{ij}$ ($a_{ij} \leq b_{ij}$) for all i, j.

Definition 3 If A is an $m \times n$ matrix and α is a scalar (real or complex number), then the elements of the $m \times n$ matrix $C \equiv \alpha A$ are given by $c_{ij} = \alpha a_{ij}$ for all i, j.

Definition 4 If A and B are two $m \times n$ matrices, then the elements of the $m \times n$ sum and difference matrices $C \equiv A \pm B$ are given by $c_{ij} = a_{ij} \pm b_{ij}$ for all i, j.

Definition 5 If A is an $m \times p$ matrix and B is a $p \times n$ matrix, then the elements of the $m \times n$ product matrix $C \equiv AB$ are given by $c_{ij} = \sum_{k=1}^{p} a_{ik} b_{kj}$ for all i, j.

Property 1
Matrix addition is both associative and commutative.

Property 2
Multiplication of matrices by scalars is associative, commutative, and distributive with respect to both scalar and matrix addition.

Property 3
Matrix multiplication is associative and distributive with respect to matrix addition, but it is *not* commutative.

Definition 6 The **complex conjugate** of a given $m \times n$ matrix A is the $m \times n$ matrix $\overline{A} \equiv [b_{ij}]$ whose elements satisfy $b_{ij} = \overline{a_{ij}}$ for all i, j.

Definition 7 The **transpose** and **conjugate transpose** of a given $m \times n$ matrix A are the $n \times m$ matrices $A' \equiv [b_{ij}]$ and $A^\star \equiv [c_{ij}]$ whose elements satisfy $b_{ij} = a_{ji}$ and $c_{ij} = \overline{a_{ji}}$, respectively, for all i, j.

Property 4

$$\begin{aligned}(A^\star)^\star &= A\\ (\alpha A)^\star &= \overline{\alpha} A^\star\\ (A \pm B)^\star &= A^\star \pm B^\star\\ (AB)^\star &= B^\star A^\star.\end{aligned}$$

Comparable relations are valid for A' and, with the exception that $\overline{(AB)} = \overline{A}\,\overline{B}$, for $\overline{A}$.

Fundamental special matrices

Definition 8 The $m \times n$ **zero matrix** is that matrix $\mathbf{0} \equiv [a_{ij}]$ whose elements satisfy $a_{ij} = 0$ for all i, j.

Definition 9 An $n \times n$ square matrix $A \equiv [a_{ij}]$ is called **upper** or **right** (**lower** or **left**) **triangular** if $a_{ij} = 0$ for all $i > j$ ($i < j$); it is termed **diagonal** if $a_{ij} = 0$ for all $i \neq j$, that is, $D \equiv [\delta_{ij} d_j]$ where δ_{ij} is the familiar Kronecker delta.

Definition 10 The special diagonal matrix $I \equiv [\delta_{ij}]$ is designated the **identity matrix**.

Property 5

$A + \mathbf{0} = A = \mathbf{0} + A$

$A - A = \mathbf{0}$

Property 6

The product of two diagonal matrices (triangular matrices of the same type) is diagonal (triangular).

Property 7

Matrix multiplication of diagonal matrices is commutative.

Property 8

$AI = A = IA$

Other special matrices

Definition 11 A square matrix A of order n with real elements such that $A' = A$ is called **symmetric**; if its elements are complex it is termed **hermitian** or **complex-symmetric*** whenever $A^\star = A$ or $A' = A$, respectively.

Definition 12 A square matrix A of order n which satisfies $A^\star = -A$ ($A' = -A$ in the real case) is called **skew-hermitian (skew-symmetric)**.

Definition 13 A matrix A such that

$$AA^\star = I = A^\star A$$

($AA' = I = A'A$ in the real case) is termed **unitary (orthogonal)**.

Definition 14 A **normal** matrix A is one which commutes with its conjugate transpose, that is, $AA^\star = A^\star A$.

Property 9
The elements on the main diagonal of a (skew) hermitian matrix are (zero) real.

Property 10
Since $A = \frac{1}{2}(A + A^\star) + \frac{1}{2}(A - A^\star)$, every square matrix can be decomposed into the sum of hermitian (symmetric) and skew-hermitian (skew-symmetric) parts.

Property 11
Hermitian, skew-hermitian, and unitary matrices are all normal.

Determinants

Definition 15 The **determinant** $\det A \equiv |a_{ij}|$ of a square matrix $A \equiv [a_{ij}]$ of order n is given by

$$\det A = \Sigma\, \epsilon_{i_1 i_2 \ldots i_n} a_{1i_1} a_{2i_2} \ldots a_{ni_n},$$

where the summation extends over all posible choices of the indices. The subscripted coefficient function ϵ takes on the values 0, 1, or -1 depend-

*Readers should be aware that complex-symmetric matrices share none of the beneficial properties regarding eigenvalues and eigenvectors of their real counterparts (see Section 1.5). As an illustrative example consider the matrix $A = \begin{bmatrix} 1 & i \\ i & 0 \end{bmatrix}$.

ing on whether any two subscripts are equal, the subscripts form an even permutation of 1,2,3, . . . , n, or the subscripts form an odd permutation of 1,2,3, . . . , n, respectively.

Property 12
$\det A' = \det A$

Property 13
If the square matrix B is formed from A by interchange of two rows (columns), then $\det B = -\det A$; if $b_{ij} = \alpha a_{ij}$ for fixed i (fixed j) and scalar α, then $\det b = \alpha \det A$; if either $b_{ij} = a_{ij} + \alpha a_{kj}$ for fixed i, k, and scalar α or $b_{ij} = a_{ij} + \alpha a_{ik}$ for fixed j, k, then $\det B = \det A$.

Property 14
If $a_{ij} = 0$ for fixed i (fixed j), then $\det A = 0$; if two rows (columns) of A are proportional, then $\det A = 0$.

Property 15
$\det AB = (\det A)(\det B)$.

Nonsingular matrices and inverses

Definition 16 Given a square matrix $A \equiv [a_{ij}]$ of order n, the **cofactor** a^{ij} of the element a_{ij} in A is defined as $(-1)^{i+j}$ times the determinant of the $(n-1)$st order submatrix obtained from A by deleting its ith row and jth column.

Property 16

$$(\det A)\delta_{ij} = \sum_{k=1}^{n} a_{ki}\, a^{kj} \qquad i, j = 1,2, \cdots, n.$$
$$\sum_{k=1}^{n} a_{ik}\, a^{jk}$$

The special case of this equality when $i = j$ is often used as an alternative definition for the determinant of square matrices.

Property 17
If $A \equiv [a_{ij}]$ is triangular, then

$$\det A = \prod_{i=0}^{n} a_{ii}\,.$$

Definition 17 A square matrix A is called **nonsingular** if $\det A \neq 0$; if $\det A = 0$, A is termed **singular**.

Definition 18 The **rank** of an arbitrary rectangular matrix A, denoted $r(A)$, is the order of the largest nonsingular submatrix.

Definition 19 Given a nonsingular matrix $A \equiv [a_{ij}]$, form its matrix of cofactors $B \equiv [a^{ij}]$. The **inverse** A^{-1} of A is then defined by

$$A^{-1} \equiv (1/\det A)\, B' \; .$$

Property 18
If A is a square matrix of order n, $r(A)$ equals n if and only if A is nonsingular.

Property 19
If A is nonsingular,

$$\begin{aligned} (A^{-1})^{-1} &= A \\ (A')^{-1} &= (A^{-1})' \\ AA^{-1} &= I = A^{-1}A \\ \det A^{-1} &= 1/\det A \end{aligned}$$

Property 20
If A and B are nonsingular matrices of the same order, then $(AB)^{-1} = B^{-1}A^{-1}$.

Property 21
The inverse of a unitary (orthogonal) matrix is its conjugate transpose (transpose).

Property 22
If $A \equiv [a_{ij}]$ is a nonsingular matrix of order n and $\mathbf{b}$ is a column vector with n elements, the set of algebraic equations

$$A\mathbf{x} = \mathbf{b}$$

has the unique solution $\mathbf{x} = A^{-1}\mathbf{b}$, that is,

$$\begin{aligned} x_i &= (1/\det A) \sum_{j=1}^{n} a^{ji}\, b_j \qquad i = 1,2, \cdots, n \\ &= (\det A_i)/(\det A) \; , \end{aligned}$$

where A_i is the $n \times n$ matrix obtained from A by replacing the ith column of A with the column vector $\mathbf{b}$. (It should be remarked that this last expression, which is commonly designated as *Cramer's Rule*, is useful primarily for theoretical purposes.)

References

Dettman, J. W. (1974): *Introduction to Linear Algebra and Differential Equations*, McGraw-Hill, New York; Chapter 2.

Finkbeiner, D. T. (1978): *Introduction to Matrices and Linear Algebra*, 3rd. ed., Freeman, San Francisco.

Noble B. and J. W. Daniel (1977): *Applied Linear Algebra*, 2nd ed., Prentice-Hall, Englewood Cliffs, N.J.

Appendix 2

Legendre polynomials and Bessel functions: basic facts

Legendre polynomials

If the method of separation of variables is applied to Laplace's equation expressed in spherical coordinates, one of the ordinary differential equations that results is essentially *Legendre's equation**:

$$(1 - x^2)y'' - 2xy' + n(n + 1)y = 0 \qquad -1 < x < 1. \tag{A2–1}$$

When n is integral (and only then; see Problem 16, Chapter 3) the equation admits a polynomial solution. Denoting by $P_n(x)$ that particular solution normalized so that $P_n(1) = 1$, we find the *Legendre polynomials* to be given by the finite summation

$$P_n(x) = \sum_{k=0}^{N} \frac{(-1)^k(2n - 2k)!}{2^n k!\,(n - k)!} \frac{x^{n-2k}}{(n - 2k)!}, \tag{A2–2}$$

where $N = n/2$ or $(n - 1)/2$, whichever is an integer. For lower-order n this formula yields

$$P_0(x) = 1, \quad P_1(x) = x, \quad P_2(x) = \tfrac{1}{2}(3x^2 - 1), \quad \text{and} \quad P_3(x) = \tfrac{1}{2}(5x^3 - 3x).$$

$P_n(x)$ is an odd or even function according as n is odd or even, and thus not only does $P_n(1) = 1$, but also $P_n(-1) = (-1)^n$ (see Figure A2.1).

The Legendre polynomials are the eigensolutions of the singular Sturm-Liouville problem consisting of equation (A2–1) on the interval $[-1,1]$ supplemented by the conditions that $y(1)$ and $y(-1)$ are both bounded. As is to be expected then, $P_n(x)$ has precisely n zeros in $(-1,1)$. The zeros are symmetrically situated in the fundamental interval and satisfy the inequalities

$$\cos \frac{2m - 1}{2n + 1}\pi \le x_{n,m} \le \cos \frac{2m}{2n + 1}\pi \qquad m = 1, 2, \cdots, n.$$

*Named for the French number and function theorist Adrien Marie Legendre (1752–1833).

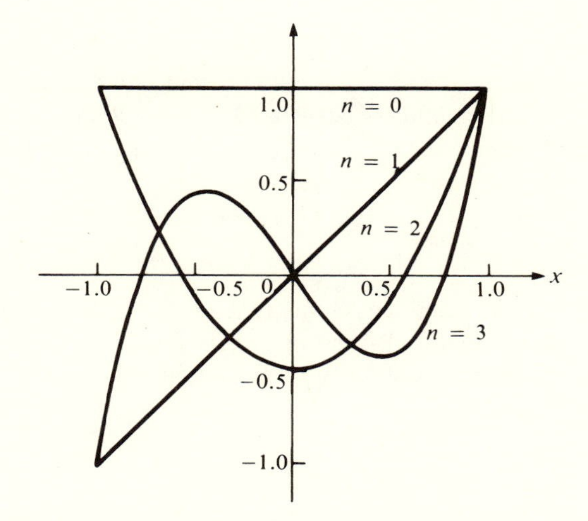

Figure A2.1 **Sketch of $P_n(x)$ as a function of x for $n = 0, 1, 2, 3$**

The functions themselves form an orthogonal set over $[-1,1]$. Indeed,

$$\int_{-1}^{1} P_n(x)\, P_m(x)\, dx = \begin{cases} 0 & m \neq n \\ 2/(2n+1) & m = n \end{cases}.$$

The Legendre polynomials are distinguished by a number of other important properties and relations. We list some of these below:

$$P_n(x) = \frac{1}{2^n n!}\frac{d^n(x^2-1)^n}{dx^n} \text{ (Rodrigues' formula);}$$

$$(n+1)P_{n+1}(x) = (2n+1)x\, P_n(x) - nP_{n-1}(x);$$

$$(x^2-1)P_n{}'(x) = nxP_n(x) - nP_{n-1}(x);$$

$$|P_n(x)| \leq 1 \qquad (-1 \leq x \leq 1);$$

$$\left|\frac{dP_n(x)}{dx}\right| \leq \tfrac{1}{2}\, n(n+1) \qquad (-1 \leq x \leq 1);$$

$$(1 - 2xt + t^2)^{-1/2} = \sum_{n=0}^{\infty} P_n(x)t^n.$$

Bessel functions

The application of separation of variables to Laplace's equation in cylindrical coordinates leads to *Bessel's equation* of order ν with parameter h*:

---

*Named after Friedrich Wilhelm Bessel (1784–1846), German astronomer and mathematician.

$$x^2y'' + xy' + (h^2x^2 - \nu^2)y = 0 \qquad x > 0. \tag{A2–3}$$

For nonnegative ν this equation has a solution bounded at the origin. When appropriately normalized, this solution is designated the *Bessel function of the first kind* of argument hx and order ν, denoted by $J_\nu(hx)$, and is given by the infinite power series

$$J_\nu(hx) = \sum_{k=0}^{\infty} \frac{(-1)^k (hx/2)^{2k+\nu}}{\Gamma(k + \nu + 1)k!}. \tag{A2–4}$$

In this expression Γ is the familiar gamma function generalization of the factorial. (See Appendix 5 for a review discussion of this function.) Plots of $J_\nu\,(hx)$ as a function of hx are given in Figure A2.2 for several familiar values of ν.

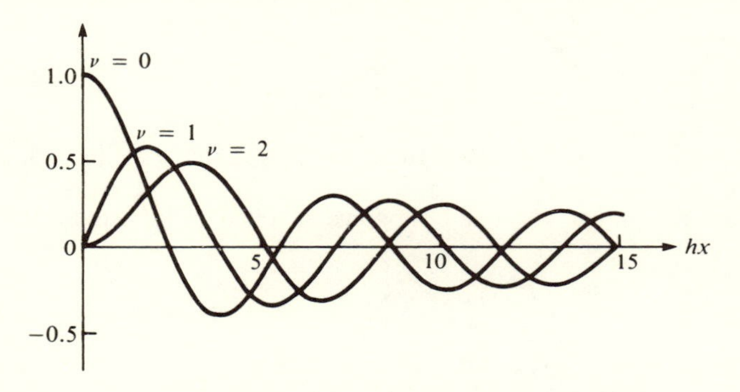

Figure A2.2 **Sketch of $J_\nu(hx)$ as a function of hx for $\nu = 0, 1, 2$**

There are a number of boundary-value problems of Sturm-Liouville type that have the Bessel function $J_\nu(hx)$ as solution. For example, if we append the boundary conditions of bounded $y(0)$ and $y(1) = 0$ to equation (A2–3), consider ν as fixed, and let h^2 play the role of the eigenparameter λ, we obtain as eigenfunctions

$$y_m(x) = J_\nu(h_{\nu,m}\, x) \qquad m = 0,1,2, \ldots,$$

where the appropriate $h_{\nu,m}$ are the (positive) zeros of $J_\nu(h)$ as a function of the argument h. When m is large, these zeros are given approximately by the expression

$$h_{\nu,m} \approx \beta - \frac{4\nu^2 - 1}{8\beta}, \tag{A2–5}$$

where $\beta = (m + 3/4 + \nu/2)\pi$. For completeness sake, we note that the zeros of $J_\nu'(h)$ (extrema of J_ν) are given approximately by

$$h'_{\nu,m} \approx \beta' - \frac{4\nu^2 + 3}{8\beta'}$$

with $\beta' = (m + 1/4 + \nu/2)\pi$ for large m. In both cases, we find that the distance between successive zeros is approximately π and tends to this value as $m \to \infty$. For different values of m, the eigenfunctions y_m turn out, as expected, to be orthogonal over $[0,1]$ with respect to the weight function x.

Of more recent importance is the variant of the problem just described in which h is held fixed and ν^2 is taken as the eigenparameter. Since this type of problem is discussed in some detail in connection with the microwave example considered in Chapter 3, we merely note here that the eigenfunctions in this situation are given by

$$y_m = J_{\nu_m}(hx), \qquad m = 0,1,2, \ldots,$$

where the appropriate ν_m are the (first real and then purely imaginary) zeros of $J_\nu(h)$ as a function of the order ν. Here the orthogonality is now with respect to the weight function $1/x$.

In many applications, an independent second solution of equation (A2–3) is needed in addition to $J_\nu(hx)$. As long as ν is nonintegral, the function $J_{-\nu}(hx)$ obtained from (A2–4) by using $-\nu$ in place of ν (note the evenness in ν of (A2–3)) suffices nicely in this regard. However, when ν is an integer, this pleasant state of affairs breaks down since, in this case,

$$J_{-n}(hx) = (-1)^n J_n(hx).$$

For this reason it is customary generally to not use $J_{-\nu}(hx)$ explicitly and to use instead, as an appropriate second solution of (A2–3), the so-called *Bessel function of the second kind* defined by the expression

$$Y_\nu(hx) \equiv \frac{\cos \nu\pi \, J_\nu(hx) - J_{-\nu}(hx)}{\sin \nu\pi}. \tag{A2–6}$$

(L'Hôpital's rule applied to (A2–5) extends this definition to integral ν.) For the Wronskians of interest it turns out that

$$W(J_\nu, J_{-\nu}) = -\frac{2 \sin \nu\pi}{\pi x} \quad \text{but} \quad W(J_\nu, Y_\nu) = \frac{2}{\pi x},$$

and thus $Y_\nu(hx)$ is well-defined and linearly independent of $J_\nu(hx)$ for all values of ν.

The Bessel functions of the first and second kinds are real-valued for h, x and ν real. However, in various practical problems, complex combinations of these functions are germane. Principal amongst these are the *Bessel functions of the third kind*, also called *Hankel functions*:

$$H_\nu^{(1)}(hx) = J_\nu(hx) + i\,Y_\nu(hx),$$
$$H_\nu^{(2)}(hx) = J_\nu(hx) - i\,Y_\nu(hx).$$

For those investigations in which purely imaginary values of x are central, the *modified Bessel functions*

$$I_\nu(hx) = e^{-i\nu\pi/2} J\nu(ihx),$$

$$K_\nu(hx) = \frac{\pi}{2} e^{i(\nu+1)\pi/2} H_\nu^{(1)}(ihx)$$

are the real-valued constructs typically employed.

As might be expected, there exists a vast array of other relations, identities, and inequalities for Bessel functions. Those of most frequent use are probably

$$\frac{d}{dx}\left[x^{\nu}y_{\nu}(hx)\right] = \begin{cases} hx^{\nu}y_{\nu-1}(hx) & (y = J,Y,I,H^{(1)},H^{(2)}) \\ -hx^{\nu}y_{\nu-1}(hx) & (y = K); \end{cases}$$

$$\frac{d}{dx}\left[x^{-\nu}y_{\nu}(hx)\right] = \begin{cases} -hx^{-\nu}y_{\nu+1}(hx) & (y = J,Y,K,H^{(1)},H^{(2)}) \\ hx^{-\nu}y_{\nu+1}(hx) & (y = I); \end{cases}$$

$$\frac{d}{dx}y_{\nu}(hx) = \frac{h}{2}\left[y_{\nu-1}(hx) - y_{\nu+1}(hx)\right] \qquad (y = J,Y,H^{(1)},H^{(2)});$$

$$\frac{d}{dx}I_{\nu}(hx) = \frac{h}{2}\left[I_{\nu-1}(hx) + I_{\nu+1}(hx)\right];$$

$$\frac{d}{dx}K_{\nu}(hx) = -\frac{h}{2}\left[K_{\nu-1}(hx) + K_{\nu+1}(hx)\right];$$

$$y_{\nu-1}(hx) + y_{\nu+1}(hx) = \frac{2\nu}{hx}y_{\nu}(hx) \qquad (y = J,Y,H^{(1)},H^{(2)});$$

$$I_{\nu-1}(hx) - I_{\nu+1}(hx) = \frac{2\nu}{hx}I_{\nu}(hx);$$

$$K_{\nu-1}(hx) - K_{\nu+1}(hx) = -\frac{2\nu}{hx}K_{\nu}(hx).$$

$$|J_{\nu}(hx)| \leq 1 \qquad (\nu \geq 0,\ hx \text{ real});$$

$$\left|\frac{d^kJ_n(hx)}{dx^k}\right| \leq |h|^k \qquad (k = 0,1,2,\ldots,\ n \text{ integral},\ hx \text{ real});$$

$$\exp\left[\frac{hx}{2}\left(t - \frac{1}{t}\right)\right] = \sum_{n=-\infty}^{\infty} t^nJ_n(hx) \qquad (t \neq 0).$$

References

Abramowitz, M. and I. A. Stegun (1965): *Handbook of Mathematical Functions*, Dover, New York; NBS Applied Math Series 55, U.S. Dept. of Commerce, Washington, D.C. (1964); Chapters 8–10, 22.

Erdélyi, A., et al. (1953): *Higher Transcendental Functions, Vols. 1, 2*, McGraw-Hill, New York; Chapters III, VII, X.

Hildebrand, F. B. (1976): *Advanced Calculus for Applications*, 2nd ed., Prentice-Hall, Englewood Cliffs, N.J.; Chapter 4.

Watson, G. N. (1958): *A Treatise on the Theory of Bessel Functions*, 2nd ed., Cambridge Univ. Press, Cambridge.

Appendix 3

Series solutions of ordinary differential equations

Many of the differential equations encountered in applied problems turn out to be homogeneous linear second-order equations with *polynomial* coefficients. A standard method of exhibiting solutions of these equations is to determine their power series representations. However, if we inflexibly seek solutions of a particular form, say,

$$y(x) = \sum_{k=0}^{\infty} c_k x^k, \tag{A3–1}$$

we should not expect our efforts to always succeed, even for relatively simple equations. For example, the equations

(I) $y'' - y = 0$,

(II) $x^2y'' + (x^2 + x)y' - y = 0$,

and

(III) $x^3y'' + y = 0$

respectively have precisely two, only one, and no nontrivial solutions with convergent power series of the form (A3–1). (Hildebrand (1976) discusses these examples in some detail.)

Ordinary points

The theory underlying series solutions of ordinary differential equations is designed to discriminate among the behaviors illustrated in the three examples above. The equation is assumed to be in the standard form*

$$y'' + f(x)y' + g(x)y = 0. \tag{A3–2}$$

The nature of various values of the independent variables are then classified according to

*All second-order linear homogeneous ordinary differential equations obviously can be so expressed.

Definition 1 A point x_o is termed an **ordinary point** of the differential equation (A3–2) if there exists an interval $|x - x_o| < R$ in which $f(x)$ and $g(x)$ have convergent power series representations

$$f(x) = \sum_{k=0}^{\infty} \alpha_k (x - x_o)^k, \qquad g(x) = \sum_{k=0}^{\infty} \beta_k (x - x_o)^k. \qquad \textbf{(A3–3)}$$

The significance of this concept is a consequence of

Theorem 1:

Let x_o be an ordinary point of the differential equation (A3–2). There then must exist *two* linearly independent solutions of (A3–2), each of which has a power series representation

$$y(x) = \sum_{k=0}^{\infty} c_k (x - x_o)^k \qquad \textbf{(A3–4)}$$

convergent in the interval $|x - x_o| < R$. □

(See Dettman (1974) or Hildebrand (1976), for example.)

For differential equations with well-behaved coefficients, almost all values of the independent variable turn out to be ordinary points. As a special case of this we observe that:

Property 1
If a second-order linear ordinary differential equation has polynomial coefficients, then all but a finite number of values of the independent variable are ordinary points.

Regular singular points

If a value of the independent variable x is not an ordinary point of a given differential equation, then it is called a *singularity* of the equation. Such is the case with the value $x = 0$ in the earlier equations II and III. As Frobenius first showed, for certain types of singularities the form of the expansion (A3–4) can be modified so that a representation for at least one solution of the equation, valid in a deleted interval $0 < |x - x_o| < R$, may be obtained. These results are customarily stated using the following terminology:

Definition 2 A point x_o is termed a **regular singular point** of the differential equation (A3–2) if there exists an interval $|x - x_o| < R$ in which $(x - x_o)f(x)$ and $(x - x_o)^2 g(x)$ have convergent power series representations

$$(x - x_o)f(x) = \sum_{k=0}^{\infty} a_k(x - x_o)^k, \qquad (x - x_o)^2 g(x) = \sum_{k=0}^{\infty} b_k(x - x_o)^k. \tag{A3–5}$$

Then

Theorem 2:

Let x_o be a regular singular point of the differential equation (A3–2). In this case there exists *at least one* solution of (A3–2) which has the representation

$$y(x) = \sum_{k=0}^{\infty} c_k(x - x_o)^{k+\alpha} \tag{A3–6}$$

convergent for $0 < |x - x_o| < R$. The constant α appearing in (A3–6) is a root of the so-called *indicial equation*

$$\alpha^2 + \alpha(a_o - 1) + b_o = 0, \tag{A3–7}$$

and the coefficients c_k are expressible in terms of the a_k, b_k of (A3–5). If both roots of the indicial equation are complex, or if both are real and do not differ by an integer, then associated with *each* there is a solution of (A3–2) representable in the form (A3–6). Moreover, the two solutions are linearly independent. □

(Again see Dettman (1974) or Hildebrand (1976), for example.)

Returning for a moment to our earlier equation II, we see that the origin $x = 0$ is a regular singular point. Moreover, the related indicial equation is $\alpha^2 - 1 = 0$, and this equation clearly has the two real roots $+1$ and -1. The solution with a convergent power series about $x = 0$ is associated with the larger root $\alpha = 1$. Even though the roots differ by an integer, using the smaller root $\alpha = -1$ in the Frobenius method leads to a second solution, namely $1/x - 1$.

As this example illustrates, when a differential equation with a regular singular point gives rise to an indicial equation with two real roots that differ by an integer, *two* linearly independent solutions of the form (A3–6) may occasionally be obtained even though Theorem 2 only guarantees the existence of one. If we are not so fortunate, as is always the situation when the two roots are the same, the second solution can only be generated by other methods, for example the "reduction of order" process. (Coddington and Levinson (1955), pp. 132ff. and Hildebrand (1976), pp. 135ff, among others, discuss these matters in considerable detail.) For our purposes it suffices to note that in all of these latter cases the independent second solution can be shown to have the form*

$$y_2(x) = By_1(x) \ln(x - x_o) + \sum_{k=0}^{\infty} C_k(x - x_o)^{k+\alpha_2}. \tag{A3–8}$$

*This is a nontrivial conclusion; interested readers may want to try their hand at a verification. (See Problem 3, Chapter 3.)

Here it is assumed that α_1 and α_2 are the real roots of the indicial equation, with α_1 the larger; $y_1(x)$ is the solution of the form (A3–6) associated with α_1 given by the Frobenius method; and B and the C_k are constants.

Whenever a second solution of a given differential equation is needed, it can often be found most easily by substituting the expression (A3–8) directly into the equation and relating recursively the unknown constants by matching coefficients of like powers of x. As an example of the efficacy of this approach, consider Bessel's equation of zero order

$$x^2y'' + xy' + x^2y = 0, \tag{A3–9}$$

for which the origin is a regular singular point with indicial roots $\alpha_1 = \alpha_2 = 0$. One solution is $y_1(x)$ given by

$$J_0(x) = \sum_{k=0}^{\infty} \frac{(-1)^k (x/2)^{2k}}{(k!)^2}. \tag{A3–10}$$

(See (A2–4), (3.6–2).) For the second solution we find, upon substituting the expression (A3–8) into equation (A3–9),

$$2\,BxJ_0'(x) + \sum_{k=0}^{\infty} C_k(k^2x^k + x^{k+2}) = 0.$$

Using the series (A3–10) and matching coefficients of like powers of x, it follows that $C_1 = C_3 = C_5 = \cdots = 0$ and

$$C_{2k} = \frac{(-1)^k}{2^{2k}\,(k!)^2}\left[C_0 - B\sum_{n=1}^{k}\frac{1}{n}\right] \qquad k \geq 1.$$

Thus

$$y_2(x) = BJ_0(x)\ln x + \sum_{k=0}^{\infty}\frac{(-1)^k}{(k!)^2}\left[C_0 - B\sum_{n=1}^{k}\frac{1}{n}\right](x/2)^{2k},$$

where we agree that the internal summation does not exist if $k = 0$. If we make the associations $B = 2/\pi$ and $C_0 = (2/\pi)(\gamma - \ln 2)$, with γ the Euler-Mascheroni constant, and recall the definition of the psi function (see equation (A5–9) of Appendix 5), this solution is identical with $Y_0(x)$ as given by (3.6–4).

Before concluding this appendix with a brief mention of other types of singularities, it is worth noting that, as expected, the singular natures of solutions and of the differential equations from which these solutions arise are inherently intertwined. For example, Theorem 1 asserts that when a solution of a differential equation exhibits discontinuous behavior in the neighborhood of a given point, that point is perforce a singularity of the equation. The converse is not universally valid for regular singular points, and Problem 4, Chapter 3 provides a case in point. On the other hand, as the elementary Cauchy-Euler equation (3.2–1) illustrates, even equations with extremely simple coefficients can be expected to lead to solutions with algebraic and perhaps logarithmic singularities.

Irregular singular points

Nonregular singularities are termed *irregular singular points*. The origin $x = 0$ in equation III above provides an example. The general study of ordinary differential equations and their solutions in the vicinity of irregular singular points is rather complicated. Existence of such a singularity usually manifests intricate behavior of the solutions, and asymptotic techniques are characteristically needed [see Coddington and Levinson (1955), Ince (1953) or Olver (1974)]. In this book we are in general not concerned with equations possessing irregular singular points, except perhaps at $x = \infty$.* (This is the situation encountered in "most" applications.) Indeed for the case of an irregular singularity at ∞, we concentrate only on those equations, such as Bessel's equation (A2–3) (3.6–1), in which this one irregular singular point has resulted from the "coalescence" of two (originally separate and distinct) regular singular points. (Ince (1953) provides a useful scheme for categorizing irregular singularities.)

References

Coddington, E. A. and N. Levinson (1955): *Theory of Ordinary Differential Equations*, McGraw-Hill, New York; Chapters 4, 5.

Dettman, J. W. (1974): *Introduction to Linear Algebra and Differential Equations*, McGraw-Hill, New York; Chapter 8.

Hildebrand, F. B. (1976): *Advanced Calculus for Applications*, 2nd ed., Prentice-Hall, Englewood Cliffs, N.J.; Chapter 4.

Ince, E. L. (1953): *Ordinary Differential Equations*, Dover, New York; Longmans, London (1927); Chapters XVI, XVII.

Olver, F. W. J. (1974): *Asymptotics and Special Functions*, Academic Press, New York; Chapters 5, 7.

*The nature of the point at infinity, with reference to a given differential equation, can be ascertained by making the change of variables $x = 1/t$ and then inspecting the behavior of the coefficients of the transformed equation in the vicinity of the origin $t = 0$.

Appendix 4

Functions of a complex variable

Complex numbers

Definition 1 A **complex number** z is an ordered pair (x,y) of real numbers x and y. Numbers of the form $(0,y)$ are called **purely imaginary numbers**. Alternatively we can write

$$z = x + iy$$

where i denotes the purely imaginary number $(0,1)$. The real numbers x,y in either of the representations are termed the **real and imaginary parts** of z, respectively, that is,

$$x = \text{Re } z, \qquad y = \text{Im } z.$$

Definition 2 Two complex numbers $a + ib$ and $c + id$ are equal if $a = c$ and $b = d$. In particular, zero is the unique complex number $z = a + ib$ with $a = 0 = b$.

Definition 3 The sum/difference of two complex numbers $z_1 = a + ib$ and $z_2 = c + id$ is the complex number

$$z_1 \pm z_2 = (a \pm c) + i(b \pm d).$$

Definition 4 The product of two complex numbers $z_1 = a + ib$ and $z_2 = c + id$ is the complex number $z_1 z_2 = (ac - bd) + i(bc + ad)$. It follows that $i^2 = -1$, and so the product of two complex numbers can be obtained by formally manipulating the terms as if they involved only real numbers and then simplifying the final result by replacing i^2 by -1 as needed.

Definition 5 The **complex conjugate** of a complex number $z = a + ib$, denoted by $\bar{z}$, is the complex number $\bar{z} = a - ib$.

Definition 6 The **reciprocal** of a nonzero complex number $z = a + ib$ is the complex number

$$\frac{1}{z} = \frac{a}{a^2 + b^2} - i\frac{b}{a^2 + b^2}.$$

This is obtained by writing the reciprocal as $1/(a + ib)$ and then multiplying both the numerator and denominator by the complex conjugate of z.

Definition 7 Division of two complex numbers $z_1 = a + ib$ and $z_2 = c + id$ is defined in terms of multiplication of the numerator by the reciprocal of the denominator. For nonzero z_1 we obtain

$$\frac{z_2}{z_1} = (c + id)\frac{1}{a + ib} = \frac{ac + bd}{a^2 + b^2} + i\frac{ad - bc}{a^2 + b^2}.$$

Property 1
The associative, commutative, and distributive properties that are valid for real numbers hold for complex numbers also.

Property 2
If $z_1 z_2 = 0$, then either $z_1 = 0$ or $z_2 = 0$ (or both).

Property 3
$\operatorname{Re} z = \frac{1}{2}(\bar{z} + z)$ $\operatorname{Im} z = \frac{i}{2}(\bar{z} - z)$,
$\operatorname{Re}(iz) = -\operatorname{Im} z$, $\operatorname{Im}(iz) = \operatorname{Re} z$,

$$(\overline{z_1 \pm z_2}) = \bar{z}_1 \pm \bar{z}_2,\ \overline{z_1 z_2} = \bar{z}_1\,\bar{z}_2,\ \overline{\left(\frac{z_1}{z_2}\right)} = \frac{\bar{z}_1}{\bar{z}_2}.$$

In view of the association of ordered pairs with points in two-space, it is convenient also to interpret complex numbers geometrically as points in the euclidean plane with axes, angles, and distances as depicted in Figure A4.1. When used for this purpose, the (*xy*)-plane is commonly termed the *complex z-plane*, with the *x* and *y* axes designated, naturally, as the *real* and *imaginary* axes, respectively.

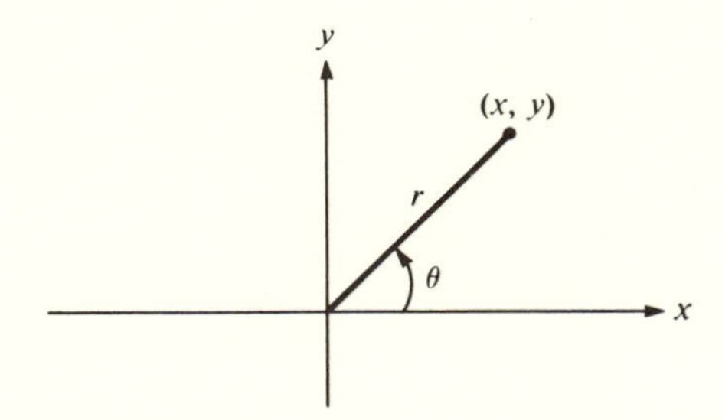

Figure A4.1

Introducing polar coordinates leads to

Definition 8 The complex number $z = x + iy$ has the alternative representation in **polar form**

$$z = r(\cos\theta + i\sin\theta)$$

or, using Euler's formula,

$$z = re^{i\theta}. \tag{A4–1}$$

Here

$$r = \sqrt{x^2 + y^2}$$

is the (positive) distance of the point (x,y) from the origin in the complex plane and

$$\theta = \tan^{-1} y/x \pm 2n\pi \qquad (n = 0,1,2, \ldots)$$

is the radian measure of the angle(s) that the directed line segment from the origin to (x,y) makes with the positive real axis. The **modulus** or **absolute value** of the complex number z, denoted by $|z|$, *is* r. The number θ is called the (an) **argument** of z and written $\theta = \arg z$.

Property 4

$|z| = r = \sqrt{(\text{Re } z)^2 + (\text{Im } z)^2} = \sqrt{z\bar{z}}$;

$|z| \geq |\text{Re } z|, \qquad |z| \geq |\text{Im } z|$.

Property 5

If $z_1 = x_1 + iy_1 = r_1 e^{i\theta_1}, \quad z_2 = x_2 + iy_2 = r_2 e^{i\theta_2}$, then

$|z_1 - z_2| = \sqrt{(x_1 - x_2)^2 + (y_1 - y_2)^2}$,

$|z_1 z_2| = r_1 r_2 = |z_1|\,|z_2|$,

$\dfrac{|z_1|}{|z_2|} = \dfrac{r_1}{r_2} = \dfrac{|z_1|}{|z_2|}$;

$\arg(z_1 z_2) = \theta_1 + \theta_2 = \arg z_1 + \arg z_2$,

$\arg\left(\dfrac{z_1}{z_2}\right) = \theta_1 - \theta_2 = \arg z_1 - \arg z_2$.

Property 6

$|z_1 + z_2| \leq |z_1| + |z_2|$ (triangle inequality), or more generally,

$$\left|\sum_{k=1}^{n} z_k\right| \leq \sum_{k=1}^{n} |z_k|,$$

$$\Big|\, |z_1| - |z_2| \,\Big| \leq |z_1 - z_2|.$$

Property 7
If α is integral

$$\begin{aligned} z^{\alpha} &= r^{\alpha}(e^{i\theta})^{\alpha} \\ &= r^{\alpha}e^{i\alpha\theta} \\ &= r^{\alpha}\cos\alpha\theta + ir^{\alpha}\sin\alpha\theta \text{ (de Moivre's Theorem).} \end{aligned}$$

The extension to rational α is straightforward.

This last result, properly interpreted, is of course valid for general complex α. However, in its present form the expression is particularly useful in determining all of the, say, nth roots of any nonzero complex number.

Complex functions

Definition 9 If $z = x + iy$ and $w = u + iv$ are two complex variables (complex numbers taking on a number of values) and if, for each value of z in some portion of the complex plane, a value of w is defined (assigned by some rule), then w is said to be a **(complex) function** of z. The notation

$$w = f(z)$$

is often employed.

If we exploit this definition, each of the real numbers u and v depends on the real numbers x and y, a fact we can express as

$$u(x,y) + i\,v(x,y) = f(z). \qquad \textbf{(A4–2)}$$

In other words, u and v satisfy

$$u(x,y) = \operatorname{Re} f(z), \qquad v(x,y) = \operatorname{Im} f(z).$$

The relation (A4–2) can also be employed in the opposite direction. Given two real-valued functions $u(x,y)$, $v(x,y)$ of the two real variables x,y, (A4–2) can be used *to define* a complex function f of the complex variable z. The simple examples

$$\text{(i) } w = (x^2 - y^2) + i(2xy) = z^2,$$

$$\text{(ii) } w = \frac{x}{x^2 + y^2} - i\frac{y}{x^2 + y^2} = \frac{1}{z},$$

$$\text{(iii) } w = x = \frac{1}{2}(z + \bar{z}),$$

serve to illustrate this point.

In working with complex functions the following notions associated with sets of complex numbers are of value:

Definition 10 Given a point z_0 in the complex z-plane and a real number $\epsilon > 0$, an **ϵ-neighborhood** or, for short, **neighborhood** of z_0, is the set of all points z such that

$$|z - z_0| < \epsilon.$$

This is obviously the set of points interior to the circle of radius ϵ about the point z_0 as center.

Definition 11 Given a set S of complex numbers, a point z_0 is said to be an **interior point** of S if there is at least one neighborhood (and hence an infinity of neighborhoods) of z_0 consisting entirely of points of S. S is called **open** if each point of the set is an interior point.

Definition 12 Given a set S of complex numbers, a point z_0 is said to be an **exterior point** of S (exterior *to* S), if there is at least one neighborhood of z_0 containing no points of S. Points that are neither interior or exterior to S are termed **boundary points** of S. A point z_0 is called a **limit** (or **accumulation) point** of S if *every* neighborhood of z_0 contains at least one point of S distinct from z_0. S is said to be **closed** if it contains all of its limit points.

Property 8
A set is open if and only if it contains *none* of its boundary points.

Property 9
Boundary points of a set are often also limit points of the set. The converse is obviously not true; just recall the nature of interior points. However, a set is closed if and only if it contains *all* of its boundary points.

Definition 13 An open set S of complex numbers is said to be **connected** if every pair of points in S can be joined by a polygonal arc, that is, a finite number of straight line segments joined end-to-end, lying entirely within the set. A **domain (region)** is an open, connected set. A **closed region** is the closed set consisting of the region together with all of its boundary (limit) points.

Definition 14 A set for which there exists a circle $|z| = R$ which surrounds it, that is, a neighborhood $|z| < R$ which contains the set in its interior, is called a **bounded** set. Sets that are not bounded are termed **unbounded**.

Definition 15 A **curve** is a set of points given by the parametric equations

$$z(t) = x(t) + i\,y(t) \qquad a \le t \le b$$

where $x(t)$, $y(t)$ are continuous. The curve is said to be **smooth** if dx/dt, dy/dt are also continuous and not both simultaneously zero. Note that a smooth curve has a continuous tangent vector for $a < t < b$. The curve is termed **piecewise** (or **sectionally**) **smooth** if it is smooth for all but a finite number of values of t. It is said to be **simple** if it is one-to-one, that is, it does not cross itself. A simple curve is occasionally called a **Jordan curve**. If it is piecewise smooth it is termed a **contour**. Curves that begin and end at the same point are said to be **closed**. Contours with this property are naturally called (simple) **closed contours**.

Property 10
A smooth curve has length

$$L = \int_a^b |z'(t)| dt$$
$$= \int_a^b \sqrt{(dx/dt)^2 + (dy/dt)^2}\, dt \, .$$

Property 11
(Jordan Curve Theorem) A simple closed curve (closed Jordan curve) C in the complex plane divides the plane into two domains. One of these domains (the interior of C) is bounded; the other (the exterior of C) is unbounded.

Definition 16 A region is **simply connected** if it has the property that every simple closed curve drawn entirely within the region encloses *only* points of the region. Regions that are not simply-connected are said to be **multiply-connected.**

Three special sets of complex-numbers, and variants thereof, occur frequently in applied problems involving complex functions, namely

$$S_1 \equiv \{z| \; |z - z_0| < r_1 \text{ with } r_1 > 0\} \, ,$$
$$S_2 \equiv \{z| \; 0 < r_1 \leq |z - z_0| < r_2\} \, ,$$
$$S_3 \equiv \{z| \; 0 < r_2 \leq |z - z_0|\} \, .$$

Readers should recognize that S_1 is a neighborhood of z_0. It is a simply-connected bounded region. S_2 is neither open nor closed; it is a bounded, multiply-connected set of complex numbers. S_3 is a closed, connected set. More precisely, it is an unbounded, multiply-connected closed region.

The earlier definition of a (complex) function permits multiple values to be assigned to $w = f(z)$, in other words multivalued functions. Normally, if it is desirable to do so, either the independent variable z or the dependent variable w can be restricted in such a way that single-valuedness results.

For single-valued functions the essential concepts such as domain, range, into, onto, one-to-one, inverse, limits, and continuity can be defined. It turns out that many of the usual results regarding real-valued continuous functions of a single real variable (for example the fact that sums, differences, products, quotients (with non-vanishing denominators), and compositions of continuous functions are continuous) carry over intact to complex functions. Two particular results are especially useful:

Property 12
If $f(z)$ is continuous at a point z_0 and $f(z_0) \neq 0$, then there is a neighborhood of z_0 throughout which $f(z) \neq 0$.

Property 13
If $f(z)$ is continuous throughout a closed, bounded region $\overline{D}$, then there exists a positive constant M such that

$$|f(z)| < M \text{ for all } z \text{ in } \overline{D}.$$

References

Churchill, R. V., J. W. Brown, and R. F. Verhey (1974): *Complex Variables and Applications*, 3rd ed., McGraw-Hill, New York.

Pennisi, L., L. I. Gordon and S. Lasher (1976): *Elements of Complex Variables*, Holt, Rinehart and Winston, New York.

Polya, G. and G. Latta (1974): *Complex Variables*, Wiley, New York.

Silverman, R. A. (1974): *Complex Analysis with Applications*, Prentice-Hall, Englewood Cliffs, N.J.

Appendix 5

The gamma and related functions

The gamma function $\Gamma(z)$ is the generalization of the familiar factorial function to nonintegral and, indeed, nonreal values of the argument z. The defining relationship is the integral representation (due to Euler)

$$\Gamma(z) = \int_0^\infty t^{z-1} e^{-t}\, dt \qquad (\mathrm{Re}\, z > 0). \tag{A5–1}$$

From this expression it follows readily, via integration by parts, that

$$\Gamma(z + 1) = z\,\Gamma(z). \tag{A5–2}$$

The recurrence relation (A5–2) can be used repeatedly to extend the definition of $\Gamma(z)$ into the left-half of the complex z-plane. It also shows that, as desired,

$$\Gamma(n + 1) = n!\,,$$

when n is a nonnegative integer.

For nonintegral z, the gamma function satisfies the useful functional equation

$$\Gamma(z)\,\Gamma(1 - z) = \frac{\pi}{\sin \pi z}. \tag{A5–3}$$

If we substitute $z = ½$ into (A5–3) we find that $\Gamma(½) = \sqrt{\pi}$. The relationship (A5–3) also shows that $\Gamma(z)$, viewed as a function of the complex variable z, is analytic for all $z \neq 0, -1, -2, \ldots$ and has simple poles for these nonpositive integral values of z.

Another functional equation that the gamma function satisfies is the multiplication formula

$$\Gamma(nz) = (2\pi)^{½(1-n)}\, n^{nz-½}\, \Gamma(z)\, \Gamma\left(z + \frac{1}{n}\right) \Gamma\left(z + \frac{2}{n}\right) \cdots \Gamma\left(z + \frac{n-1}{n}\right).$$

The special case with $n = 2$, namely

$$\Gamma(2z) = (2\pi)^{-½}\, 2^{2z-½}\, \Gamma(z)\, \Gamma(z + ½), \tag{A5–4}$$

is called the *Legendre duplication formula*.

For larger values of z the gamma function has the asymptotic expansion

$$\Gamma(z) \sim \sqrt{2\pi}\, e^{(z-½)\ln z - z} \left[1 + \frac{1}{12z} + \frac{1}{288z^2} - \frac{139}{51840z^3} + \cdots\right] \tag{A5–5}$$

as $z \to \infty$, $|\arg z| < \pi$. Although (A5–5) is customarily termed *Stirling's formula*, the result actually derived by Stirling in 1730 was an asymptotic representation for the natural logarithm of the gamma (factorial) function. In modern notation we would have

$$\ln \Gamma(z+1) \sim (z + ½) \ln z - z + ½ \ln (2\pi) + \frac{1}{12z} - \frac{1}{360z^3} + \frac{1}{1260z^5} - \frac{1}{1680z^7} + \cdots \qquad \textbf{(A5–6)}$$

as $z \to \infty$, $|\arg z| < \pi$. Utilizing either (A5–5) or (A5–6) we can establish the useful fact that

$$\lim_{z\to\infty} \frac{\Gamma(z)\, z^a}{\Gamma(z+a)} = 1 \qquad |\arg z| < \pi.$$

Two functions closely related to the gamma function are the beta function

$$B(z,w) \equiv \frac{\Gamma(z)\,\Gamma(w)}{\Gamma(z+w)}$$

and the psi (or digamma) function

$$\psi(z) \equiv \frac{d(\ln \Gamma(z))}{dz} = \frac{\Gamma'(z)}{\Gamma(z)}.$$

The beta function has the well-known integral representation

$$B(z,w) = \int_0^1 t^{z-1}(1-t)^{w-1}\,dt. \qquad \textbf{(A5–7)}$$

For $z \neq -1, -2, -3, \ldots$, the psi function can be expressed by the convergent series

$$\psi(1+z) = -\gamma + \sum_{k=1}^{\infty} \frac{z}{k(k+z)}, \qquad \textbf{(A5–8)}$$

where

$$\gamma = \lim_{N\to\infty}\left[\sum_{k=1}^{N}\frac{1}{k} - \ln N\right] \doteq 0.57722$$

is the Euler-Mascheroni constant. In the special case of a positive integer n we find $\psi(1) = -\gamma$ and

$$\psi(n) = -\gamma + \sum_{k=1}^{\infty}\frac{n-1}{k(k+n-1)} = -\gamma + \sum_{k=1}^{\infty}\left[\frac{1}{k} - \frac{1}{k+n-1}\right] = -\gamma + \sum_{k=1}^{n-1}\frac{1}{k} \qquad n \geq 2. \qquad \textbf{(A5–9)}$$

References

Abramowitz, M. and I. A. Stegun (1965): *Handbook of Mathematical Functions*, Dover, New York; NBS Applied Math Series 55, U.S. Dept. of Commerce, Washington, D.C. (1964); Chapter 6.

Erdélyi, A., et al. (1953): *Higher Transcendental Functions, Vol. I,* McGraw-Hill, New York; Chapter I.

Hildebrand, F. B. (1976): *Advanced Calculus for Applications*, 2nd ed., Prentice-Hall, Englewood Cliffs, N.J.; Chapter 1.

Appendix 6

Fourier and Laplace transforms

The Fourier transform

The (complex) Fourier transform and its inversion formula are given by

$$\mathscr{F}[f(t)] \equiv F(\omega) \equiv \frac{1}{\sqrt{2\pi}} \int_{-\infty}^{\infty} f(t)\, e^{i\omega t}\, dt$$

and

$$f(t) = \frac{1}{\sqrt{2\pi}} \int_{-\infty}^{\infty} F(\omega)\, e^{-i\omega t}\, d\omega,$$

respectively.

Closely related are the Fourier cosine transform pair

$$F_c(\omega) \equiv \sqrt{\frac{2}{\pi}} \int_0^{\infty} f(t) \cos \omega t\, dt,$$

$$f(t) \equiv \sqrt{\frac{2}{\pi}} \int_0^{\infty} F_c(\omega) \cos \omega t\, d\omega,$$

and the sine transform pair

$$F_s(\omega) \equiv \sqrt{\frac{2}{\pi}} \int_0^{\infty} f(t) \sin \omega t\, dt,$$

$$f(t) \equiv \sqrt{\frac{2}{\pi}} \int_0^{\infty} F_s(\omega) \sin \omega t\, d\omega.$$

For even functions $f(t) = f(-t)$, it follows that

$$F(\omega) = F_c(\omega).$$

Similarly, for odd functions $f(t) = -f(-t)$ we have

$$F(\omega) = iF_s(\omega).$$

More generally:

Property 1
$$\mathscr{F}[f(t/a+b)] = ae^{-iab\omega}F(a\omega) \qquad (a>0)$$

Property 2
$$\mathscr{F}[f(at)e^{ibt}] = \frac{1}{a}F\left(\frac{\omega+b}{a}\right) \qquad (a>0)$$

Property 3
$$\mathscr{F}[t f(t)] = -i\frac{d}{d\omega}F(\omega)$$

Property 4
$$\mathscr{F}\left[\frac{df(t)}{dt}\right] = -i\omega F(\omega).$$

Selected examples of Fourier transforms

Example 1 If $f(t) = \begin{cases} 1 & |t|<1 \\ 0 & |t|>1 \end{cases}$,

then
$$F(\omega) = \sqrt{2/\pi}\,\frac{\sin\omega}{\omega}. \qquad \square$$

Example 2 If $f(t) = \dfrac{1}{1+t^2}$,

then
$$F(\omega) = \sqrt{\tfrac{1}{2}\pi}\, e^{-|\omega|}. \qquad \square$$

Example 3 If $f(t) = |t|^{-\alpha}$
with $0 < \text{Re}\,\alpha < 1$, then
$$F(\omega) = \sqrt{2/\pi}\,\Gamma(1-\alpha)\sin(\tfrac{1}{2}\pi\alpha)\,|\omega|^{\alpha-1} \qquad \square$$

Example 4 If
$$f(t) = (\alpha + it)^{-\beta}$$
with $\text{Re}\,\alpha > 0$, $\text{Re}\,\beta > 0$, then
$$F(\omega) = \begin{cases} \dfrac{\sqrt{2}\,\pi}{\Gamma(\beta)}\,\omega^{\beta-1}e^{-\alpha\omega} & \omega>0 \\ 0 & \omega<0. \end{cases} \qquad \square$$

Example 5 If

$$f(t) = \begin{cases} (1 - t)^{\alpha-1}(1 + t)^{\beta-1} & |t| < 1 \\ 0 & |t| > 1 \end{cases}$$

with Re $\alpha > 0$, Re $\beta > 0$, then

$$F(\omega) = \frac{2^{\alpha+\beta-3/2}}{\sqrt{\pi}} B(\alpha,\beta)e^{-i\omega} M(\alpha;\alpha + \beta;2i\omega),$$

where $\beta(\alpha,\beta)$ is the standard beta function (see Appendix 5) and $M(a;c;x)$ designates the confluent hypergeometric function discussed in Chapter 3. The special case when $\alpha = \beta$ is of interest since then the result can be reexpressed in terms of Bessel functions (see (3.2–4)):

$$F(\omega) = 2^{\beta-1}\,\omega^{1/2-\beta}\,\Gamma(\beta)\,J_{\beta-1/2}(\omega). \qquad \square$$

Example 6 If

$$f(t) = e^{-\alpha|t|}\sin\beta t$$

with Re $\alpha > |\text{Im}\,\beta|$, then

$$F(\omega) = \frac{4i\alpha\beta\omega/\sqrt{2\pi}}{(\alpha^2 + \beta^2 + \omega^2)^2 - 4\beta^2\omega^2}. \qquad \square$$

Example 7 If

$$f(t) = e^{-\alpha t^2}$$

with Re $\alpha > 0$, then

$$F(\omega) = \frac{1}{\sqrt{2\alpha}} e^{-\omega^2/4\alpha}. \qquad \square$$

Example 8 If

$$f(t) = \tan^{-1}(2/t^2),$$

then

$$F(\omega) = \frac{\sqrt{2\pi}}{\omega} e^{-\omega}\sin\omega. \qquad \square$$

For extensive lists of Fourier transforms see Oberhettinger (1957), (1973), and Erdélyi et al (1954).

The Laplace transform

The Laplace transform and its inversion formula are given by

$$L[f(t)] \equiv \mathscr{L}(s) \equiv \int_0^\infty f(t)\,e^{-st}\,dt \qquad \text{Re}\,s > a$$

and

$$f(t) = \frac{1}{2\pi i}\int_{\sigma-i\infty}^{\sigma+i\infty} \mathscr{L}(s)\, e^{st}\, ds,$$

respectively. In analogy with the Fourier transform we have the basic

Property 1

$$L[f(t/a)] = a\, \mathscr{L}(\text{as})\ (a > 0)$$

Property 2

$$L[f(at)e^{-bt}] = \frac{1}{a}\mathscr{L}\left(\frac{s+b}{a}\right) \qquad (a > 0)$$

Property 3

$$L[tf(t)] = -\frac{d}{ds}\mathscr{L}(s)$$

Property 4

$$L\left[\frac{df(t)}{dt}\right] = s\,\mathscr{L}(s) - f(0).$$

It also easily follows from the above definition that

Property 5

$$L\left[\int_0^t f(\tau)d\tau\right] = \mathscr{L}(s)/s$$

Property 6

$$L\left[\int_0^t f(\tau)\, g(t-\tau)d\tau\right] = \mathscr{L}_f(s)\, \mathscr{L}_g(s).$$

Selected examples of Laplace transforms

Example 1 If

$$f(t) = t^{\alpha}$$

with Re $\alpha > -1$, then

$$\mathscr{L}(s) = \frac{\Gamma(\alpha+1)}{s^{\alpha+1}} \qquad \text{Re}\, s > 0.$$

□

Example 2 If

$$f(t) = (t^2 + 2\alpha t)^{\beta - 1/2}$$

with $|\arg \alpha| < \pi$ and Re $\beta > -1/2$, then

$$\mathscr{L}(s) = \frac{1}{\sqrt{\pi}} \Gamma(\beta + 1/2)\,(2\alpha/s)^{\beta}\, e^{\alpha s} K_{\beta}(\alpha s) \qquad \text{Re } s > 0,$$

where K_{β} is the modified Bessel function of the third kind (see Appendix 2 and Chapter 3). □

Example 3 If

$$f(t) = e^{-\alpha t},$$

then

$$\mathscr{L}(s) = \frac{1}{s + \alpha} \qquad \text{Re } (s + \alpha) > 0.$$ □

Example 4 If

$$f(t) = (1 - e^{-t})^{\alpha - 1} (1 - xe^{-t})^{-\beta}$$

with Re $\alpha > 0$, $|\arg(1 - x)| < \pi$, then

$$\mathscr{L}(s) = B(\alpha,s)\,{}_2F_1(\beta,s;\alpha + s; x) \qquad \text{Re } s > 0,$$

where $B(\alpha,s)$ is the standard beta function (see Appendix 5). When $x = 1$, the hypergeometric function can be summed in terms of gamma functions and

$$\mathscr{L}(s) = B(\alpha - \beta, s) \qquad \text{Re } \alpha > \text{Re } \beta \quad \text{Re } s > 0.$$ □

Example 5 If

$$f(t) = \frac{1}{\sqrt{t}} e^{-\alpha/4t}$$

with Re $\alpha \geq 0$, then

$$\mathscr{L}(s) = \sqrt{\pi/s}\, e^{-\sqrt{\alpha s}} \qquad \text{Re } s > 0.$$ □

Example 6 If

$$f(t) = \sin(2\sqrt{\alpha t}),$$

then

$$\mathscr{L}(s) = \frac{\sqrt{\pi\alpha}}{s^{3/2}} e^{-\alpha/s} \qquad \text{Re } s > 0.$$ □

Example 7 If

$$f(t) = \tanh t,$$

then

$$\mathscr{L}(s) = \tfrac{1}{2}\,\psi(\tfrac{1}{4}s + \tfrac{1}{2}) - \tfrac{1}{2}\,\psi(\tfrac{1}{4}s) - s^{-1} \qquad \operatorname{Re} s > 0,$$

where ψ is the psi (or digamma) function (see Appendix 5). □

Example 8 If

$$f(t) = t^{\alpha-1} \ln t$$

with $\operatorname{Re} \alpha > 0$, then

$$\mathscr{L}(s) = \frac{\Gamma(\alpha)}{s^{\alpha}}\,[\psi(\alpha) - \ln s] \qquad \operatorname{Re} s > 0.$$ □

For extensive lists of Laplace transforms see Oberhettinger and Badii (1973) or Erdélyi (1954).

References

Erdélyi, A., et al. (1954): *Tables of Integral Transforms, Vol. 1,* McGraw-Hill, New York; Chapters I–V.

Oberhettinger, F. (1957): *Tabellen zur Fourier Transformation*, Springer-Verlag, Berlin.

——— (1973): *Fourier Transforms of Distributions and Their Inverses*, Academic Press, New York.

——— and L. Badii (1973): *Tables of Laplace Transforms*, Springer-Verlag, New York.

Index